On the Frontier
My Life in Science

On the Frontier
My Life in Science

Frederick Seitz

President Emeritus
The Rockefeller University

American Institute of Physics   New York

AIP Press
American Institute of Physics
500 Sunnyside Boulevard
Woodbury, NY 11797-2999

Library of Congress Cataloging-in-Publication Data
Seitz, Fredrick, 1911–
    On the frontier : my life in science / by Fredrick Seitz.
        p. cm.
    Includes index.
    ISBN 1-56396-197-0 (alk. paper)
    1. Seitz, Fredrick, 1911– . 2. Research--United States--History.
3. Physicists--United States--Biography. I. Title.
QC16.S36A3 1994   94-17625
530'.092--dc20   CIP
[B]

10 9 8 7 6 5 4 3 2 1

Cover design by: Marilyn R. Horowitz and Linda Katz

To Kathleen Foley, M.D.

*She and her staff made it possible
for my beloved Elizabeth
to leave this world with dignity.*

Contents

Preface

This book came into being almost accidentally; it is the result of a longstanding, routine request from the National Academy of Sciences to its membership for biographical material for its archives. My wife's illness, starting in 1991, left me with long hours in hospital waiting rooms and by her bedside as she underwent treatment, and provided an opportunity, unhappy though it was, to make a beginning. With her help the project grew over the months into its present form. She enjoyed discussing the events of our lives described in the book and together we pored over our collection of photographs of the friends and associates who meant so much to us. Although we had substantial source material for guidance, some of the story had to be reconstructed from memory and as a result may be partly flawed, however I believe most of the reporting is reliable. While I have never regarded myself as having a particularly remarkable memory, I am continually astonished at the large number of experiences extending over many decades one can, with reasonable patience, withdraw from one's memory bank.

Some colleagues and friends who have read the book in draft form have pointed out that it is not a conventional autobiography, focussing as it often does on the activities of others who were colleagues of mine, or on the environment in which I found myself. This focus is intentional. I have always enjoyed observing my fellow scientists and feel that this aspect of the book is very much a reflection of my own interest in the ever-evolving scientific scene. I have had a great deal of personal satisfaction through following the results of research in which I was not immediately involved, and through associating with individuals in various areas of science. This interest predates by many years my period as President of the National Academy of Sciences.

Several individuals who have been closely involved with this work deserve much credit. First is Mrs. Florence Arwade who manages my office and kept the manuscript and all correspondence flowing as needed. Mrs. Lucie Sewell Marshall, my wife's sister-in-law, and Leonard Lee Bacon, my old friend, both provided valuable comments on the entire book and helped eliminate many flaws in the text. John S. Coleman, with whom I worked closely at the National Academy of Sciences, read both Chapter 13 which deals with the Academy and Chapter 9 which tells of the work carried out during World War II. Dr. Dillon E. Mapother, a former colleague at the University of Illinois, generously read through and provided substantial help with Chapter 11 centering on the years we spent at the University. He in turn referred some points to Dr. George A. Russell, a former colleague who is now chancellor of the University of Missouri. Similarly Dr. Maclyn McCarty and Mr. Rodney W. Nichols did much to strengthen and add detail to Chapter 14, dealing with The Rockefeller University. Mrs. Ruth Carney Brown, my high school classmate, made constructive comments on the early chapters. I am also indebted to Professor Daniel Kevles, who kindly provided information relevant to my days at the California Institute of Technology.

Ms. Barbara Sullivan, the text editor engaged by the publisher, made many cogent comments and did a great deal to improve the coherence, clarity, and flow of language in the book, for which I am most grateful.

Dr. Alexander G. Bearn and Mrs. Sonya Wohl Mirsky, the former librarian of The Rockefeller University, also provided much valuable guidance and encouragement along the way.

I would like to express gratitude to Dr. Lillian Hoddeson for discussions of many facets of topics appearing in this book. We are also grateful to the publishing staff of the American Institute of Physics for its warm and efficient cooperation.

Assembling the photographs was a labor of love. Many are from the Emilio Segrè collection of the American Institute of Physics; there Mr. Douglas Egan was very helpful. Some are from the National Academy of Sciences where we owe much to Ms. Mary H. Brown and her colleagues of the Office of the Home Secretary and especially to Ms. Janice Goldblum of the Archives. Ms. Rene Mastrocco of the Archive Center of The Rockefeller University provided similar help in relation to university photographs. Several friends at the University of Illinois, particularly Dr. James S. Koehler and Dillon Mapother, came to my aid in connection with photographs from that institution. Princeton University provided photographs of Harry Hess and the Great Hall of the Graduate College. Mr. Rolfe Neill of the *Charlotte Observer* in North Carolina and Mr. Bradford Bollinger of the Santa Rosa, California, *Press Democrat* provided valuable advice on items that might be found in special press sources. Dr. J. M. Cadiou, the Assistant Secretary General for Scientific and Environmental Affairs of NATO, and Mrs. Rudolf Schrader, the wife of my former deputy at NATO, furnished photographs from that institution. Mrs. Pierre Aigrain found good photographs of Jean Monnet and Raymond Aron in Paris. George Washington University generously provided a copy of the assembled group attending the meeting at the University in January of 1939 where Niels Bohr announced the discovery of fission. I am particularly grateful to Professor Marina von Neumann Whitman for the photograph of her father — one which revives warm memories of the highly stimulating time spent in the presence of this remarkable man. Also, I should express thanks to Canadian friends as well as to the staff of the Canadian Academy of Sciences for help in obtaining a photograph of the late J. Tuzo Wilson. Not least, I am grateful to the staff at Lick-Wilmerding High School in my native San Francisco for some of the photographs in Chapter 3. Headmaster Adams provided much friendly support. Finally, a number of friends and institutions, mentioned in captions, delved into personal or institutional files to provide valuable photographs from their own collections.

The book could never have been started or finished without the magnificent support I have received from the administration and board of The Rockefeller University. Here my gratitude is overwhelming, not least to Dr. Attallah Kappas and to his successors as Physician-in-Chief of the University Hospital, Drs. Jan Breslow and Jules Hirsch.

Added note: In preparing the name index for this book, I note the absence of many individuals, including friends, students, postdoctoral fellows, colleagues and others, who did much to enrich our professional and personal lives. The constraints of a single volume make limitations necessary; however, the memory of all such individuals is contained in spirit within these covers.

Preamble

To the extent our existence has meaning in this strange world in which we find ourselves, there should be much more to life than providing the everyday needs of food, shelter, and creature comforts for ourselves and our families. Traditionally, both religion and art have been regarded as essential adjuncts to life and will undoubtedly remain so in some form for most individuals. Science as a systematic form of human endeavor has fulfilled another great need, by providing a deeper understanding of the details of the natural world in which we live—details which sometimes seem to run counter to what we like to think of as "common sense."

For some, science is absorbed indirectly as they follow advances at the moving frontier. For others, the ingestion happens much more indirectly, as if by osmosis. Today only a few rational individuals in advanced countries believe that the earth is flat or that emotion and reasoning are sited in a bodily organ other than the brain—issues which were once subjects of serious debate. In fact, the enlightening discoveries of science have engendered in those who appreciate them a combined sense of pride and humility—pride in the ingenuity and brilliance of accomplishments, humility at the realization that our own sun-centered planetary system is but a minute speck in a vast universe. This realization makes our own status in the world quite unclear; are we and our biological relatives on earth, as living beings, a special miracle of natural creation or a commonplace and insignificant phenomenon when viewed within the universe as a whole?

This brief century has brought us an irreversible revolution in technology as a result of scientific discoveries, a revolution which has been hard-earned and which is an important adjunct to the advance of science. Unfortunately, there is a tendency for the leadership in our somewhat populist, democratic societies to attempt to shape popular demand as it sees fit, and to value science more for its immediately practical products rather than as a manifestation of the free wills and imaginations of its creators, especially when they are engaged in the most basic research. Many leaders behave as if the basic and the applied could be somehow miraculously separated by government edict, neglecting the first and stimulating the second. This attitude could well prove fatal for science, both basic and applied, in the coming century. In the bygone days when aristocrats had a strong voice in the running of society, the support of art and scholarship was regarded as one of the fundamental responsibilities of leadership. The results of that attitude can be seen in the great monuments which we travel around the world to see, as well as in the universities which take pride in their ancient traditions and their venerable libraries. It would be a pity if the granting of political freedom to the people at large led to the exclusive cultivation of the mundane. We would inevitably destroy that part of the scientific adventure intended to expand our minds, and everyone would be the loser.

President John F. Kennedy was one leader who appreciated the larger issues. He exhorted us to walk on the surface of the moon and to inspect Mars with landed instruments—both great scientific as well as technical adventures. Have we as a society lost our taste for such adventures, or will we, in spite of our various domestic troubles, take further dramatic steps, such as establishing an observatory on the moon? Will we show that there is more to human destiny than distributing our wealth broadly in small packages (an endeavor which need not be appreciably hindered by the support of great science)? Is the centuries-long struggle to create open and free societies culminating in the triumph of the ordinary? Are we at the end of the human quest to comprehend the stars and the structure of matter, or at the beginning of a glorious new chapter?

From my childhood, I was intrigued by science and its accomplishments, and decided early on that I would devote my life to some field of research if circumstances and my own abilities permitted. My experience at a fine high school in San Francisco, which had an excellent mathematics and physical sciences staff, combined with the support of indulgent parents, strongly encouraged me to pursue the goal. The entire world of science began to open up to me during my happy and productive undergraduate years at Stanford and the California Institute of Technology, followed by graduate school at Princeton. At Princeton I not only had the good fortune to work with Edward Condon and Eugene Wigner, but encountered many more of the most gifted scientists of the century—an inspiration in itself.

Had World War II not intervened, the chances are very great that I would have settled into a faculty position somewhere and spent my life devoted to research, teaching, and the cultivation of students, perhaps the most satisfactory career one could hope for in the best of times. As fate would have it, I became involved part-time in administration during World War II and was never again a truly free agent. Fortunately, administrative work can have its own satisfactions if one is willing and able to suppress one's ego sufficiently in order to help others flourish. I found it feasible and rewarding to do so in the exhilarating environment of rapidly developing science which obtained both in the United States and abroad during the twentieth century. Along the way I had the good fortune to meet and work with a wide range of dedicated, creative individuals from the various walks of life that intersect with science in one way or another.

My judgment is possibly wrong, but I have the feeling that for better or worse it would be far more difficult at the present time to recreate anything like the path I followed. It is not only that the fields of science are more crowded and more highly specialized. Much of science is beset by complex pressures originating both outside and inside the scientific community which make administrative work more difficult. As the instrumentation needed to advance scientific frontiers has become more complex, it has also become more expensive. While the support provided to science by the federal government managed to keep up with the growing need in the years immediately following World War II, that is no longer the case, and science now faces great financial stresses after years of plenty, along with uncertainties concerning future government policies.

Then too, many members of the scientific community tend to be much more intensely involved today in issues considered to have a high degree of ideological content, such as arms control and the status of the environment, on which scientific information has a bearing. As a result, discussions of matters such as the technical feasibility of anti-ballistic missile defense, the likelihood of global warming, or the long-range effect of genetic research, all of which are only marginally understood, and which once could be faced constructively in the spirit of civility and good humor, are now more apt to be the focus of heated debate between inflexible adversaries. It would now be much more difficult to become involved in *both* basic and applied science, as I was. Fortunately, large areas of science are still free of ideological content and continue to make remarkable strides within the traditional framework. May they remain so!

In a few words, the present-day world of science is different in many ways from that which I entered as a young student, but it is still a fascinating place to be.

Chapter One
Early Years

I was born in San Francisco in a flat at 50 Santa Rosa Avenue, just a half-block from its start at Mission Street. Mission Street, which ambles north/south through the city, is part of the old Camino Real, the royal Spanish road along which the missions of Father Junipero Serra were located, starting in San Diego and ending in Sonoma. The date was July 4, 1911, the year in which Rutherford discovered the atomic nucleus, Kamerlingh Onnes discovered superconductivity, and Sun Yat-sen overthrew the Chinese monarchy and established a republic.

It must have been a very noisy day, if my memories of subsequent birthdays are any indication. At that time there were no restrictions on the use of fireworks in the city, and the merchants of Chinatown each year provided patriotic celebrants with a virtually unlimited supply. On a typical July Fourth, the crackling and booming of fireworks everywhere competed with the clanging of fire engines, which were hard at work all day. As a child, I usually ended the day badly singed and partly deaf—but highly contented.

The coast range mountains of northern California are now believed to have been created when one plate of the earth's crust slid beneath its continental neighbor which scraped off a layer of debris in the process. That same abrasion resulted in the hills and valleys of San Francisco. Each valley has its own microclimate, depending upon the extent to which the surrounding hills block the inflow of moist sea air. The downtown area and the so-called Mission District are shielded to a great extent by Twin Peaks, and are therefore relatively free of daytime sea fog, while the Richmond District along Golden Gate Straits, being much more exposed, has a great deal of fog. The southernmost valley, where I grew up, is shielded relatively little because of a breach in the coastal hills, created in part by the San Andreas fault, which leaves the land just south of the city limits and joins it again in Marin County to the north, creating Bolinas and Tomales Bays. As a result of this geological quirk, the fog has always had its way with the area, and we knew many gray summer days when other valleys enjoyed the daytime sun. I have never been enamored of a northern coastal climate since.

In my childhood the various valleys also possessed what might be termed cultural identity since neighborhood relationships were more important then. I do not know to what extent something of the kind exists today and, if it does, whether the degree is comparable.

1

In that southern part of the city where Santa Rosa Avenue was located, Mission Street ran along relatively high ground on the eastern side of the valley, which was bordered on the west by thousand-foot-high Mount Davidson and on the east by a low range of coastal hills that edged along the Bay. Santa Rosa Avenue cut across the valley from east to west.

The Roman Catholic Church possessed a large complex of buildings at the western, or lower, end of the block where we lived. Here was Corpus Christi, the largest Catholic church in the district. The complex was a major social hub as well as a religious center. Attached to the church was a large enclosed churchyard; this was a wonderful haven for children of all sizes since it was away from traffic and big enough to allow for a variety of games.

At that time the church was bordered by a long dirt road parallel to Mission Street; this was an access road serving what remained of the old truck farms, which dated from the early days of the city. Most were owned and operated by Italian immigrants or their descendants, but a few were run by Chinese farmers who carried on as they had in old China, performing much of their labor by hand.

An entrepreneurial group had once constructed a scenic railroad line, The Ocean Shore Railroad, along that dirt road, leading from a point near the center of downtown San Francisco to Halfmoon Bay some twenty five miles south along the coast. It was, however, a failure from the start since most of the route was shrouded in a dense mist whenever fog prevailed, which happened most often during the summer. As children we placed pennies and nails on the rails to see them flattened by the train wheels, just for the fun of it. The rails were torn up soon after World War I, and the road has since been replaced by a high speed north/south highway, Alemany Freeway, which has taken most of the through traffic away from Mission Street. It also has an extension going east to the Bayshore Highway.

Five years or so before I was born, my father, after whom I was named, had opened a fancy bake shop, or patisserie, close to home in the main shopping district on Mission Street. He chose the site partly because it was within his means and partly because the area had a growing residential population.

He was born on January 23, 1876. He and his five siblings had come into the world near Heidelberg in the Palatinate, in a well-to-do family with all the promises of a fine education and the other benefits that go with material comfort. Unfortunately, his father, George, apparently a poor business-man, risked his fortune on a plan for the industrialization of flour mills, lost everything, and died soon thereafter when my father was in his early teens.

One of the few interesting anecdotes concerning my father's father I have been able to glean from European relatives is that, as a student, he had been very active in a rebellion going with the ill-fated social revolutions that swept over the German states prior to unification in 1870, and which were echoes of the great revolution of 1848. He eluded an extensive police search, which probably would have resulted in a prison term, only because his parents managed to hide him skillfully in the attic of the family home while denying that they knew of his whereabouts.

The extended family provided help to my father's widowed mother, Elizabeth, and to his older brother, who was given a technical education. My father, however, was apprenticed at the age of fourteen to the Viennese-style pastry baker (Konditor) in Heidelberg for three years. There followed two more years of apprenticeship at a similar bakery in Innsbruck, Austria, a major resort area. At last in 1895, at the age of nineteen, he set out from Austria for New York City, where he obtained a good job as a full-fledged pastry baker at Cushman's bakery, one of the leading specialty shops in the city. The young man set himself three goals in his new life: to help his mother and sisters, to absorb as much education as circumstances would allow, and to save enough to eventually go into business for himself. His sisters, who no longer had expectations of dowries, would need to prepare to become teachers. At that time the opportunities available in New York City to working individuals for serious spare time study were excellent as a result of the presence of many immigrants anxious to become part of their new world. Those, like my father, who were sufficiently earnest to take full advantage of such resources, probably did at least as well as the average college student of today.

Ten years later he was in San Francisco, a year before the Great Earthquake of 1906, making plans to become established.

My father's life might have turned out very differently. Before he entered his baker's apprenticeship, his mother was approached by the local Jesuit leaders, who offered to give him a scholar's education through their seminaries if she would allow him to become a member of the Catholic Church. The family had a Protestant tradition at that time, strongly reinforced by intermarriage with Huguenots, and the offer was refused. Today it probably would be accepted since religious links are no longer drawn so sharply.

My father was not a tall man, perhaps five foot six inches, but he was well proportioned, muscular, and very active. In appearance he was basically the alpine type, although I suspect that strains from both eastern and western Europe had intermingled in the family. There was, for example, the aforementioned intermarriage with the Huguenots, and it would be surprising if there were not some Slavic component as well, since the family name is particularly prominent in Austria, where there has been much ethnic mixing.

Whatever the different strands of my father's lineage, together they made for strong family bonds. He was deeply devoted to his mother during her lifetime, corresponding regularly and offering whatever help he could. She was devout, and instilled in him the highest principles of honesty and integrity. His standards were exacting in both business and personal relationships. Years later, an individual who had been one of my childhood friends said to me: "We all regarded your father as one of the fairest men in the community."

My father's personality was also strongly shaped by the fact that a combination of independence and responsibility was thrust upon him at a very early age. He never carried any life insurance or belonged to any of the benevolent organizations common to those times. He did join a workers' union in San Francisco, but only because he had to belong in order to gain employment before going into business, since it was what was called a "closed-shop" city, that is, strongly unionized. I do not know what subsidiary benefits union membership carried, if any, but I doubt if he would have attached much importance to them.

Had fate been kinder, my father would have preferred a highly technical profession, and probably would, for example, have made an excellent chemical engineer. While still a bachelor, he spent his spare time reading semi-popular scientific and technical literature, and retained a surprising amount of the knowledge gleaned from his studies.

In retrospect, I believe that my knowledge at a relatively early age of the extent to which my father had been denied a truly formal education provided me with a strong incentive to go as far as circumstances would permit me to.

Somewhere along the way, he developed a considerable fascination with gold mining and spent a number of months as an amateur prospector in the Panamint Valley area of California before settling down to the bakery business. One consequence of this adventure was his purchase, with several partners, of a slightly developed gold mine in the mother lode area near Sonora. The mine, which was then worked through tunnelling, more or less broke even until the Depression of the 1930s caused some of the partners to suspend the operations, feeling the business was too risky. My father was seldom happier than when prowling about the mining area with his geologist's pick, looking for material to pan or to send off for assay. Fifty years later my brother and I reaped a modest profit when efficient open-pit methods were used to exploit the ore.

For reasons I have never understood, the region around Boise, Idaho had long held an attraction for members of the Seitz family. My father's older half-brother (the son of my grandfather's deceased first wife), his younger brother, Louis, and his first cousin, Fred Baust, all settled there. My father himself made an extended visit there on his way west, but went on since he wanted to enter business in a major city. I suspect a family friend had migrated there in the 1850s and must have recommended it as an attractive part of the American frontier.

My mother, Emily Charlotte (Figure 1.1), was born in San Francisco in 1883 and was raised in the Potrero District, which was mainly residential prior to the earthquake of 1906. Her father was a Civil War veteran named Henkel, originally from Missouri, who went to California soon after the war. There he met and married a young woman, Elena von Schwartzenhorn, who had recently come to the United States from the Halberstadt-Quedlinburg area of Saxony, where her family had lived for a number of generations. Someone in her father's lineage had apparently been a courtier and was knighted at some point for special service to the royal family in Saxony. Judging from her mother's advanced age at the time Elena was born, namely thirty seven, it would seem that Elena was a younger daughter in what was then a relatively large family, and that she was both daring and venturesome enough to want to see what the United States had to offer. While my mother's father was a fine man in other respects, he turned out to be a poor provider, and Elena eventually left him. Somewhat later she married a Norwegian named Holmberg, the captain of a sailing ship called the *John Smith*, which transported redwood lumber up and down the West Coast. I never knew either of these men, but my older brother had the privilege of meeting our genetic maternal grandfather in the full regalia of the Grand Army of the Republic sometime around 1915 when the latter was living at a retirement home in the Presidio in northwest San Francisco. Unfortunately, all records of vital statistics in San Francisco prior to 1906 were destroyed in the great earthquake and fire of that year.

My mother was a tall, handsome woman with a light complexion and light brown hair. In spite of the hardships her mother faced from time to time as a young woman, my mother was able to enjoy the gaiety of San Francisco life. She loved music and the opera, and participated in several of the social lodges which were an important part of life in her day. My father, with his serious outlook (Figure 1.2), might have seemed an unlikely husband for Emily, but her mother saw his virtues clearly and undoubtedly exerted her influence. In fact, my father was very kind to his mother-in-law, who died prematurely of a heart condition in 1911.

My mother had an extended family in the greater Bay Area, suggesting that someone in her family in the previous generation, which I never knew about, or rather never had the interest to ask about, had first come to California and had formed a base of family support for those who followed. There were apparently some aunts who had lived in San Francisco and who died before my time, and a living aunt, Mrs. Mastretti, in Bodie, California. There were also several cousins of ours in nearby towns. My father once told me that one of my mother's uncles, who had been a purveyor to the Nevada silver mines, was killed by an Indian while plying his trade. He may well have been Mrs. Mastretti's first husband since she had two sons, Joseph and George Langrell, who owned and operated the hotel in Bodie. In its time, Bodie was one of the communities linked to the silver mines;

Figure 1.1. *(Left) My mother as I knew her as a child. We had a most happy relationship.* **Figure 1.2.** *(Right) My father in his early twenties. Deprived by accident of the opportunities in life he would have preferred, he shouldered his responsibilities imaginatively and without complaint. He had an excellent amateur knowledge of science.*

today it is a ghost town and a tourist attraction (Figure 1.3). Joseph was clearly the business manager while his brother seems to have operated a more or less non-stop card game in the hotel. The former also had the more social gift of rattling off popular tunes on the piano by ear. Whatever their familial links to European culture may have been, in their looks, language, and behavior, the two sons were pure products of their frontier upbringing.

My mother had spent a year or so in Bodie with her aunt as a child, and she spoke of the cold winters and, more fondly, of the warm hearts of the people she encountered.

She had a younger half-brother, Conrad, who was quite prosperous in his younger days and enjoyed San Francisco life to the full. He had a lively, quick-witted sense of humor. Perhaps not surprisingly, one of his grandchildren became a popular stand-up comedian in the Bay Area. Emily also had a younger half-sister, Teresa, who married a pleasant, warm-hearted man who worked for the Standard Oil Company. Tragically, he died of throat cancer within a few years of their marriage. Teresa then lived for a year or two in a flat in the same building on Santa Rosa Avenue where I was born until she married a widower, Rudolph Bruhns, who had three children and owned a popular downtown restaurant. I formed close friendships with these step-cousins. Prior to my aunt's second marriage, most of

Figure 1.3. The ghost gold mining town of Hodson neighboring my father's mining property. I spent many hours prowling through the abandoned buildings in the 1920s. The scene is typical of what one experiences along the mother-lode country of California. (Photograph provided by Alexander Cereghino.)

our family social life centered around old San Francisco friends of my mother. Those relationships waned thereafter, and were replaced in large part by purely family affairs.

The nature of my personal relationship with my mother can be described by a minor anecdote. One of her childhood friends, who remained a life-long intimate of hers, married a midwesterner, Charles Cowan, who worked for a food company in the East Bay area. They had a son, Russell, slightly younger than I, whose birthday occurred in mid-February and was celebrated on Washington's birthday. At one of Russell's parties, all the children were given papier-mache hatchets loaded with beaded candy as favors. Much to his sorrow, Russell broke his through misuse. My mother took me quietly aside and said: "Please trade yours with him. I will make it up to you." I responded good-naturedly. A month later, long after I had forgotten the incident, she presented me with a toy locomotive—loaded with much better candy! She taught me the importance of kindness, and reinforced the lesson with her own generosity.

Most of the families living in our block on Santa Rosa Avenue were first- or second-generation Italians, who had come almost exclusively from the Genoa area. Two exceptions were the O'Briens and the Gallaghers. A third was the Heckstrom family, of Swedish descent. Mr. Heckstrom was a policeman whose beat was in Chinatown. Each July 4th he brought home a large sack of fireworks which kept us busy and happy for a good part of the day.

There was one relatively wealthy family on our block, the DePaolis, who lived across the street from us. They had a large mansion with an exotic cylindrical tower on a half-acre estate surrounded by a tall cypress hedge. The family had made its money in real estate and truck gardening. The then head of the family, probably second generation, happened to die on the day I was born, a coincidence which my mother was inclined to regard as having some sort of significance with respect to cosmic balance. The DePaoli's son, Louis, in his mid-twenties when I was a child, was well-liked by the neighbors. He was the first person in the neighborhood to acquire an automobile, and on several memorable occasions took a group of us children along when he made business calls, driving us to nearby Colma on the southern side of the city, and treating us to ice cream cones while we waited for him in the car. He eventually became a prominent real estate agent. His widowed mother, however, was not quite as fond of us. During my childhood, she would sit at a window in the tower of her mansion while we played in the street, to make certain that no one trespassed on her property. Her efforts were only partly successful, particularly if a fly ball went over the hedge and landed inside.

The LaFerreras were another private, reserved family in the neighborhood, although at the time they were not as wealthy as the DePaolis. The children of the family seemed to spend all of their out-of-school hours practicing on musical instruments. However the youngsters may have felt then about their loss of play-time, the sacrifice paid off for one of them, at least. Vincent LaFerrera, who was perhaps fifteen years older than I, became a celebrated violinist and orchestra leader. His group gave concerts in the Bay Area and played at hotel dances.

Fairly early in my life I was fortunate enough to find Silvio Delvecchio; we were lifelong friends until he passed away in the early 1980s. He was slightly younger, but was similar to me physically. He lived on Mission Street, but liked to come and join the children on Santa Rosa Avenue. Except for a brief period during World War II when he was stationed at an army camp in Utah, Silvio spent his entire life in the city, working as an accountant at a local insurance company. He maintained an encyclopedic knowledge of almost everyone who lived in what we then termed "our district."

Silvio and I were frequent, welcome visitors in each other's homes, moving in and out freely and thoroughly enjoying the associations. His family, like most Italian families in the neighborhood, originally came from the vicinity of Genoa. His parents and other relatives had spent some time in New York City on their way to California and had a memento of their stay there—in the form of a brilliant red table cloth on which representations of the Brooklyn Bridge and the Woolworth Building were boldly crocheted in white. The wonderful cloth was used on holiday occasions and made my mother's white linens seem quite uninspired by comparison.

On reaching California, the Delvecchio family settled for a while in the gold mining area near Jackson, in the mother lode country. Two of Mrs. Delvecchio's brothers had operated a saloon there. One of the relics of this period was an electric-powered, air-driven, coin-operated harmonium that could play several dozen popular tunes from those earlier times. It was invariably put into operation at the more boisterous Delvecchio family parties, and never failed to enhance the fun.

Mrs. Delvecchio was a fine cook in the Italian style. Her ravioli was a particular favorite of mine, and the occasions on which she served the dish are among the highlights of my many fond memories of this warm-hearted family.

Like many similar families, the Delvecchio clan had one member, a grandparent, who never learned English. As a result, the primary language at home, and that with which the children were raised, was Italian. So many of my playmates, who had been born right in the neighborhood, spoke English with a discernible Italian accent.

Every autumn, all hands in the Delvecchio family, along with a few friends, turned to making wine. They milled the grapes, stems and all, into a huge vat and let it ferment, pouring it into barrels when the fermentation was completed. This wine was never destined for the Tastevin Society since the fermentation was carried out crudely, the presence of stems in the mix guaranteeing that it was both sour and bitter. Nevertheless, among the neighbors it had appreciative consumers.

The city of San Francisco never took to Prohibition very well, even though the state had passed its own Prohibition Law during World War I, prior to the enactment of the Volstead Act. Whenever the state law came up for reconsideration on the ballot, the city of San Francisco voted overwhelmingly for its repeal. As I recall, the state law actually was repealed prior to the Volstead Act.

When I was seven, we moved to a more luxurious flat on Mission Street, but little else changed since the distance moved was short. The move did encourage me to expand my horizons by exploring life in the street just north of and parallel to Santa Rosa Avenue, which was called Francis Street. It was dominated by Irish families, although again there were exceptions. Street play was more dynamic there than on Santa Rosa Avenue, but the risk of being involved in a brawl was much greater. The young priests from the church, who were also mostly of Irish descent, made it their business to keep an eye out for such troubles and often emerged to break up altercations. The combatants were then conducted into the church for a period of repentant prayer.

There was one striking difference between life on the two streets. On

Santa Rosa Avenue the nature of the street play was invariably adjusted to include girls if they wanted to join in the fun. This never occurred on Francis Street, the girls playing games of their own choosing.

One or two of my memories of daily life when I was a child still have a particular resonance. In those days, the coaches attending funerals and weddings at the local church were mostly horse-drawn carriages, with coachmen wearing knee britches and top hats. In fact, about half of all traffic in the city then was still horse drawn. I recall standing with my father on Mission Street about the year 1917 as the traffic flowed by. He watched the mingling autos and horses, and commented: "I guess we are going to get rid of the horse manure but heaven knows what will follow."

The relative simplicity of life at that time is symbolized for me by the fact that the postman, the policeman who helped school children cross Mission Street, and even the man who broom-swept the streets were all well-known and well-respected individuals. They were regarded as an integral part of the community and were courteously referred to by their last names.

My own memory in that early period of life goes back to my third year, 1914. I recall quite vividly the excitement caused by the outbreak of the war in Europe in August of that year. We heard the news while on vacation in Guerneville on the Russian River, and I remember being profoundly moved, although I did not, of course, know precisely what "war" involved.

I also remember being taken to the Marina District to see the early preparations for the Panama Pacific International Exposition which was to be opened the next year, that is, 1915. That preview trip was followed the next spring by visits to the actual great exposition. I was awed by the giant ferris wheels and the airplane acrobatics of Arthur Smith, one of the early barnstormers of the day. Many other wonders such as the Tower of Jewels and the Palace of Fine Arts, which still remains as a memento of the occasion, are deeply imprinted in my mind. My father said at the time that because of World War I, the exhibition was a pale image of like ones that had preceded it in Saint Louis and Chicago, but to me, it was all that was glamorous and thrilling.

One of the notable events of this childhood period was the arrival of my father's younger brother, Louis (Figures 1.4 and 1.5), from Idaho. He stayed with us for a month or so when I was four years old and accompanied us on our visits to the great Panama Pacific Exposition of 1915. At that time he was foreman in a prominent Idaho lumber company, Boise-Payette. Unlike my father, Louis was a very powerful man, with a hearty laugh and a love of fun. I am reminded of him when I occasionally see the motion picture "I Remember Mama," since he looked and acted very much like Uncle Chris as played by Oscar Homolka, walrus mustache and all! Uncle Louis' arrival was preceded, like that of some exotic foreign potentate, by a large crate of fur rugs made of assorted animal pelts which he had acquired on various hunting trips. I do not know what my mother thought of the gift, since the

Figure 1.4. (Left) My uncle, Louis, in the early 1920s, dressed in his Sunday best. Probably a photograph he sent to his far off mother. Figure 1.5. (Right) My father's mother at the time of her marriage in 1872. He adored her.

rugs were more appropriate for the decor of a hunting lodge, but they did eventually become part of the household. My brother still had a somewhat moth-eaten bear rug from the collection hanging in his garage in the early 1990s.

Our uncle's visit was a special treat for my brother, Charles, who was four years older than I, since he was permitted to accompany my uncle to the fairground. They went frequently, always returning with a number of trophies from the amusement section of the fair. These consisted of such valuable items as plaster of Paris dolls and large boxes of mediocre grades of candy. Even at that early age I was able to discriminate between the best and the not-so-good of the confectioner's art since my father's profession meant we usually had lots of the former available at home.

It was a sad day for my brother and me when we finally saw Uncle Louis off back to Idaho on the Northern Pacific Railroad. He returned once again ten years or so later with the idea that he might buy a fruit farm in the Peninsula Area, but nothing came of it. Instead, he bought a farm in Idaho in the 1930s, and managed it exceedingly well, closely following the advice provided by the various government agencies on the best methods of irrigation. He became a source of sound counsel to some of his less literate neighbors.

Not long after Uncle Louis' visit in 1915, my mother discovered how to have a quiet Saturday afternoon; her method had a profound effect on my life. For two years or so my brother Charles, then eight years old, had been permitted to attend the Saturday afternoon showing at the local moving picture house, which was about three blocks away on Mission Street, life being simpler and safer then. It was a store-front affair with hard wooden chairs, no air circulation, and a pretentious name, the Panama Theatre. Admission for children was five cents. My mother decided that it would be a fine idea if my brother took me along on these outings, leaving her in relative tranquility for three luxurious hours or so. While he did not complain openly, my brother undoubtedly felt the job to be demeaning, since serving as a baby-sitter diminished the air of mature independence that he was anxious to cultivate before his eight-year-old friends. While on the way to the theater he did not dare hit or kick me because of the inevitable retribution he could expect at home as soon as I came within earshot of our mother. His companions, however, had no such incentive to restrain themselves. I tagged along at a respectable distance.

Once in the theater he would sit me down in an aisle seat, stare me in the face and order, "Stay there until it is over and I come to get you—or else!" Duty done, he then joined his friends elsewhere in the house for the noisy fun. I had no objection to being alone since, once the show started, I was in a special world. Those were the days of westerns, slap-stick comedies, and fifteen-week serials. I knew all the principal actors of those genres from Bill Hart to Pearl White to Charlie Chaplin in his earliest comedies. I was particularly fond of Buster Keaton and became a great annoyance to my friends by declaring argumentatively that he was vastly superior to Chaplin. These early pleasures proved addictive, and I was a steady movie-goer until the responsibilities of later life deprived me of the spare time. Since the days of Keaton and Chaplin, the technology of film making and displaying has improved beyond all imagining—unfortunately, it seems that at the same time the standards of the scripting of films have been greatly cheapened through the introduction of apparently limitless and irreversible vulgarity and violence.

The Panama Theatre was eventually replaced in the early 1920s by a much more palatial edifice with plush seats and a stage large enough to accommodate so-called four-a-day vaudeville acts on Sunday. Nothing approaching the quality of Shakespearean drama ever appeared on that stage.

Once, when a bad cold prevented me from accompanying my brother to the Saturday matinee, he returned breathlessly to tell me of the astonishing wonder he had witnessed on the screen that day—a cattle stampede! I listened with a combination of awe and depression, certain that I had missed a once-in-a-lifetime opportunity. Little did I realize that I would see several dozen stampedes before half my life was spent.

Perhaps I should add that in maturity my brother and I became very close friends, occasionally enjoying not only the movies but many other activities together.

My family had acquired the usual hand-cranked Victrola record player before I was born. By the time I was old enough to manipulate it we had perhaps one hundred or more discs of various kinds. Some of the older ones were actually recorded only on one side. About half were arias from operas sung by the stars of the day, from Caruso and McCormack to Farrer and Schumann-Heink. My mother enjoyed light operas too, so there were records of excerpts from Victor Herbert and various Viennese light operas. We also had recordings of popular patriotic tunes like the "Star Spangled Banner" and "Stars and Stripes Forever," the latter as played by Sousa's band.

Included in our record collection were about a dozen so-called comic ethnic records such as "Hilda from Holland," "Affairs at Pumpkin Hollow," "The Italian Fruit Vendor," "Cohen on the Telephone," "Casey at the Dentist," and "Scenes at a Hungarian Restaurant." These would of course be taboo today. It is interesting that such records were generally acceptable when we looked upon our country as a melting pot, but would not be now when we emphasize ethnic diversity and ethnic groups frequently advertise their ethnic humor. I assume at least one Ph.D. thesis could be written on this profound topic.

For most of the thirty-odd years he was in business, my father did not take a vacation beyond an occasional day in the country during the week. He had a dozen or so employees and a number of daily schedules to maintain. My mother, however, took us children on a vacation for two or three weeks each summer. When I was a very young child we usually went to stay at a boarding-style ranch in the Russian River area north of San Francisco, near Guerneville. My brother and I revelled in these journeys. I can still call up a poignant memory of the odors of the dry summer fields and of the luxurious green growth along the river.

Following the influenza epidemic of 1918–19, which generated a combination of fear and death in the population, my mother and I went for several weeks to a resort in Santa Cruz along the coast south of San Francisco, spending several hours a day on the beach when it was sunny. My mother and I had a great deal of fun together, particularly on vacations, since we could do trivial things we both enjoyed without causing my father to look askance at our frivolous waste of time. I recall one vacation when we managed to attend the movies every night for a full week!

During the great influenza epidemic, my father displayed his perception and wisdom in an interesting way. He noted on observing our neighbors that, when taken ill, many seemed to recover and become ambulatory within a week or so. If they promptly resumed a full schedule of normal activities, however, they were apt to be struck down, often fatally, with a second severe illness. When, therefore, members of our family became ill, my father insisted that we remain at home and at rest for two to three weeks *after* the initial illness had passed. Richard Shope later demonstrated at the Rockefeller Institute that the first illness was the result of the viral infection responsible for the epidemic, whereas the second, severe phase, was the result of bacterial pneumonia for which the victims of the virus were particularly susceptible. My father's perspicacity may well have saved our lives. The grade schools were closed for a number of weeks during the epidemic. When we returned, a half-dozen desks were vacant, including that of a close friend, Arni.

Later, in the 1920s, my mother changed our vacation venue; we spent several weeks each summer at Camp Curry in Yosemite valley, with excursions to such places as Glacier Point and Wawona. It was truly a paradise then, with a small population and a very relaxed atmosphere. My mother trusted me sufficiently to allow me to roam freely both on the valley floor and on the numerous hiking trails, which led to such places as Nevada Falls and Sentinel Dome. I sigh when I see what has happened to that magnificent valley since, but it was unavoidable with the growth of population and the ease of travel.

My mother had a remarkable, and perhaps understandable, reflexive reaction toward the Chinese community in San Francisco which was originally established during the Gold Rush that started in 1849. A large number of Chinese immigrants were engaged in essential, productive work, building the great railroads eastward until they were completed in the early 1870s. The roads across the Sierras undoubtedly would not have been finished as rapidly as they were had it not been for the extraordinary labors of these people. Once that work was over, however, the people faced sudden and almost universal unemployment, and many entered into illicit activities in a wholesale way. Competing gangs or tongs were formed which were frequently at war with one another.

This turmoil diminished considerably, or at least became more regulated, after the Great Earthquake and Fire of 1906, when the entire downtown community had to join forces in order to rebuild the city. Thereafter Chinatown became more normally commercial. In any event, the area was regarded as a dangerous part of the city in the 1880s and 1890s when my mother was growing up. As a consequence of her old fears, she was denied the pleasures of a Chinese dinner until 1939, when my wife and I took her to Mr. King's Shanghai Gardens Restaurant in Philadelphia. Her approval was evident.

Chapter Two
Elementary School

In September of 1917, not long after my sixth birthday, my mother took me to Monroe elementary school a few blocks from home. I had been to a preschool, then called a kindergarten, but it involved no more than pastime activities. That decade was characterized by a great expansion of the population of children in San Francisco and a number of other cities, the result of the great wave of immigration from Europe that had taken place in the decades just before World War I. Both parents of at least half the children in the school were foreign-born, Italy or Ireland being perhaps the most heavily represented countries. Of the remaining children, most, like me, had one foreign-born parent.

Wooden shacks had to be built hastily to take care of the school's overflow of baby boomers, and my first two years were spent in such buildings while a second permanent building was being erected nearby. Fortunately, the mild San Francisco climate made heating the sheds unnecessary. We were dressed for the weather, wearing woolen underwear in winter.

We were perhaps fifty in a class and discipline was strictly maintained while we were being drilled in the three Rs. Each morning we lined up in the schoolyard to salute the flag, recite the pledge of allegiance, and receive a few disciplinary comments from the principal, Miss Agnes Haggerty, before marching in order to class. She was a no-nonsense woman, highly conscious of her responsibilities and, as I learned to my sorrow, inclined to regard an accused student as guilty until proven innocent—sometimes even to the point of disregarding the evidence. The teachers were highly competent and, on the whole, warm-heartedly sympathetic to the students, who responded in a disciplined and respectful way. Those teaching the upper grades had specialized in some field such as American history or English literature in college.

Taken as a whole, the classes were directed toward hard-core subjects, such as arithmetic, spelling, American history, geography, and literature. We diagrammed sentences, memorized some classical poems, and, in the upper grades, read several plays by Shakespeare including *The Merchant of Venice* and *Julius Caesar*. There were some extracurricular activities such as freehand drawing, woodworking, and singing. Handball was the principal school-time sport.

Freehand drawing in fact occupied a fair amount of time in the earlier years. I was once criticized by my school first-grade teacher for outlining

15

individual objects in still-life drawings with a thin dark line. Apparently I was quite sensitive at that stage of my visual development to the so-called Mach effect, whereby the optical system in the eye and brain gives special emphasis to sharp transitions in contrast — such as between objects in a still-life — a process that is apparently very important in giving clear visual definition to letters of print. We discussed my practice, to which she objected, and she of course ultimately won the argument by edict. Actually, I did achieve a certain vindication since the contrasting features in the drawing itself undoubtedly displayed the Mach effect to a degree even without my additions.

Our first-year teacher, Miss Raleigh, among our many other lessons, led us through the alphabet and into simple reading. It may be commonplace at that level of instruction, but she succeeded in impressing me in a very special way with an understanding of how each word should be looked upon as a pattern unto itself, so that one reads words as such, and not a string of letters. It seems many students at that level become lost in the thicket of letters in the alphabet, and the development of their reading skills is somewhat delayed as a result. With this start, I soon developed my own form of speed reading of light material, although I never used it when reading books I really loved, such as *Huckleberry Finn*. The process of reading quickly became a simple matter of word recognition rather than deciphering.

There was one aberration in the school system of the time, which presumably would not be found at present in a comparable school. Some students were regarded as "difficult," either because they were particularly slow to learn or because they were of an uncertain temperament; these worrisome cases were apt to be herded together in one class assigned to a special teacher. Once, when our regular teacher was ill and a substitute was not available on short notice, the members of my class were distributed to other classrooms and for several days I found myself a silent and amazed witness to the style in the "special" classroom. Sitting at the very back of the room, I watched as the special teacher managed to maintain what appeared to be a perpetual state of rage from one end of the day to the other, and kept her regular students cowering at their desks. Should one become obstinate or otherwise objectionable in her eyes, she would not hesitate to pummel the offender while delivering intense invective. Since she sustained this style day in and day out, she must have had some deep-seated source of frustration that, among other consequences, provided her with a virtually inexhaustible supply of nervous energy. I shrank down at my desk with the others, trying to remain as inconspicuous as possible, and was hugely relieved when my own class reassembled.

At appropriate times during the school year, an entire class would be taken on a special outing to Golden Gate Park or to hear the San Francisco Symphony Orchestra, then under the direction of Alfred Hertz, a remarkably talented musician. In retrospect, I marvel at the ability of the teachers involved to maintain a reasonable degree of coordination within the group as we climbed on and off streetcars and wove excitedly through the crowds.

There are several events which stand out sharply in my memories of school days. Fairly early in the year my third-grade teacher, a Miss Mahr, noticed that I became listless and bored during the endless repetition of basic subjects — such as spelling and multiplication — that was needed by some of the slower students. The teachers, of course, were duty-bound to pass students on to the next grade only if they were reasonably well-qualified but it was clear to Miss Mahr that I found these drills trying. So one day, during one of the study periods devoted to review, she quietly escorted me to the back of the classroom, opened a book cabinet and handed me a book, saying, "Come here and take a book to read whenever you feel like it." The book I held was *The Swiss Family Robinson*, and it opened up a whole new world of imaginative adventure to me. It is not that we did not have a great deal of reading matter at home, but the choice available there was usually oriented toward the practical rather than the dashing. For example, we had the Encyclopedia Americana as well as a simpler version written for young children, and at Christmas time or as a birthday present, I was apt to receive a book with a title such as "The Boy Naturalist in the Azores," which I would read dutifully but without great zeal. Miss Mahr's books, on the other hand, were thrilling and I rapidly became an avid reader. Soon after, I met another boy, Emile Freeman Ford, who was a fellow bookworm and with whom I traded exciting new discoveries. While my parents were somewhat dubious about the direction I was taking, they had the good judgment to let me find my own ways.

Most of our teachers, while working in the Mission District, lived nearer to the center of the city where they could more easily avail themselves of cultural activities and live more private lives. Some, however, such as Miss Mahr, were part of the local community. My fifth-grade teacher, Miss Clyde, was active in the local Catholic Church, and she would frequently entertain a neighborhood audience during a church social evening. She generally performed a monologue in which she would play one or more roles with a dramatic flair. I was always welcome to attend these events with my Catholic friends and never failed to do so, particularly if ice cream and cake were part of the fare.

When I was in the fifth grade, the members of my class were one day shepherded into a large room where we could be well separated, and were given a version of the Terman aptitude test, newly devised by the psychologist Professor Lewis M. Terman at Stanford University. Shortly thereafter my father received a letter from Stanford asking if my parents would be willing to take me into the city to be registered as part of a group. My father wisely decided to do nothing. He told me some time later that he had felt that the last thing I needed at that stage of development was to be pigeonholed into any special category — life itself would do the sorting soon enough.

Luis Alvarez, the brilliantly creative experimental physicist—one of the greatest of this century—was in the San Francisco public school system at the same time as I, but apparently missed being tested, perhaps because he was absent from school on the appropriate day. Years later he asked me whether I had ever taken the Terman test, as though he may have lost out on some kind of opportunity, and seemed to be bothered by the fact that he had missed it. I told him my father's view of the matter and reassured him that life, on its own, had done a fairly impressive job of sorting him out.

For our weekly singing class, in the fifth or sixth year, we were provided with songbooks furnished by the city. Browsing through one of them, I saw that certain pages had been systematically cut out of it. There was, however, an index which allowed one to discover just what had been excised. The offending numbers turned out to be German folk songs, some by Beethoven, Mendelssohn, Schubert, and Schumann, which had been removed at the height of patriotic fervor in World War I. One had escaped, however—Schubert's "The Linden Tree" in English translation. Apparently the song was a favorite of our teacher, Miss van Arsdale, perhaps because the level of dissonance produced by the students while singing it was somewhat less than that inflicted on other pieces. While I enjoyed singing this piece, my real favorite was the more boisterous "Men of Harlech!" I had little idea then what a Saxon was, and even less of an inkling that by stretching things I might be considered one.

Looking back, I know that I owe a great debt to all those dedicated teachers. In addition to Miss Mahr, mentioned earlier, I gratefully remember Miss Means (Figure 2.1), who carried us through the final two elementary grades—the seventh and eighth. She was graceful, intelligent, well-informed, and had had an excellent education herself. If she were alive today, she would be an asset to any of the best high schools or junior colleges. Above all, while a respected disciplinarian, she was prepared to credit us with as high a degree of maturity as we deserved. She eased our transition between childhood and the greater responsibilities that were to come with high school. To show our appreciation and admiration, a group of the boys in the class spent overtime in the school wood shop making her a well-varnished cedar chest which we delivered to her downtown apartment as a surprise in a borrowed delivery truck. She seemed deeply pleased. I have often wondered since how she managed with the bulky thing.

My father's cousin, Fred Baust, had settled as a pharmacist in Boise, Idaho, when he arrived from Europe soon after 1900. Early in the 1920s, he moved with his family to California, and they stayed with us in San Francisco for six weeks or so while finalizing their plans. The two children, Walter and Irma, were born in the United States and were, respectively, about the same age as my brother and I. They impressed upon me once again the rapidity and completeness with which the children of immigrants placed at the American frontier could become an intimate part of the local American scene.

Figure 2.1. *My graduating grammar school class of December 1924. Our much-admired teacher Miss Means is on the right. I am kneeling, essentially alone to the left of center in the second row. My good friend Alfons Castelli is off to the right alone in the same row.*

These young people spoke with an authentic frontier twang and had little knowledge of or interest in — or so it seemed to me — in their European roots. The family eventually decided to establish themselves in Los Angeles; unfortunately, I lost track of them when my educational pursuits took me to the East Coast.

In my later years of elementary school, about 1922 or 1923, my friends and I began to be fascinated by the technological wonders of radio broadcasting, as stations like KPO and KGO came on the air. As the fever grew in me, I began stringing up antennas at an alarming rate, and acquiring as many crystal sets and earphones as I could get my hands on. Soon, a high school graduate in the neighborhood, George Munk, began assembling vacuum tube sets at his home and offering them for sale. His home and workshop became a mecca for a small group of enthusiasts. Along the way, I developed a new close friendship with a boy named Alfons Castelli (Figure 2.1), who shared my enthusiasm for this new form of technology. It is interesting to note in passing that a simple triode vacuum tube retailed at that time for five dollars, equivalent to about one hundred dollars in the 1990s. My father was highly sympathetic to the interest and funded me with reasonable generosity.

Our family acquired its first real radio from George Munk in 1925. It was a clumsy but effective set with a Brandes loudspeaker and three variable condensers with large dials that had to be manipulated independently to tune in to any one of the local stations. All stations in the San Francisco time zone remained silent between 7:30 and 8:00 pm during part of that decade in order to allow devotees to see if they could tune in stations in another time zone. It was always a great thrill when we managed to pick up faint voices from Chicago or Pittsburgh amid the static noise (Figure 2.2).

By the end of the decade, every middle class home in the district had a radio complete with a loudspeaker sitting prominently displayed in its living room. During the period when radio programs were becoming universally popular, a private foundation provided funds that made it possible for the Coolidge String Quartet to give radio concerts around the country. The quartet spent several winter periods in San Francisco, broadcasting on one or another of the area's three stations. Although I had had some previous exposure to chamber music, the radio concerts provided a much richer and more complete experience. I became, and remained, deeply devoted to the art form. I believe that many composers use this elegant medium to reveal the innermost regions of their psyches, just as great painters often use charcoal drawings and etchings as a form of artistic liberation. Its comparative simplicity I find profoundly moving.

The publications of Hugo Gernsbach and his colleagues, who were based in New York City, had names like *The Radio Experimenter*, *Science and Inven-*

Figure 2.2. *The future Professor H. Richard Crane at the console of his radio broadcasting system. The photograph probably dates from the early 1920s, making him a pioneer among the amateur radio hams. (AIP Emilio Segrè Visual Archives.)*

tion, and *Amazing Stories*, and were required reading for all radio buffs. These tomes did much to add spice to this already exciting period of rising amateur interest in science and technology through radio. In retrospect, it is clear that the publications contained their share of nonsense, but they satisfied an important need at the time and were read by a remarkably wide audience of teenagers, many of whom — perhaps despite those readings! — later became prominent scientists and engineers. Later in my life I met several colleagues in the National Academy of Sciences, such as Lloyd Berkner, who had been Gernsbach addicts.

Another source of excitement during this period was the so-called Scopes Trial which concerned the right of a secondary-school teacher to teach Darwin's principles of evolution. The trial took place in Tennessee but of course the issue was of intense interest to the whole country, polarizing communities everywhere. Our local Hearst papers were full of reports of the trial, but, in this case at least, managed to avoid actually taking sides. Having done a reasonable amount of reading, I was a committed evolutionist, and had little difficulty in stirring up arguments among my friends. At grammar school the issue was brought dramatically into focus by our geography teacher, Miss O'Flaherty. She was a tall, stately woman, undoubtedly with ancestral ties to Brian Boru, and the evolution question gave her not a moment's pause. She informed us quite unequivocally: "Mr. Darwin is English and therefore clearly descended from monkeys. That is not true of any of *my* ancestors."

During Lent, just as baseball season was getting underway, my Catholic friends were always torn between the temptation to play baseball and their obligation to study the catechism in the church. The lure of the game was generally strong enough for them to risk imperiling their souls — but there were other, more immediate dangers. The best ball grounds were at Balboa Park, a fifteen- or twenty-minute walk from the neighborhood. Unfortunately, the route to the park passed right by the church where the priests would often lie in wait for their young, truant parishioners during this season. One of my duties was to go ahead of the group past the church to make sure the coast was clear. On the few occasions when a priest did confront me, he retreated hastily before my declaration of faith and the Catholic boys skipped past while his back was turned. The bonds of friendship in our group were such that they were never marred by religious disputes. My friend Silvio Delvecchio did once comment to me, in an almost matter-of-fact way: "They tell us in church that you guys worship the devil." I let it pass, deciding not to make an issue of the sale of indulgences or Martin Luther's famous edicts for reform. Baseball came first.

Chapter Three
High School

In the 1920s there were three excellent public high schools in San Francisco with good pre-college programs: Lowell, Polytechnic, and Galileo High. There were also good public high schools in Berkeley, Redwood City, and Palo Alto. Those in San Francisco were all located in residential districts of the city that were inconveniently far removed from my home. There was a public high school in a neighboring district that I could have reached with a ten-minute street-car ride, but it did not have a comparable academic reputation.

In addition to the public schools there were three well-known private high schools. One, Sacred Heart, catered mainly to parochial school students. Another, Cogswell High, featured a utilitarian education and a modest pre-college program. The third private high school was Lick-Wilmerding, then in the Potrero District which had been a mixed residential and industrial area before the 1906 earthquake, and was about a half-hour away by street-car.

Long before leaving elementary school, I had decided that Lick-Wilmerding was my choice. For one thing, my hero, George Munk, the expert radio amateur, had gone there and my brother had been a student at Lick, as it was known in short, for several years, so I had a number of his old yearbooks which I pored through with a sense of enchantment.

The funds for the creation of the original school had been provided in 1875 by the great San Francisco philanthropist James Lick (1796–1876), a Pennsylvanian by birth, who had been an apprentice in piano factories in Baltimore and New York and who eventually opened several similar factories in South America (Figure 3.1). There he learned, with the aid of armed henchmen, how to retain his holdings amid revolutions and other violent upheavals. He was located in Lima, Peru in 1848 when the United States acquired California, and he promptly liquidated all his holdings in South America for thirty thousand dollars in gold—a sum equivalent to more than two million dollars in 1990. Lick moved to San Francisco with his fortune, and he also persuaded a good friend of his in Lima, who was in the chocolate business and whose name was Domingo Ghirardelli, to move at the same time. Neither had cause to regret his decision.

On arriving in San Francisco, Lick bought up as much available land, both in the heart of the new community and elsewhere, as his resources

Figure 3.1. James Lick, the wealthiest man in California in his day and a remarkable philanthropist. Joseph Henry persuaded him to create the Lick Observatory. Lick is buried under the great telescope. In a biography of him, his great grandniece, Rosemary Lick, referred to him as a "generous miser." (Courtesy of Lick-Wilmerding High School.)

permitted. The Gold Rush began a year later and he was soon the wealthiest man in the city, as well as one of its most prominent citizens. In his later years, he established a number of philanthropies, most of which still endure. Among his legacies is the Lick Observatory on Mount Hamilton east of San Jose. Like many individuals he was fascinated by the heavens and was persuaded to invest in a great telescope by Joseph Henry, the President of the National Academy of Sciences and Secretary (director) of the Smithsonian Institution when the latter visited California in the 1870s. Lick is buried under the large telescope. The private high school that bears his name legally carries the formidable title of The California School of Mechanical Arts, indicative of Lick's interest in supporting apprenticeships. The school opened its doors in 1894 under the leadership of one George A. Merrill, then twenty eight years old—and about whom more is told below.

Two other private schools were founded along more modest lines by other San Francisco philanthropists. These schools, Wilmerding High and Lux High, a girls school, became closely linked with Lick High to provide economy of operation. Lick and Wilmerding High schools essentially fused in the process. Lux remained independent, but girls and boys shared upper, college-oriented classes. Formally, Wilmerding had as its official title: The Wilmerding School of Industrial Arts. The Wilmerding and Lux families, although also pioneers, earned their fortunes in less dramatic ways. They were, however, no less public spirited.

The girls at Lux came from middle-class families, as did the boys. The girls dressed in the normal conservative dresses of the time; the boys wore suits and neckties. Our dates for class dances and picnics, such as they were, were almost exclusively with girls from Lux. Had I stayed in the Bay Area, the chances were great that I would have ended up marrying a classmate.

The original Lick-Wilmerding High School buildings, which remained in use until after World War II, had been built by the students under the supervision of the shop faculty while classes were held in temporary quarters. They were sturdily utilitarian, and not architectural masterpieces.

At Lick a student pursued one of two types of program: the first was for college-bound students, and the second was for those desiring practical training for a trade. Vocational courses included automobile repair, sheet-metal working, mechanical drawing, commercial art, and cabinet making, or one could train to be a chemical technician.

During the first two years, all students followed the same course, taking classes five days a week in basic algebra, elementary chemistry, physics, and English literature. In addition, we all took both mechanical and freehand drawing. A typical shop program during this period placed emphasis on masonry, tinsmithing, and woodworking of various kinds.

Those of us going on to college found ourselves, during our second two years of high school, in classes with girls from Lux High. The program was a continuation of the basic science and English literature classes with the addition of plain and solid geometry and trigonometry. The students who specialized in the shops were not required to take other courses. Our physics program took us through electromagnetism while our chemistry courses focussed on inorganic chemistry, with the major emphasis being placed on electrochemistry, the periodic chart, and mineralogy. I took a shop course in automobile repair in my third year, in order to gain practical knowledge of the art. The only language available at the time was a two-year course in the basics of Latin, taught fairly perfunctorily five days a week. It was required for entrance to the University of California at Berkeley. The only real lack in the program was that we had no courses in history, but I compensated for this with a great deal of outside reading in the field — and studying history for pleasure became a lifelong addiction.

While a program of this type was acceptable to West Coast colleges and universities, it would have been regarded as inadequate at one of the elite eastern private universities. I learned only later, after being accepted as a graduate student in physics at Princeton, that the fact that I had attended a high school where shop was practiced and where Latin was taught for a mere two years had counted strongly against me. I believe that the continual downgrading of the status of hands-on technology at those institutions, with the admitted exception of computer use and programming, may provide additional signs of a form of national decay. Along the way the requirements in foreign languages have also diminished in many schools. This is clearly a national disaster for which we will pay a price at both practical and cultural levels.

My experience in the school's auto repair shop taught me the satisfaction of technical competence, and brought some bonus rewards as well. One of my classmates, Gene Mires, had somewhere acquired a 1923 Buick touring car. We worked on it lovingly in the shop until it was in excellent condition, and in it, a group of us went camping in Yosemite valley for several weeks one summer. At that time there was an almost unlimited number of designated places in the valley where you could park your car, pitch a tent, and lead a pleasant, uncrowded existence for as long as you wished — or at least until school started.

George A. Merrill, the young Director mentioned earlier, had been an engineering major as an undergraduate at the University of California at Berkeley (Figures 3.2 and 3.3). His primary interest though was in physics, and he had in fact written the physics text that we used in high school. The physics instructor during those years, Ralph H. Britton (Figures 3.4, 3.5, and 3.6), had a similar background, and he was a special source of inspiration to me. He was wonderfully well-informed on the many aspects of physical science, and he delivered his information in a terse, intellectually satisfying style that I have rarely since seen surpassed. I corresponded on and off with Mr. Britton until his death at the age of 96. Both he and Mr. Merrill had the innate ability to have become prominent professional physicists in today's environment, although I believe Mr. Britton was much the more imaginative of the two. Mr. Merrill, who lived in Redwood City and commuted to school by train, somehow found the time and energy to serve as Mayor of that city during my days in high school. Mr. Britton eventually succeeded him as Director for a period and oversaw the move of the institution to another part of the city.

The chemistry instructor was Mr. Sidney A. Tibbets, who was not only a

Figure 3.2. George Merrill in his early days as head of the high school. (Courtesy of Lick-Wilmerding High School.)

Figure 3.3. George Merrill standing by the workbench that James Lick brought with him from South America. (Courtesy of Lick-Wilmerding High School.)

Figure 3.4. My much admired physics teacher, Ralph Britton. (Courtesy of Lick-Wilmerding High School.)

Figure 3.5. *Ralph Britton fifty years later at the time of my class reunion in 1978.*

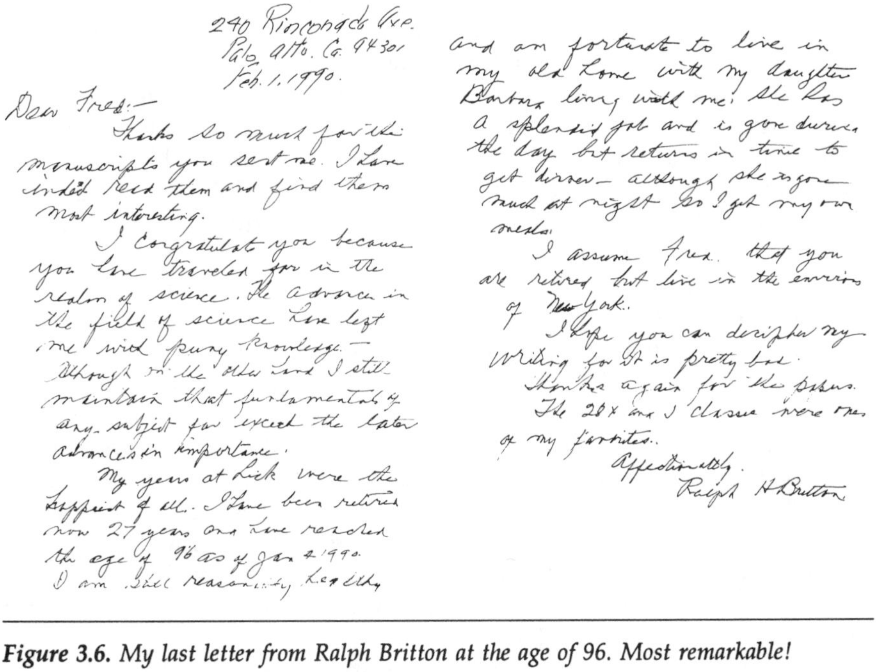

[Handwritten letter]

290 Rincanada Ave.
Palo Alto. Ca. 94301
Feb. 1, 1990.

Dear Fred:—

Thanks so much for the manuscripts you sent me. I have indeed read them and find them most interesting.

I congratulate you because you have traveled far in the realm of science. The advance in the field of science have left me with puny knowledge.— Although in the older land I still maintain that fundamentals of any subject far exceed the later advances in importance.

My years at Lick were the happiest of all. I have been retired now 27 years and have reached the age of 96 as of Jan 2 1990. I am still reasonably healthy

and am fortunate to live in my old home with my daughter Barbara living with me. She has a splendid job and is gone during the day but returns in time to get dinner — although she is gone much at night so I get my own meals.

I assume Fred that you are retired but live in the environs of New York.

I hope you can decipher my writing for it is pretty bad.

Thanks again for the papers. The 28× and I classes were one of my favorites.

Affectionately.
Ralph H Britton

Figure 3.6. *My last letter from Ralph Britton at the age of 96. Most remarkable!*

good teacher but a consultant as well. Mr. Tibbets ran a thriving business in which he involved his senior students destined to be chemical technicians, giving them invaluable practical experience. His style of teaching was more dogmatic and impassioned than Mr. Britton's, but thoroughly admirable. I incurred his wrath one day when I arrived early and was caught fiddling idly with the equipment on the lecture table. He reached for a meter stick with the intention of batting me on the arm and I compounded my offense by jerking my arm away so that he scattered his demonstration over half the room. It took a great many correct answers to the questions on his periodic examinations before I could even partially redeem myself.

The senior mathematics instructor, E. R. Booker, clearly had the ambition, as well as the ability, to become a college teacher, and took summer courses at Stanford to enhance his qualifications. Several years later when I was a student at Stanford myself, we found one another in the same summer class in integral calculus, and I had the remarkable pleasure of doing homework with a former teacher.

George Merrill was a stern disciplinarian, as I learned once to my cost when he, for his own good reasons, singled me out for disciplinary action. Our class, of which I was President, had caused a rumpus during a study period when the instructor was absent. As the head of the class, I was held responsible even though I was in no sense the originator of the disorder and could have done little to quell it. Such measures caused my father what I regarded as unnecessary distress. By 1990s standards, our excesses were no more than playful exuberance, but Mr. Merrill held us to Victorian standards of conduct. Having been deeply involved in the founding of the institution, he felt a great sense of responsibility for what went on in it. Moreover, it must be admitted that the various shops around the school contained equipment that was inherently dangerous so that a high degree of discipline had to be maintained. Doubtless Mr. Merrill's overall concerns about student behavior in all other situations were strongly affected by his anxieties regarding these potential hazards.

In any case, one unfortunate result of the incident was that he apparently became reluctant to recommend me for admission to Stanford University, in spite of the fact that my grades were acceptable. Fortunately for me, I did well on the upgraded Terman aptitude test required for admission to Stanford and I had additional help from the District Attorney of San Francisco, Matthew Brady, a close friend of the family, who wrote a strong letter of support. Only one other of my classmates, Gene Mires, of auto shop fame, entered Stanford. We were close friends up until college, but Gene was intent on a commercial career and our interests quickly diverged. Some years later my father apologized to me over the anger he displayed toward me as a result of Mr. Merrill's actions. He had decided that the man had reacted too severely to a minor incident.

At the suggestion of my cousin, Henry Bruhns, who lived in a different part of the city, I joined the downtown YMCA soon after entering high school to take part in the Saturday morning sports activities there. In this way, I gained a whole new spectrum of friends my own age from the city at large. Most of them went to other high schools, and we shared with each other a considerable amount of cross-cultural information. In several cases these friendships were greatly strengthened by mutual attendance at the YMCA summer camp at Pinecrest in the Sierra Nevada mountains, near Lake Strawberry on the road to Sonora Pass. One of the boys with whom I formed a lifetime bond was John Purcell. He was a year or so younger than I but physically mature for his age. As a result of our friendship John entered Lick High a couple of years behind me.

Although John's family name was of Irish origin, his parents were actually Hungarians from Budapest. Apparently the name went back to an Irish adventurer who had been an officer in the Hungarian army several generations before. John also followed me at Stanford, majoring in civil engineering. We remained close friends. On graduating, he joined the Caltex Oil Company and became involved in worldwide oil exploration. Our paths crossed again late in life when he had retired to take up residence in Cape Town, South Africa.

One of the great pleasures of the period for me and my classmates and our other friends was a weekend hiking expedition to Muir Woods or Mt. Tamalpais in Marin County. Armed with sleeping bags and knapsacks we first took the ferry boat to Sausalito — an artist's paradise at the time — boarded a Northern Pacific electric train to Mill Valley, or to a similar point such as Manzanita, where the hiking could begin. Actually, there was a very crooked railroad line to the top of Mt. Tamalpais but that was only for children and the elderly. I do not recall any of my Italian or Irish neighborhood friends being involved in these expeditions. Apparently it was something that went back into the older traditions of San Francisco.

One of the dramatic events which occurred soon after our graduation was the murder, in broad daylight, of our talented English literature teacher, Clara Boeke, by a mentally disturbed man whom she had tried to befriend. He was apparently homicidally inclined since he stabbed her while strolling in a park. The tabloids made much of it. Students past and present were genuinely grieved.

My greatest claim to distinction in high school was my position as co-editor of the class yearbook, a job I shared with Ruth Carney, an attractive, highly intelligent classmate. Here too, however, I managed to run afoul of the authorities. The printer, through an error of his own, had an under-run in the number of volumes produced and had to set up the plates for it twice. The mistake was laid upon my shoulders instead of the printer's and I found myself once again the cause of a great deal of consternation in Mr. Merrill's office because extra charges were levied. My stock in Mr. Merrill's office dropped even lower in spite of my remonstrations of innocence.

Figure 3.7. My graduation picture at Lick High in 1928.

Lick High School was relocated from the Potrero District after World War II with the increased industrialization of the area, and it currently occupies a site less than a mile away from my original home. It is now truly coeducational and retains an outstanding reputation in the Bay Area. In the meantime Lux closed its doors. Lick-Wilmerding alumni are well-represented in admissions to the University of California at Berkeley—a great mark of distinction at present.

My own class, which graduated in December of 1928 (Figure 3.7), in keeping with a now abandoned tradition of having two admissions and graduations during the school year, held a fiftieth reunion in San Francisco in 1978 with about half of the graduates attending. Inasmuch as a great depression and a great war, among other incidents, had occurred since many of us had last met, it would have been an excellent opportunity for us to reminisce and inquire about each other's various adventures, as well as to learn the fates of some of our absent classmates. Unfortunately, the high school alumni managers apparently thought it most important to have a jazz band dominate the event, so conversation was all but impossible. I fear this is an all-too-common occurrence on these occasions.

We had a final reunion fifteen years later to which all of the remaining 1928 graduates were invited, but only a tiny group was able to attend.

There are many individuals who, having lived reasonably happy and productive lives, would gladly contemplate reincarnation. I would consider it seriously only if I could somehow arrange to skip the high school years. Even though I enjoyed my friends and the class work at Lick High, I would not want to relive the emotional turmoil of adolescence. One is neither grown up nor still a carefree child. One has a great desire to take wing but realizes that one is really not prepared to do so. One has a sense of growing responsibility without knowing how to cope with it.

Chapter Four
Undergraduate Days

STANFORD

Having graduated from high school in December 1928, I entered college in January of 1929 at the awkward mid-year point. Most of my classmates had already formed bonds of friendship, and I found myself somewhat detached from the group. As a result, I never developed much class spirit and went my own way, knowing that sooner or later I would have to decide what lead to follow. From the start, I treasured the great freedom to find one's own way that college life offered. My two roommates at Encina Hall, Charles Coffen and Donald Slocum (Figure 4.1), were both from medical families in Oregon and were themselves headed toward medical school. Slocum, a capable individual, eventually gained national distinction as a knee surgeon. They were both being sought out by fraternities for their sophomore year, but I, the latecomer, enjoyed benign neglect without regret.

The President of Stanford University at that time, Ray Lyman Wilbur, was on leave in Washington when I arrived, serving as Secretary of the Interior in President Hoover's administration. The Acting President was Robert E. Swain, previously a member of the chemistry department faculty.

My father, ever the wise and practical person, knew from the start that I would have to find my own place in life and asked only that I spend some of my time deciding upon a career. Knowing of my interest in science, he suggested that I consider the field of biochemistry; with remarkable prescience, he felt that it was certain to become a major field. Inasmuch as Lick High had lacked a basic biology program, I welcomed the opportunity to become familiar with the field and spent much of my first year absorbed in class work and auxiliary reading of biological texts. The instructors were excellent, and the general study stood me in good stead in later years when I became closely associated with leaders in the field of cellular and molecular biology, even granting that great revolutionary advances had occurred in the meantime. One of the very fine young lecturers in general biology was Douglas Wittaker who eventually joined the staff of The Rockefeller University as a leading member of Detlev W. Bronk's support staff.

Regarding chemistry, I focussed most of my attention on physical chemistry although there was an excellent group at Stanford working on organic

Figure 4.1. The future surgeon, Donald Slocum (right), and me outside the freshman dormitory at Stanford (Encina Hall) in the spring of 1929.

chemistry under the leadership of Francis W. Bergstrom. One of the outstanding chemists in the department was Edward C. Franklin, who had done extensive research in the field of ammonia chemistry as an analog of aqueous chemistry.

In the course of class work, I met a particularly interesting student, Harlan Hess, who had been raised on a farm in Nebraska and had a profound knowledge of the interplay of living systems as they can be observed in the everyday natural world. I have been reminded of him often in reading the remarkable essays of Loren Eiseley who also grew up on a midwestern farm and had a keen eye for the natural world about him. Harlan and I made many excursions in the open areas of the Stanford estate as well as along the seacoast and I learned much from him. He had the ambition to become a college biology teacher. The world depression which set in during the autumn of 1929 placed impediments in his path and I never learned of his fate since we eventually became separated by the width of the continent and different areas of interest.

In fact, the advent of the depression cast a great pall over all of campus life. By the autumn of 1930 many students were compelled to leave for lack of funds and the jubilation that usually accompanied the year's big Stanford-Berkeley football game had all but vanished. Tragedy also struck several of the fine Italian families in my neighborhood in San Francisco; many had gambled heavily on shares in the Bank of America, previously known as the Bank of Italy, and were wiped out by the crash. Fortunately for my family, my father had cleared himself of all debt several decades earlier, thanks to his prosperous business, and had never borrowed money for his mining investments, but had used available cash. As a result, I enjoyed a sheltered position during the Depression that was shared by very few.

To broaden my knowledge of the earth, I took an introductory course in geology with Elliot Blackwelder, a leading geologist of the day. In the lab we found inspiration in a framed, topographic drawing bearing the signature "Bert Hoover" — a product of the future president's student years.

It is remarkable to recall the strange lengths the geologists of that day would go to explain mountain-building in the absence of our contemporary knowledge of plate tectonics. I imagine that Blackwelder responded to the new knowledge rapidly if he retained the acuity he displayed in his lectures, although he was in his eighties at the time continental drift was verified. In contrast, a leading government geologist in Washington, whom I came to know well during my days as President of the National Academy of Sciences, was emphatically denying the existence of continental drift as late as four years after the crucial revelations by Harry Hess, Frederick Vine, and J. Tuzo Wilson.

In the spring of 1929, I formed what would be a lifelong friendship with a fellow student in a mathematics class named Alexander de Bretteville, Jr. (Figure 4.2). My talks with him helped me make many important decisions in my undergraduate years. Alex came from a distinguished San Francisco family which traced its name back to Norman nobility. One of his aunts had married the heir to the Spreckels sugar fortune, and his mother was a member of a prominent English family. Alex's younger brother, Charles, also a student at Stanford, eventually became President of the California Bank.

Figure 4.2. Alexander de Bretteville, Jr. during a trip to the Sierra Nevada Mountains in the early 1930s.

Alex had attended Galileo high school and was intensely interested in both physics and astronomy. We spent the summer of 1929 touring southern California and the Grand Canyon area in a car my father gave me, visiting whatever scientific institutions came within driving distance, including the Mount Wilson Observatory. Alex's purposefulness inspired me to focus much more closely on my own interests.

Stanford offered the student a far greater range of choices of classes than the eastern universities with which I later became familiar; a feature that had both bad and good consequences. Taken as a whole, I might have benefitted had I been subject to more discipline. On the other hand, I had the opportunity to enjoy some remarkable humanities courses; I recall being enthralled by Professor Edward M. Hulme's lectures on Italian Renaissance history and art. He emphasized that any piece of great art must be viewed in a multi-dimensional context, not least as a major element of history. While the concept was not new with him, it helped nurture in me what was already beginning to be a lifelong interest in history.

Stanford had a fine gymnasium which I used a great deal during my early years. I took up boxing at one point, but perhaps fortunately have found little cause to employ such pugilistic arts in the intervening years. Later on, I began jogging through the Stanford hills—this sport, still in its infancy at that time, appealed to me because it required no fixed schedule. My friend Alex, who was a really fine athlete, assured me that I displayed no remarkable style while running.

I also belonged, with John Purcell, to an informal beach club near Mussel Rock on the coast just south of the border between San Francisco and San Mateo counties. Part of our pleasure herein was the half-hour walk to the club along the shore, which was then almost deserted. I understand that the area is now unsafe for strolling. The water was normally much too cold for swimming, but we tried.

John and I often went into the city on weekends, and spent Saturday evenings going to a movie and roaming through various parts of the city more or less at random late at night—a pursuit no longer considered wise. During these walks we sometimes passed the Bohemian Club on Taylor Street and did our best to peer through the front door, as the club had a considerable reputation as an organization devoted to hedonistic debauchery. Years later, when I became a member, I learned that it was actually composed of home-loving, middle-aged individuals devoted to their families, who enjoyed occasional convivial gatherings that emphasized the pleasures of talking and arguing, as well as participating in member-generated drama and music of remarkably high quality.

By the spring of 1930 it had become clear to me that my interest lay in physics and mathematics, and I began to cultivate the two departments in depth. Both had some remarkable faculty members. The Head of the Physics Department was David L. Webster (Figure 4.3), an x-ray physicist who had studied at Harvard and who reportedly had come west for the sake of

Figure 4.3. Professor David Locke Webster, the head of the Stanford physics department and a New Englander to the core. (AIP Gallery of Member Society Presidents.)

the health of one of his children. He was a large, friendly man, of an aristocratic New England family, somewhat shy and not given to small talk.

While a graduate student at Harvard, Webster had made one of the early independent measurements of Planck's constant by determining the x-ray frequency associated with the upper edge of the continuum spectrum as a function of voltage. He was also very much an expert in the field of classical electromagnetic theory, and was widely respected for this by the most advanced electrical engineers in the community. He became an expert on flight dynamics in World War I and helped develop a technique for pulling a stalled airplane out of a spin. We last met during World War II at Aberdeen Proving Ground in Maryland where he was studying rocket dynamics. I was always somewhat awed by him and regret that I never saw more of him in later life.

Another x-ray specialist in the department was Perley Ason Ross, who was celebrated for his development of a set of monochromatic filters for the x-ray spectrum, as well as for early confirmation of the details of the Compton effect. He and Mrs. Ross were fine hosts; those of us in the department enjoyed frequent informal evenings at their home. One of their daughters later married my friend William Hansen.

There was also a spectroscopist, George R. Harrison, on the senior staff. He ultimately left Stanford to head the Eastman Spectroscopic Laboratories at M.I.T. He gave a very dramatic series of lectures with excellent demonstration apparatus, and left behind a reasonably well-equipped spectroscopy laboratory with a large Rowland optical grating. De Bretteville and I spent many hours there testing the equipment in an amateur way.

I formed close friendships with two members of the graduate staff. One was William Webster Hansen (Figure 4.4) who later invented the klystron and was a major pioneer in the field of what were then regarded as microwaves. He was a burly Norseman with the strength of a giant who had grown up in Fresno, California. His remarkable abilities were discovered early by Webster who gave him a free rein in the department, which was run informally to say the least. Unfortunately, Hansen died prematurely at the age of forty after a brilliant but all-too-brief career, presumably as a result of a genetic defect leading to emphysema.

Hansen took considerable interest in my own development when he realized that I had a serious commitment to physics. He encouraged me to attend his lectures in quantum mechanics and electromagnetic theory and, in many other ways, played the role of a surrogate older brother. He had a profound influence on my attitude toward involvement in research. Through his indoctrination, I received an early exposure to wave mechanics at a reasonably sophisticated level.

Another close friend was John C. Clark (Figures 4.5 and 4.6) who had spent a period at the University of Chicago as a laboratory assistant to Albert A. Michelson, concerning whom he had some interesting tales including one memorable experience. During Clark's tenure, Michelson wrote a small book called "Experiments in Optics." Clark had worked hard to assemble the equipment needed for the photographs of diffraction patterns which

Figure 4.4. My mentor, William Webster Hansen, the inventor of the Klystron. (AIP Niels Bohr Library.)

Figure 4.5. (Left) Dr. John C. Clark in the early 1930s. The intensely serious look is quite out of character. (Courtesy of J. C. Clark.) Figure 4.6. (Right) Dr. John C. Clark at Lake George, New York, in 1989. (Courtesy of J. C. Clark.)

were featured in the book and, indeed, had taken and prepared the pictures himself. When the book appeared, Clark bought a copy and asked Michelson if he would be willing to autograph it—the master refused with a growl. Such are the ways of genius.

Others in the department included William R. Hewlett, usually to be found in the library, and the Varian brothers, Russell and Sigurd, who later established an important corporation at nearby Cupertino. Hewlett's future partner in business, David Packard, was a major football hero on campus.

The members of the Physics Department tended to be practical jokers with a heavy-handed style—David Webster occasionally participating. Professor Lewis M. Terman's celebrated Psychology Department was located on a mezzanine in the same building as physics and was one, but by no means the only, target of some of the practical jokes. One of my close friends, Bourne Eaton, the nephew of the founder of Forest Lawn Cemetery in Los Angeles, had been given a miniature Austin sedan as a present. A favorite diversion enjoyed by the staff was to hand-carry this miniature car up the stairs onto the mezzanine, without Eaton's knowledge, and leave it in front of Professor Terman's office door. My high school director, Mr. Merrill, should have been gratified to learn that I avoided joining in these adventures.

The Head of the Mathematics Department was a distinguished Danish mathematician, Hans Frederik Blichfeldt (Figure 4.7), a specialist in analysis and a kindly man. One of his brilliant colleagues was Professor V. Uspenski, a Russian refugee who was also a specialist in analysis as well as in classical probability theory. One of the more transient members was Harold Hotelling, a specialist in more modern mathematical statistics of the Fisher school, a field considered crude by some of the old-line mathematicians of the time. He transferred first to Columbia University and then to Duke, where I had the pleasure of meeting him years later in the 1960s.

The Mathematics Department was run somewhat more formally than the physics department, but still constituted what might be termed a close, happy family. I was given free access to the library and spent many hours in that quiet haven. As a matter of convenience and to maintain good relations with the departmental secretary, I majored in mathematics rather than physics, and Professor Blichfeldt gave a small luncheon party for me when I went off to graduate school. Unfortunately he became ill during World War II and died near the end of it so I never had an opportunity to pay my respects to him again.

Foreign scientists invited to tour the United States enjoyed visiting California and being part-time guests at the University. For example, H. A. Kramers, the brilliant Dutch physicist, gave a series of lectures on quantum mechanics in the spring of 1930. Everyone in the department turned out for the lectures, as much to meet the great man as to learn the newest developments in quantum physics.

Similarly, George de Hevesy gave several lectures on the techniques of radioactive dating with the use of the natural radioactive elements. At that

Figure 4.7. Professor Hans Frederik Blichfeldt, head of the Stanford Mathematics Department. (Courtesy of the National Academy of Sciences.)

time he appeared fiercely formidable to me. When I met him again years later, he turned out to be a genial, witty man.

Somewhat later Harald Bohr, the mathematician brother of Niels Bohr, was in residence for six months, lecturing among other things on his work on nearly periodic functions. To the mathematicians he was *The* Bohr. His classroom lectures were models of clarity and dramatic style. He employed at least three colors of chalk on the boards to stress various points — a practice that impressed me a great deal. The proof of Kurt Goedel's theorem on the limits to the use of the axioms in a given field of mathematics for proving theorems in that field was causing much excitement among mathematicians at this time, particularly those in Europe. Harald Bohr spent a good part of an hour illustrating the principles involved in the theorem by emphasizing that one had to go to a higher level of mathematics, such as to the field of calculus in the complex plane, to prove the fundamental theorem of algebra, that is, that an algebraic equation of the nth degree has n roots. The visits to the Bay Area of individuals of this level of distinction were usually noted in the San Francisco newspapers, the guests being referred to as "savants."

The Physics Department at Stanford usually had a guest lecturer in the summertime — in 1930 it was the twenty eight year old theoretical physicist, Edward U. Condon (Figure 4.8). He had completed graduate research earlier at Berkeley with Raymond T. Birge, a molecular spectroscopist; he had spent two years in Europe at Goettingen and Munich, served on the staff of the Bell Telephone Laboratories, and was now at Princeton University. He and Ronald W. Gurney had recently provided the explanation of radioactive alpha particle decay in terms of quantum mechanical tunnelling. He was brilliant and

Figure 4.8. Edward U. Condon as he appeared in the 1940s. I owe him a great debt on many scores. (AIP Gallery of Member Society Presidents.)

friendly, and possessed a wonderful gift of exposition, thanks in part to the fact that he had earned his way through college as a newspaper reporter. Having stiff, unruly hair, he favored crewcuts his entire life.

Condon and I became good friends when he realized that I had a serious interest in physics, and he suggested that I consider applying to Princeton for graduate study. Between his good word and the fact that the Head of the Stanford Physics Department, David Webster, was a close friend of his fellow physicist, Augustus Trowbridge, the Dean of Graduate Studies at Princeton, I eventually made the grade and was due to be admitted in January of 1932.

In mid-summer of 1930, during his lecture period at Stanford, Condon visited Berkeley to see his old friends on campus. On returning to Palo Alto he told us of meeting with Ernest O. Lawrence, a new member of the Berkeley physics faculty (Figure 6.2), who was developing what he called a resonance accelerator—the cyclotron. Condon spent a good portion of a classroom hour describing the theory as well as the potential and possible limitations of the device. A new era in nuclear physics was on its way. Condon also mentioned that a brilliant American theoretical physicist he had met in Goettingen, named J. Robert Oppenheimer (Figure 4.9), had just joined the Berkeley staff.

Figure 4.9. J. Robert Oppenheimer in the early 1930s. (AIP Niels Bohr Library, Physics Today Collection.)

CALTECH

During the summer of 1930 I came to feel that it would be to my advantage to transfer to the California Institute of Technology in Pasadena, which was clearly the best technical school in the west. I took an extensive admissions examination and was accepted for my junior year which began that autumn. My experience there was magnificent in all respects but one. At that time the vast majority of the undergraduate students were from the Los Angeles area, and with few exceptions left the campus at noon. There were no undergraduate dormitory facilities and the only on-campus dining service available to undergraduates was light lunch served in a primitive wooden shed. The facilities used by the graduate and postdoctoral staff, on the other hand, were excellent since the Athenaeum was available to them. Making the best of things, I found a comfortable attic room in a private home on Hudson Street near to the campus, and focussed all of my attention on class work.

Both the lectures and resources at Caltech were excellent. The lectures in mathematical physics were given by William V. Houston who had spent several postdoctorate years in Europe, first with Arnold Sommerfeld and then with Werner Heisenberg; in these connections he had contributed greatly to the understanding of electrical conductivity in metals. Houston later became the President of Rice University and was active in the work of the National Academy of Sciences.

I was befriended in a generous way by a professor of physical chemistry who specialized in electrochemistry, S. Jeffrey Bates. He appreciated my somewhat unusual situation as a transfer student and provided me relatively full run of the department. Among other things, he arranged for me to sit in on Linus Pauling's lecture-seminar on chemical bonding, which was establishing a major new tradition in aspects of theoretical chemistry. In reading a recent history of the Institute, I noted that Bates was one of the first members of the chemistry faculty.

As future events demonstrated, Pauling (Figure 4.10), then still in his twenties, was a truly remarkable chemist, possessing a deeply intuitive understanding of molecular and crystalline structures which he elucidated and amplified by available instrumentation, not least the methods of x-ray diffraction. An early association with newly developing wave mechanics provided him with a practical understanding of the way in which patterns of electron distributions between atoms that are favorable for bonding can be obtained by the superposition of monatomic patterns — when nature cooperates. This knowledge, guided by intuition, enabled him to develop, at a critical time, a new way of discussing the properties of chemical bonds. In the process, he provided succeeding generations of chemists with a useful framework, and its associated language, in which to deal with molecular structures — methods that could be made semi-quantitative with the use of empirical parameters.

Figure 4.10. Professor Linus Pauling in 1988, nearly sixty years after I first had the privilege of meeting him. (Courtesy of the National Academy of Sciences.)

The procedures used in deriving this formalism lack the type of precision traditionally followed in advances in the frontier fields of physics—but as Condon once observed, physicists have special freedoms in selecting the systems they study in order to simplify their problems, while chemists have a professional obligation to deal with what physicists would regard as hopelessly murky ones—or, as Condon put it: "They are willing to wade deep out into the muck." The framework created by Pauling, and expanded by others over the years, has proven remarkably useful, advancing chemistry in many new directions.

The more-or-less weekly physics colloquium was always a treat since the lecturer was usually a major figure in physics, astronomy, or cosmology. One highly memorable set of lectures was given by the colorful Dutch physicist Paul Ehrenfest, who had inspired several generations of continental physicists. His histrionic abilities, used to emphasize major points in his lectures, were most remarkable. I had the privilege of having lunch at his table on several occasions, in the wooden shed mentioned above, and appreciated the special magic that emanated from him. Tragically, he committed suicide a few years later, despondent over the fate of one of his children who was hopelessly ill.

Other notables at Caltech during this time included Paul S. Epstein, the leading theoretical physicist; Charles Lauritsen, who was developing a high voltage generator for nuclear and other studies, and Fritz Zwicky, who was in the midst of his investigations of novae and supernovae. Robert A. Millikan was a familiar figure as he strode about the campus attending to his various responsibilities—always dynamically impressive. In his many speeches of praise for the institution, he enjoyed declaiming that "the Massachusetts

Institute of Technology is the greatest Institute of its kind in the United States—*east of the Rocky Mountains!"*

Perhaps above all, Caltech had an esprit-de-corps which made one feel that there one was experiencing the wave of the future. The Depression which lay ahead, however, would prove very difficult for the school; it was only after World War II, when Lee A. DuBridge became President, that it resumed its upward surge begun under Millikan.

William S. Shockley was a fellow student of mine in several classes. We had met previously in Palo Alto since his widowed mother was a good friend of Professor Ross's family who had entertained us so often. We had a lifelong association which was professionally constructive in the early years, but which became difficult in the 1960s when Shockley abandoned his interest in the physical sciences and became deeply absorbed in controversial approaches to studies of the relative levels of intelligence of different ethnic groups.

STANFORD AGAIN

During the winter of 1930–31, I decided that it would be wisest to finish up my undergraduate work at Stanford, where I had accumulated sufficient credits to graduate by December of 1931, and move on to graduate study. The fact that student life at Stanford was much pleasanter undoubtedly played a significant role in my decision. Nevertheless, the time I spent at Caltech made an indelible impression on me, and I was always grateful later on for that enriching experience.

The summer lecturer at Stanford in 1931 was Professor Floyd K. Richtmyer of Cornell University, an x-ray physicist well-known to the group at Stanford. I was particularly anxious to meet him because he had written an inspiring book, *Introduction to Modern Physics*, which not only gave a very lucid account of recent developments in physics, but also contained an introductory chapter that reviewed the key historical developments in the history of physics since Newton's time. It provided a historical perspective that I very much appreciated at that stage of my development. Richtmyer was also editor at that time of the McGraw-Hill "Green" series in modern physics, and was largely responsible for giving the series the luster it enjoyed for several decades.

LEAVING

When I left California in January 1932 for graduate school in the East, my thoughts were so completely focussed on the immediate future that it never occurred to me in any significant way that I might be making a permanent move, and would never return to the state except as a visitor. Nevertheless,

I retained an underlying feeling of what might be called homesickness for a number of years. I was finally cured of it once we acquired a summer place at Lake George in the Adirondacks in 1937. Meanwhile, the state of California has undergone such profound changes that I am sure that moving back would now feel like moving to a foreign country.

When I left the state, the city of San Francisco was still a powerful constituency, but Los Angeles, following new attitudes and policies linked to the influx of dynamic Midwesterners and fewer controls from labor unions, was moving ahead as rapidly as the Great Depression permitted. Today the greater Los Angeles area, extending from Santa Barbara to San Diego, dominates both the politics and economics of the state, whereas the city of San Francisco seems to have followed a policy of minimal growth. In fact, the Bay Area would probably have been even more static were it not for the expansion of the electronics industry in the Peninsula — the region now called Silicon Valley. In my student days the area was known for its fruit farms, meadows, and bedroom communities.

In visiting Los Angeles in recent years, I have been amused to note that many people there unconsciously regard Yosemite Valley as lying in the southern part of the state even though it is directly east of foggy San Francisco, which is looked upon by the same people as a strange community well apart from the south. In my day the border between north and south was clearly understood to be the Tehachapi Mountain Range running east-west just north of Santa Barbara.

When my mother was born in 1883, the population of California was approximately one million. Since then, it essentially doubled every twenty years, a growth rate which was maintained until about 1960 when the population reached sixteen million. By the 1990s it had grown to nearly thirty million. My father had predicted in the 1920s that the state population could be expected to continue its growth until it reached approximately sixty million, at which time it would exhibit some of the types of population pressure and regulatory problems found in European countries. This to me remains a reasonably reliable estimate, granting that the time scale involved will be influenced by many factors.

As I contemplate the status of California at present, I must admit that it is very difficult to envision what it will be like when it finally achieves what may be termed mature growth in a half century or more. Most new arrivals from other parts of the country or abroad have the feeling that they are entering a brave new world where older traditions should be muted. California's agriculture, which was such an important basic resource after the Gold Rush, is transforming as the region begins to face substantial difficulties with its water supply. On top of this, the state has yet to feel the full effect of immigration from Mexico and Asia, although it is becoming increasingly conscious of both. While a large border state such as Texas faces some similar issues, that state has preserved some well-defined traditions, tied to the period when it gained independence from Mexico, while California has abandoned most of those which defined its pioneering character up to World War I, and which were, in the main, associated with developments in the northern area in the decades following the Gold Rush. One tradition which fortunately has not been lost is California's strong support of excellent educational facilities. If preserved, these, indeed, will be a source of strength in the next century. In opposition, however, is the very questionable influence of Hollywood which, to a significant degree, has flooded the minds of both young and old with superficial, if not downright dangerous, attitudes.

Alongside all of this, the ruthless exploitation of previously undisturbed terrain by the first post-war generation of real estate developers had the unfortunate consequence of generating an over-zealous backlash on the part of some environmentalists. Somehow a compromise between progressive development and the preservation of a natural environment must be found.

To gain control of its future, California will need to revive its dedication to work, both creative and routine, which was weakened in the relatively easy years between 1945 and 1965 when our country had little competition from abroad. It will also need renewed emphasis on the early years of education, re-instituting some forms of discipline which have been neglected along the way.

It would be my personal guess that economic problems, not least those arising from severe international competition and the way the state responds to it, will eventually determine the outcome.

Chapter Five

Princeton Years

While planning my departure for Princeton, I had a surfeit of advice from relatives and friends on the best route to take to New Jersey—most of it of little value. I finally decided to travel on the Santa Fe railroad because of the Harvey restaurant system, the train stopping at special stations during mealtimes so that the passengers could leave the cars and eat at well-prepared station restaurants. Later on, the Harvey system and the women who served the meals were the subjects of an overblown but colorful motion picture. One of my relatives dourly predicted that I probably would never return from the East alive. In defiance of this self-styled Cassandra, I started out for my new life in early January of 1932.

One of the most poignant moments in my life occurred as the ferry boat taking me to the train in Oakland pulled away from the San Francisco pier. My parents had come to see me off and here I took leave of them. They stood disconsolately on the pier watching the boat move off; all three of us knew that our relationship was about to be permanently changed. What had been to me a long twenty years of growing up had been to them an almost incredibly brief and busy time, doing their best with an active but good-natured child. They anxiously hoped that I knew what I was doing, but were glad that I had found a mission.

On my way east I stopped in Chicago for several days and was welcomed by the Gamma Alpha fraternity house of the university, the arrangements having been made by my Stanford friend John Clark. My guest status made it possible for me to have an interesting tour through Arthur H. Compton's laboratory and to see the planetarium at the lake shore, which was then a great novelty.

Arriving in New Jersey in January as an inexperienced Californian, I was subconsciously surprised to see so many "dead" woodlands. The awakening of nature that occurred, however, once the eastern spring arrived, is still treasured in my memory as an almost miraculous experience.

Princeton was a graduate student's paradise. Not only were the routine burdens of the students minimal, but there were fewer than 150 graduate students in the entire university, only about fifteen of whom were in physics. In 1932, the Physics Department was housed in a substantial, high-ceilinged building, Palmer Laboratory, which dated from the previous century; the Physics and Mathematics Departments were closely linked by a short bridge. Among the senior members of the Physics Department at that time were E. P. Adams, a classical physicist of the Cambridge University school; Allen G. Shenstone, a spectroscopist doing a careful investigation of the spectrum of copper; Henry D. Smyth (Figure 5.1) who became celebrated for his work with the Manhattan District, recorded in his postwar treatise known as the "Smyth Report;" Rudolf Ladenburg (Figure 5.2), a senior German physicist recently arrived in the United States; and Howard P. Robertson who, among his other accomplishments, was a brilliant theoretical cosmologist.

Ladenburg had arrived in the United States in 1931, abandoning a distinguished post in Berlin. He had apparently, like so many others, become concerned about the rising strength of the National Socialists, and the consequences for his family. Shenstone, who was a member of a prominent Canadian family, had served as an officer in the British Army in World War I. He and Ladenburg discovered that fate had placed them close to each other but on opposites sides of the same front during the latter part of the war. One of Ladenburg's remarkable achievements at Princeton was to excite a rare gas (neon) to a state in which most of the atoms were in an upper metastable level so that the so-called anomalous dispersion associated with the optical transition was reversed. Two decades later it would be recognized that he had, in the process, developed the basis for laser excitation.

Robertson (Figure 9.13), then in his early thirties, had been a protégé of

Figure 5.1. (Left) Henry DeWolf Smyth, head of the Princeton Physics Department. (AIP Emilio Segrè Visual Archives.) Figure 5.2. (Right) Professor Rudolf Ladenburg. (AIP Emilio Segrè Visual Archives. Photograph by Alan W. Richards.)

Richard Tolman at the California Institute of Technology. His contributions to relativistic cosmology are legends in the lore of the field. He was then very much a purist in the field of mathematical physics, and emphasized mathematical elegance in everything he did. At that time, he was translating into English Hermann Weyl's book *Group Theory and Quantum Mechanics*—a relatively exotic work with a special appeal to mathematicians. Wigner's book on the same subject, covering his path-breaking work in the field and to be discussed later, addressed itself much more directly to the everyday working problems of physics. Robertson's lectures on general relativity became a benchmark of elegance and clarity and were well attended by both theoretical physicists and mathematicians. The comparatively gritty theoretical work that was the basis of ongoing experiments held little appeal for him at this stage of his career. Fate would decree, however, that, like it or not, he would be plunged into a welter of applied problems when, in World War II, he became a key member of the group of American operation analysts in England. On the surface he seemed the hail-fellow-well-met type, but during my own wartime associations with him, I found that he was actually highly sensitive and introspective.

Also on the faculty were Louis A. Turner (Figure 5.3), an excellent lecturer with a fine analytical mind; Gaylord P. Harnwell (Figure 8.1) who eventually went to the University of Pennsylvania; and Walker Bleakney (Figure 5.4) who spent his entire professional career at Princeton. Bleakney later did outstanding research in the field of ballistics and shock wave generation in World War II. Turner, who followed the advancing frontier of atomic and nuclear physics quite closely, predicted immediately after the

Figure 5.3. Professor Louis A. Turner, a remarkably gifted physicist. (Argonne National Laboratory and the AIP Emilio Segrè Visual Archives.)

Figure 5.4. Professor Walker Bleakney, a young member of the department when I arrived. (Ulli Steltzer and the AIP Emilio Segrè Visual Archives.)

announcement of nuclear fission in 1939 that the elements higher in the periodic series produced by the addition of neutrons to uranium of mass 238 would probably be fissionable and could be separated from uranium chemically; subsequent measurements demonstrated just that. He became a division director at the radiation laboratory at M.I.T. during World War II. Years later, while I was at the University of Illinois, Turner became Head of the Physics Division at Argonne National Laboratory near Chicago. Unfortunately it was a time when relations between the laboratory and the state universities were somewhat strained, for reasons unrelated to Turner; as a result, he did not have a happy time there.

Turner, in my opinion, had the potential to be a much more celebrated scientist than he ultimately was. It is possible that the Princeton community did not provide him the right environment at an early stage of his career. Segments of the faculty, to which Turner and his wife belonged, led a complex social life which could be distracting professionally.

The Princeton Mathematics Department had an equally interesting faculty. Luther P. Eisenhart and Oswald Veblen were the senior members. The former became the Dean of Graduate Studies, replacing Trowbridge, soon after my arrival. The latter played a major role in the development of the Institute for Advanced Study, which was initially founded as a haven for European refugee scientists.

The math department was housed in a new building that had only opened the previous year (Figure 5.5). It had been built with funds provided by the family of the mathematician, H. B. Fine. It was an architectural gem, with richly carved wood panelling and stained glass windows depicting famous

Figure 5.5. *The entrance to what was Fine Hall in my day. It has been renamed Jones Hall. The photograph, taken in the summer of 1992, shows my granddaughter, Jennifer, and me. To me, the building is full of ghosts.*

equations from both mathematics and physics. Whenever I entered this precious building, I had the feeling of walking into a sanctuary or shrine. It suffered greatly during World War II from over-use and the introduction of crude temporary partitions to create extra "private" desk space, although it was partially restored subsequently. During my years at Princeton, various offices in the building were occupied by such notables as Albert Einstein, John von Neumann, Eugene P. Wigner, H. P. Robertson, Oswald Veblen, Luther Eisenhart, and Edward Condon. To me the building remains full of the ghosts of these geniuses.

Most of the mathematics students from my perspective, at least, seemed to make up a rather insular group, with its own special concerns and pleasures, but one or two mixed more freely with the physics community. One of these was Abraham H. Taub, who had a variety of interests, including general relativity and cosmology. He was also close to von Neumann. Our paths were to cross frequently in the future, both during World War II and then again at the University of Illinois.

During my time there, Princeton had very strong programs in biology, geology, astronomy, and chemistry as well as in physics and math. The school had been systematically cultivating the sciences ever since the appointment of Joseph Henry, the most distinguished American physicist of his time, in 1832. The career of this great scientist is traced out in detail in the *Papers of Joseph Henry* being published volume by volume under the auspices of the Smithsonian Institution. The program was started with the leadership of the historian Nathaniel Reingold and has been continued by Marc Rothenberg.

One of my fellow physics graduate students was Joseph O. Hirschfelder, who had joined the department from Yale the previous autumn. Soon after I arrived, he took me in tow to help me orient myself and avoid the more obvious pitfalls. He was a generous-hearted, imaginative person with a lively sense of humor. His father was on the medical faculty of the University of Minnesota. Our paths crossed frequently in later life. I also became friends with three chemists, all of whom had been at the University of California and had homes in the state. We sat together frequently in the graduate college at meal times.[1] One of them, Albert Sherman, was a National Research Council Fellow working on the theory of chemical rates. The other two, Arthur A. Frost and Everett Gorin, were both, like me, in their first year of graduate study. Another comrade of mine was one of the biology students, Herbert Shapiro, who was working with the Department Head, Professor E. Newton Harvey, on the physical properties of cell membranes as determined by the amount of cell deformation generated during centrifuging. But my closest friend during my graduate years was John P. Blewett (Figure 5.6), a physics student who had been at the University of Toronto. He subsequently joined the General Electric Research Laboratory and remained there until Brookhaven Laboratory was created at the end of World War II.

Edwin M. McMillan, later the co-discoverer of plutonium with Glenn T. Seaborg, was the senior graduate student in physics that year. He went on to join E. O. Lawrence at Berkeley. He shared a Nobel Prize in 1951 for his work with plutonium early in World War II and became director of the laboratory in 1971.

I became thoroughly familiar with the Chemistry Department through my friendships with the graduate students there. The program was, in the main, devoted to physical chemistry and was headed by Professor Hugh S. Taylor, an English physical chemist. One of the stars on the faculty was Henry Eyring, a friendly, prolifically creative scientist who had worked on the theory of reaction rates with Eugene Wigner and Michael Polanyi in Berlin. Eyring not only generalized this early work into a well-developed formal theory, but demonstrated its application to many important chemi-

---

[1] During vacations and other periods when the graduate student dining hall was shut down, I loved to feast at Griggs Imperial Restaurant in the heart of a substantial Afro-American community which then occupied the section of Witherspoon Street in Princeton, just north of Nassau Street—a welcome change from the routine college fare. I still treasure memories of breakfast involving delicious hot apple and custard pies. Legend had it that the core of the population consisted of the descendents of individuals who had been brought to Princeton as man-servants by students from the southern states prior to the Civil War and were manumitted when the student graduated.

Figure 5.6. John P. Blewett, one of my best friends during graduate years. (AIP Niels Bohr Library.)

cal problems. A member of a distinguished Mormon family, he eventually accepted a position at the University of Utah, as he had a special sense of commitment to the community there. His departure was greatly regretted by those he left behind. In fact, Hugh Taylor, a devout Roman Catholic, did his best to persuade Henry Eyring to stay. After an extended private meeting with Eyring he reported, "I failed completely to persuade him, but he nearly converted me to Mormonism."

In the autumn of 1933, I was granted a special fellowship, one of the privileges of which was living accommodations in one wing of a duplex apartment at the Graduate College (Figure 5.7). The other wing was shared by a chemistry student, John Turkevich, who would spend the remainder of his career at Princeton. His father had been the bishop of the Orthodox Russian church in New York City, and was abruptly stranded there in 1917 by the Communist Revolution.

John was not only a remarkably good chemist but, naturally enough, kept himself well-informed on affairs in the Soviet Union during his career. Once, during the Khrushchev era, he headed up a scientific delegation associated with an exhibit of American products our country was showing in Moscow. Khrushchev appeared, and when he discovered that Turkevich spoke Russian, he informed John that this exhibit meant nothing, and that in ten years the Soviet Union would bury the United States in consumer technology. Turkevich offered a good-humored rebuttal, which led to a more or less humorous bantering debate between the two that was aired on television in the United States. As it happened, Turkevich's perceptions of the Soviet Union were remarkably prescient. When I saw him after Mikhail Gorbachev had

Figure 5.7. The hall where we dined in splendor and in ragged academic gowns. (Courtesy of Princeton University.)

been in office for about a year, he told me quite earnestly that the communist system was going to crash and that the leadership would have to find an entirely new route to follow. His comments were hard to believe at the time but, knowing him, I was inclined to trust his judgment.

John von Neumann (Figure 5.8) and Eugene Wigner (Figure 5.9, also Figure 9.10), both in their late twenties, were, in 1931–32, in their second year of sharing a professorship at Princeton.[2] They spent the autumn semester at Princeton and the remainder of the year in Europe at Goettingen. Since in those days the autumn semester ran until nearly the end of January, I had the pleasure, upon arriving at Princeton, of catching the last part of Wigner's lectures that term on elementary quantum mechanics.

Wigner and von Neumann were clearly amused by some of the intellectual posturing which was then common in the Princeton community both on and off campus. Such behavior later led Albert Einstein to comment, "Some of the people in this community gain stature by walking on stilts." The exchange of knowing winks between Wigner and von Neumann on some occasions did not escape the notice of the students.

In addition to (indeed, because of) its resident population of great scien-

[2] Biographies of von Neumann and Wigner appeared in 1992: Norman Macrae, *John von Neumann* (Pantheon Press, New York, 1992); Andrew Szanton, *The Recollections of Eugene P. Wigner* (Plenum Press, New York, 1992).

Figure 5.8. (Left) John von Neumann, the universal genius. (Courtesy of his daughter, Professor Marina von Neumann Whitman.) Figure 5.9. (Right) Eugene Paul Wigner to whom I owe much. (AIP Niels Bohr Library, Physics Today Collection.)

tists, Princeton attracted many distinguished visitors at that time. From abroad came luminaries like R. H. Fowler, Paul A. M. Dirac (Figure 5.10), Erwin Schroedinger, and Abbé le Maître, a Belgian Cosmologist. Clarence M. Zener (Figure 5.11), with whom I became close friends, was a visiting National Research Fellow working prematurely to try to solve the mysteries of superconductivity.

All the theoretical students at Princeton idolized Dirac. Our admiration was not only for the brilliance that he demonstrated in developing a linear first order relativistic equation for the electron and his role in the discovery of what are called Fermi-Dirac statistics, but also for his crystal-clear and highly formal book on quantum mechanics. So devoted were we students that we became like the rabbinical scholars who could tell you exactly at what part of what page in the sacred text a given phrase occurred.

We were also amazed at Dirac's style of teaching, for he had another singular talent. He lectured without any notes, and yet one could have taken down his words and printed them verbatim; his deliveries were always letter-perfect. The most remarkable display of this ability that I witnessed occurred many years later, in 1983, at the summer school at Erice at the western end of Sicily. The Soviet government had allowed Peter Kapitza to attend the meeting and Dirac, who had known and admired the great Russian scientist from their days at Cambridge, delivered a forty-minute enco-

Figure 5.10. *(Left) Paul A. M. Dirac as he was during the Princeton years. (AIP Emilio Segrè Visual Archives, E. Scott Barr Collection.)* *Figure 5.11.* *(Right) Clarence M. Zener. (AIP Niels Bohr Library,* Physics Today *Collection.)*

mium to his renowned colleague. Again, he spoke from memory, without notes. It could be transcribed without any significant alteration.

Once a month or so on a Saturday, I. I. Rabi (Figures 5.12 and 12.7) would bring his team from Columbia University to spend the day with Condon going over the latest results of their work with molecular beams. He was often accompanied by his remarkable and devoted wife Helen, who came to enjoy the campus and visit with friends. Condon was one of the few individuals on the East Coast at that time who had a full working knowledge of quantum mechanics of the type that was immediately useful to Rabi. As you can imagine, I was thrilled to be able to be a silent observer of this profound and exciting work.

Emilie Condon, who had grown up at Klamath Falls near the California-Oregon border, met her husband on the Berkeley campus. She was as warm-hearted and friendly as he was, and a generous hostess to the students. The three Condon children, who led very productive lives, grew up in an active intellectual atmosphere, as both parents were well read and enjoyed vigorous debates on a variety of topics.

Mrs. Condon's parents were immigrants from Czechoslovakia, and she naturally shared their deep concern about the frightening events that were taking place in their home country as World War II approached. She tolerated me in spite of my background, but, like many others, was not immune

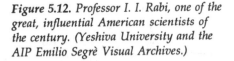
Figure 5.12. Professor I. I. Rabi, one of the great, influential American scientists of the century. (Yeshiva University and the AIP Emilio Segrè Visual Archives.)

to conflicting feelings. She once declared with great earnestness, "I do not at all understand ethnic prejudice. The only people I hate are the Germans and the Hungarians."

When I arrived at Princeton early in 1932, Condon was working with George H. Shortley, an advanced student, on their classical treatise, "The Theory of Atomic Spectra." I helped check some of the text and made a few minor suggestions based on my reading of Wigner's newly published book on group theory. However, in discussing my own future with Condon, he expressed the view that until some new experimental technology came along to open up new vistas, the opportunities for exciting new discoveries in atomic and molecular spectroscopy were probably exhausted, and it might be a good time to look to other areas of research. In particular, he suggested looking for ways in which one could use quantum mechanics to explore the properties of crystalline solids while trying to be quantitative if possible. During the spring, we made what might be called a symbolic start together by examining the possibility that one might observe double refraction of light in cubic crystals at sufficiently short wavelengths. Mostly, however, I spent a number of those months exploring the relevant literature.

Apart from the classical work on the crystalline symmetry groups and on the macroscopic physical properties of single crystals, including much that dated well back into the nineteenth century, some very stimulating news was emerging just then from abroad. P. Drude's original free electron theory of metals, formulated at the turn of the century using Boltzmann statistics for the hypothetical electron gas, had generated more in the way of paradoxes than enlightenment. Now, the advent of wave mechanics and quantum statistics began clearing the way for resolution of some of Drude's mysteries, particularly those related to electronic specific heats and the variations in electric and magnetic properties. Arnold Sommerfeld and Werner Heisenberg, working with young colleagues such as Rudolf Peierls, Hans Bethe, Nathaniel Frank, Felix Bloch, and William Houston, opened up the field in amazing ways. In addition, A.H. Wilson, R. Peierls, and Leon Brillouin wrote enlightening general accounts of this work which were particularly accessible to the student. Rudolf Hilsch of Pohl's laboratory in Goettingen wrote a good review of the work on color centers there, and J. H. Van Vleck's book on electric and magnetic susceptibilities appeared at this time. The famous Bethe-Sommerfeld treatise on the theory of metals was still to come; nevertheless, there was a great deal to absorb. I forged on, absorbed in literature that was mostly in French or German.

It was during the spring of 1932 that James Chadwick finally discovered the neutron for which he had searched in vain just after the end of World War I and had missed by a fluke. Until that discovery, the interior of the nucleus had been a mysterious place. Suddenly a momentous shift in the course of physics was underway. The field of nuclear physics received new attention throughout the world and quickly became the central topic of advanced research in physics and chemistry.

When the autumn of 1932 rolled around, Condon and Shortley were so immersed in the final stages of their book that Condon generously determined that it would be more profitable for me to work with Eugene Wigner, and broached the matter with him when he returned from Europe for the autumn semester. In a later interview, Wigner comments that European students of that period were far better prepared than those in the United States. I fear I provided him with a prime example of American backwardness, although he seemed pleased that I had absorbed his book on group theory, and that I was thoroughly infected with what was then called the *Gruppenpest*—or "disease" related to using the mathematics of group theory in physics. This form of expertise eventually made it possible for me to develop the irreducible representations of the crystalline space groups. General recognition of the importance of the methods of group theory spread very slowly among theoretical physicists at that time.

In any event, Wigner accepted me as his first American graduate student and suggested that we see if we could develop some realistic wave functions for a simple crystalline solid.

It should be emphasized that Wigner was following many pursuits at that time, including a growing interest in nuclear physics. In fact, it was during that same autumn that he wrote the paper on nuclear binding energies that gave rise to the term "Wigner forces." He saw clearly that the large difference in the binding energy of the deuteron and the alpha particle implied that nuclear forces were of short range.

Working with him was, of course, one of the most remarkable experiences of my life. At that time, he was far from world famous and was, on the whole, probably much better known to the physical chemists than to the general physics community because of his work on crystal structures and his use of quantum mechanics to provide a basic understanding of the factors which determine chemical reaction rates. At his father's behest, he had gone dutifully from Budapest to Berlin to study chemical engineering, but he spent most of his very considerable spare intellectual energy absorbing nutrients from the rich atmosphere of theoretical physics to be found in Berlin.

Wigner had a great natural gift for mathematics, and had become close friends with John von Neumann while a teenager in Budapest, as a result of their mutual love of the field. The friendship remained steadfast until the latter's death in 1957, and each benefitted from it in his own way. Actually, the two men had very different personalities. Von Neumann enjoyed fast automobiles and noisy intellectual cocktail parties although he drank little himself. Wigner was far more reserved and preferred smaller discussion groups. The papers he wrote on his own were often highly individualistic, and used sophisticated mathematics. They were not readily comprehensible at the time to the average physicist, unlike, for example, the papers of Enrico Fermi, who prided himself on his gift for using simple mathematics to achieve results. It was only later that it became clear to the entire physics community that Wigner was one of the great pathfinders of our century and that the mathematics he used was essential for future developments in physics.

I do not believe that at this stage of their careers either von Neumann or Wigner anticipated that the United States would become their permanent homes so soon. Wigner did not adapt easily to a strange environment, particularly one as strange as the United States must have seemed to him at that time, since he was so deeply imbued with central European culture. He and I spent much of our spare time discussing what he referred to as the "customs of your country." I am not certain how clear an understanding of Americana I succeeded in imparting to him, but I probably developed a trace of a Hungarian accent in the process. Both von Neumann and Wigner soon became devoted citizens and probably appreciated the virtues of our country even more than most of us who grew up here do.

Wigner's original guiding thought for our work was based on the supposition that the wave functions of the electrons in occupied states in the solid should be much smoother than those in the free atom because of the relaxation of boundary conditions and that the main source of cohesion for non-ionic solids such as metals might well lie in a net lowering of kinetic energy. We searched for a suitable candidate to provide a test of this concept.

A literature search revealed that a Russian physicist, W. Prokofiew, had developed a core potential field for the sodium atom that generated valence electron levels that were good to about a percent. We decided to focus on metallic sodium. In brief, we spent the autumn combining and modifying the atomic wave functions of the free atom going with the Prokofiew field in order to see if we could justify and indeed quantify to a degree the lowering of the kinetic energy. Mounds of paper accumulated on my desk but nothing quantitatively interesting emerged. By the beginning of December we had begun to wonder if we really had a good problem.

The campus was virtually deserted over the Christmas holidays and I had the physics department to myself. I had no relatives in the East and my home in San Francisco was a long way off in the days before airplane travel, so I stayed on quietly in Princeton. As I ruminated on the work of the autumn, it occurred to me that we had been approaching the problem from the wrong end. What was important for our particular inquiry was the Prokofiew field itself and not the wave functions of the free atom. After a happy week or so of integration by the method of finite differences and the use of a Monroe calculator, the cellular method of deriving solid state wave functions was born, and when Wigner returned at the end of the holidays we carried on with increasing excitement. Our key paper was published the following May.

While the ground state wave function found in this way was not bad and demonstrated that both the kinetic and potential energies of the valence electron were lower than for the free atom, the full derivation of a sensible value of the cohesive energy of the metal required a treatment of the electron correlation energy, and Wigner provided this in one of his numerous seminal papers during the following spring and summer. We were, incidentally, fortunate in our choice of sodium as a model since the effective electron mass is close to unity for the occupied levels. Even more important than our immediate results was the doorway that our discovery had opened for quantitative expansion of the field.

In view of the levels of precision which have been achieved by others over the intervening years with the use of sophisticated pseudo-potentials, clever expansion methods, modern computers, and much ingenuity, I fully appreciate the fact that our original work now seems antediluvian. That, however, is the way of science.

When Wigner was packing his bags to return to Germany in early February of 1933, the news broke that President von Hindenburg had appointed Adolf Hitler the Chancellor in Germany. I happened to be with him when he heard the news. He sighed and said, "Ah, Weh!" He returned to Budapest instead.

The number of visitors from abroad swelled to flood-like levels after Hitler came to power and began doing his best to destroy Germany's cultural life. The creation of the Institute for Advanced Study and the arrival in Princeton of Albert Einstein in 1934 were additional positive attractions.

Einstein (Figure 5.13) soon became a familiar figure about the campus and in town, usually dressed quite informally, a fashion statement well ahead

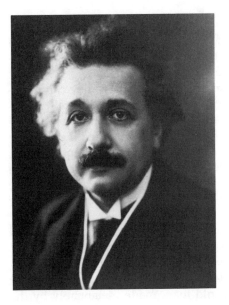

Figure 5.13. My favorite photograph of Albert Einstein taken in his early forties at the peak of his career. (AIP Niels Bohr Library.)

of its time. To many in the scientific community, and to visitors, he was a delightful curiosity—a legendary symbol of brilliant science. To those such as Wigner, who had known him in Berlin, he was a combination of a beloved elder statesman of science and an inspiring model. Even though by this time Einstein's greatest years of creativity were behind him, one still listened with a sense of awe to the accounts of his critical contributions to the development of physics. He was, finally, a scientist's scientist, whose true worth could best be appreciated by those professionals to whom he had provided leadership at periods of confusion and doubt.

During my student period, Einstein had two great scientific interests: he was attempting to find a unified basis for all the fundamental concepts of physics as they were understood then—a goal still beyond our grasp—and he was looking for fundamental inconsistencies in wave mechanics as it had been formulated in the second half of the 1920s, with so many apparently inborn paradoxes. While thus engaged, he frequently carried on spirited duels with his friend and esteemed colleague, Niels Bohr, attempting to dislodge the accepted foundations of the structure. We are indebted to Einstein for this, as his skepticism led to many critical reviews of those foundations. So far, however, testing has only served to place the theory on ever-firmer ground, in spite of what are sometimes termed its "spooky" predictions—predictions which are less spooky to those who grew up with the field. The fact that the electron combines the properties of both a wave and particle leads to results that seem at variance with everyday "common sense."

As students we were somewhat puzzled by Einstein's attacks on wave mechanics, without losing a trace of our respect and admiration for him. For we knew he had spent his younger years revelling in the production of paradoxical structures to solve some of the great earlier riddles. Apparently we all have our limits, even the Einsteins among us.

Rabi once said that most good scientists are fortunate if they can maintain their highest quality of productivity for fifteen years. Einstein maintained his for a quarter century. He made his last truly memorable contribution in the year 1924–25 in the field of quantum statistics dealing with the distribution of energy among a group of particles in thermal equilibrium. He was inspired by the work of an Indian physicist, S. N. Bose, and here again, Einstein pointed the way to a new understanding. Three years later, in 1928, he was very seriously ill and upon recovering assumed a rather different role.

Einstein, of course, was always a popular subject for photographers. His classic portrait, seen frequently at present—long, untrimmed white hair, drooping bushy mustache, mournful expression—has come to be the symbol of science; the universally recognized icon of intellect. But in this most familiar picture, Einstein is seen as an old man, made melancholy by the follies and cruelties of humanity. To a scientist, however, the really stirring Einstein pictures are those taken during his most creative days, before 1925, when the special inner light of his genius could be clearly seen in his every feature.

Another great scientist who joined the faculty of the Institute for Advanced Study at that time was the German mathematician, Hermann Weyl (Figure 5.14). Having a Jewish wife and no admiration for Hitler, he left his

Figure 5.14. Herman Weyl during his later Princeton days. (Courtesy of Martha Weyl Kenworthy.)

position in Goettingen as soon as he feasibly could. Von Neumann had spent a critical period with him in Zurich as a student. The story goes that it was Weyl who had convinced von Neumann's parents that it was more important for John to follow his career as a mathematician than for him to join the family bank in Budapest, even though he was the eldest son. Weyl appeared to be somewhat disdainful of American scholarship when he first came to the United States, but he mellowed considerably with the passage of time. Some fifteen years after his arrival, he spent a number of weeks in Pittsburgh lecturing at the Carnegie Institute of Technology where I was at the time. He proved to be a wonderfully affable and friendly guest.

Weyl's son Joachim (Figure 5.15), known to his many friends as "Jo," followed in his father's footsteps, acquiring a doctorate in mathematics at Princeton in 1939. He spent a number of years as a civilian research mathematician in the navy, starting in World War II and rising to the rank of chief scientist before returning to academic work in the 1960s. He and his wife Martha (popularly known as Sonja) were always the center of a large circle of admiring friends and colleagues.

In retrospect, one of the major events of my graduate school years was the advent of John Bardeen[3] (Figure 5.16) during the winter semester in

Figure 5.15. (Left) Herman Weyl, right, with his son, Joachim and daughter-in-law, Martha (Sonja). (Courtesy of Martha Weyl Kenworthy.) Figure 5.16. (Right) John Bardeen as I first knew him. (Courtesy of Mrs. Bardeen.)

[3] Much biographical material for John Bardeen is available in *Physics Today*, Volume 45, No. 4 (April 1992), organized with the substantial cooperation of David Pines.

early 1933. He was to be the most illustrious of our group of students. As I recall, he enrolled in the physics department as a self-supporting student at start. On the day of his arrival, one of the senior faculty members who had interviewed him introduced us and asked if I would take him to the graduate college for lunch. John had just resigned from a highly promising career at Gulf Research Laboratories in Pittsburgh where he had spent three years working on theoretical aspects of techniques for oil prospecting. He mentioned off-handedly a special method he had developed for using magnetic field measurements to determine underlying geological structures—an invention whose details I learned later were held rigidly as a company secret. His apparently phlegmatic, or matter-of-fact, demeanor masked one of the most powerful and determined analytical minds of our generation. A fellow graduate student who had joined us for lunch, and who was having considerable trouble with his exams, took me aside later and asked if someone should warn the seemingly naive Bardeen of the difficulties that graduate work at Princeton posed. My friend need not have worried.

Bardeen's knowledge and wisdom quickly won him the respect and admiration of everyone who came in contact with him. He doled out his talents like precious nuggets in a seemingly parsimonious way, characteristic of his manner. He had many gifts, including a willingness to tackle complex problems with persistence and patience. When most of us were still attempting to understand the intrinsic properties of perfect or nearly perfect solids, he turned his attention with considerable success to the behavior in the transition layer between the surface of a metal and the vacuum—work which served him very well later, when he and Walter H. Brattain worked on the point-contact transistor. He recognized the importance of solving the mystery of superconductivity early on, and worked tenaciously on the problem for twenty years until he achieved success. His approach to problem-solving did not have the elegance of mathematical style which characterized Eugene Wigner's work, but it was highly effective. In many ways, his method of selecting and attacking problems resembled that of Irving Langmuir in a different field and another time.

There were aspects of Bardeen that I did not appreciate until many years later. For example, he enjoyed golf, which is unusual among academic physicists, but I assumed he liked the exercise and the pleasure of the open air amid green surroundings. When eventually I learned that he had been quite a competitive athlete while growing up, it became clear that his passion for golf was simply a strong desire to get as low a score as possible. He did not stop at half-measures and strove for the hole-in-one every time.

In June of 1934 Bardeen and I drove to Chicago together in his car—we had both just passed our preliminary physics examinations for the physics department. We stopped on the way to spend several days relaxing in Pittsburgh. While working at the Gulf Research Laboratories, John had lived in one of the fraternity houses associated with what was then the Carnegie

Institute of Technology, and we were greeted there noisily and as honored guests. The night before we left, the fraternity threw a big party for John and, without too much encouragement, he soon became the life and soul of the boisterous gathering. That night I saw John in a rare departure from his usual sober demeanor, a treat that few of his later friends would experience in quite the way I did. I might add that I was put quietly to bed long before John was ready to quit.

A much greater boon came to John as a result of his connection with the Carnegie Institute of Technology. His charming, spirited future wife, Jane Maxwell (Figure 5.17), was a student there, and there it was that they met.

Once a start was made, several of Wigner's other students were attracted to associated problems in solid state physics; in addition to John Bardeen, were, most notably, Hillard B. Huntington, Conyers W. Herring, and Roman Smoluchowski, with consequences then and later that need no embellishment here. We maintained loose but lifelong connections.

As a result of the new approach to solids, two major groups at other universities moved into the field in the 1930s. The first was headed by John C. Slater (Figure 5.18), then a key figure at M.I.T. I like to remind others that in 1923 Slater, as a young postdoctoral investigator in Europe, was the first

Figure 5.17. (Left) Jane and John Bardeen at the time of their marriage in 1938. (Courtesy of Mrs. Bardeen.) Figure 5.18. (Right) John Clarke Slater as he appeared in the 1930s. (Massachusetts Institute of Technology and the AIP Emilio Segrè Visual Archives.)

individual to my knowledge to propose that the wave/particle dilemma might be resolved if the waves could, in some sense, be looked upon as statistical entities guiding the particles—a concept that was quickly seized upon by Bohr and Kramers just prior to the development of wave mechanics. The three published a paper with which Slater was not entirely pleased as his partners took control and introduced concepts with which he was not sympathetic. In the meantime he had become a leading figure in theoretical atomic and molecular physics.

Not only did Slater personally make several valuable contributions to solid state theory in the 1930s, but a number of his students carried out interesting band calculations. One of those students was William Shockley whom, as mentioned in Chapter 4, I had known since undergraduate days. Using an empty lattice model, Shockley demonstrated that the precise determination of band gaps for real crystals by means of the cellular method would require highly evolved techniques. He went on to join the Bell Telephone Laboratories where, along with Bardeen and Brattain, he played a crucial role in the invention of the transistor after World War II.

A group led by Nevill F. Mott (Figure 7.9), then at the University of Bristol, was also focussing on solids. Mott had previously gained fame for his work on what is called Mott scattering and, with Harold S. W. Massey, had written a book on the scattering of electrons by atoms. He quickly became a major force in the field of solids, publishing a number of highly illuminating papers and books, both alone and with colleagues. I have always felt that Mott deserved the Nobel Prize fully as much for his work prior to World War II as for what he achieved later. His work with Ronald W. Gurney in 1937 and 1938, on the explanation of the latent photographic image is, for example, a major classic. Mott was a frequent visitor to the United States during the second half of the 1930s.

One of the special benefits of being at Princeton, I quickly learned, was its close proximity to New York City and Philadelphia with their myriad delights. My first opportunity to enjoy the Metropolitan Opera occurred during those years, and I have particularly vivid memories of hearing Kirsten Flagstad and Lauritz Melchior in their prime. Meanwhile, in Pennsylvania, Leopold Stokowski was conductor at the Philadelphia Academy of Music, and I joined the throng of worshippers with fellow students for many Saturday night concerts.

I was also able to enjoy both exciting work and recreational pleasures when I was welcomed back to the Stanford campus during the three summers of my graduate years, and renewed some old friendships. A notable addition to the department in 1934 was Felix Bloch, the brilliant Swiss physicist who had spent several years in Leipzig with Werner Heisenberg before the disaster of Hitler struck. Alex de Bretteville, John Clark, and I took advantage of these reunions to go backpacking in the higher reaches of the Sierra Nevada mountains with a rented horse to help haul equipment. The mountain trails were almost empty in those years, even in summer, and we found we could easily go for an entire day or two without seeing another party.

In August of 1932 I was beginning to make plans to return to New Jersey from Stanford, when Mrs. Ross, the wife of Professor Ross of the physics department, brought me some news from her friend, Mrs. Shockley, in Hollywood. It seemed that Mrs. Shockley's son William had received an appointment at M.I.T. and was planning to drive east in his DeSoto convertible. Would I care to drive with him and share the costs? I eagerly agreed, and embarked on one of the most carefree two-week periods of my life.

Shockley, I quickly realized, was strongly influenced by the Hollywood culture of the day, fancying himself a cross between Douglas Fairbanks, Sr. and Bulldog Drummond, with perhaps a dash of Ronald Colman. He received the pronouncements of Hollywood stars on political, social, and economic issues with the same degree of seriousness that I would have taken those of Governor James Rolph or President Herbert Hoover. More disturbing was the fact that to enhance his assumed image, he kept a loaded pistol in the glove compartment of his car. I was handy with a rifle at that time, but still looked askance at travelling thousands of miles in the company of a loaded pistol. Shockley's swashbuckling air and his pistol eventually brought him to grief as he drove through Newark, New Jersey after dropping me off at Princeton. The Newark police pegged him as a suspicious character and upon discovering his pistol, asked him to come along quietly. He never gave me the details but he evidently had a rather difficult interview with a Newark judge who made imaginative use of the English language.

We chose the southern route cross-country through Arizona (Figures 5.19 and 5.20), New Mexico, Texas, and Arkansas, eventually reaching the Lincoln Highway in Ohio. We encountered torrential rains after leaving Carlsbad Caverns. The rains were so heavy that the highway was obliterated so we parked the car off the highway at a rise and spent the night wandering through the desert. This gave Shockley an opportunity to fire his pistol several times, his purpose being to ward off attack by a group of coyotes we could hear howling in the distance. It also caused a gas station attendant at a nearby store to excitedly inform us the next morning that the local police had been alerted that two desperadoes were loose in the area. He enlisted our aid in keeping a look-out for the outlaws.

Along the way we visited the Kentucky Caves, including the then-famous Floyd Collins Crystal Cave. Collins, the cave's discoverer, had been trapped and killed by a cave-in during a later exploration. Rescuers' vain attempts to reach him before he died were featured for days in the national press. His supposed body was on exhibit in a glass covered coffin at the Crystal Cave, although I must confess that to me the object looked surprisingly like a dummy from a cheap clothing store. Our guide, however, assured us of its authenticity, reverently declaring, "He was, as you can see, a very handsome man."

Figure 5.19. (Left) William Shockley quenching his thirst in the Arizona desert at a desperate moment in 1932. (AIP Emilio Segrè Visual Archives.) Figure 5.20. (Right) William Shockley in a more formal pose. (AIP Niels Bohr Library.)

We toyed with a plan to visit the long and justly famous Mammoth Caverns, which have been the object of much further exploration since, but were disuaded by the staff at the Collins Cave with the comment: "Don't waste your time boys. They've all been smoked up by torches!"

Our careers, not to mention our lives, were almost cut short in a harrowing incident that occurred as we were traversing the hills of Kentucky one early evening on a narrow two-lane, two-way road. I was driving and had a sheer drop-off on my right. As we rounded a curve we saw hurtling down toward us two trucks, clearly racing one another, taking up both lanes. Apparently chasing each other up and down the mountains was a popular sport among the spirited, young locals. By the grace of the Lord, I had just enough room on the shoulder to squeeze by the oncoming truck in our lane with perhaps an inch to spare. The trucks rattled on their way, leaving us behind, shaken but relieved. To the best of my knowledge I have never been closer to instant death than in those few seconds.

One might ask if any of the character traits for which Shockley would be both admired and denounced later in his life were evident at this early stage of his career. It was clear from the start, of course, that he was unusually intelligent. His later fame, and notoriety, grew in large measure from two dominant characteristics. First and foremost was his ability to pinpoint the core issues in a scientific problem and bring them to the surface in a dramatically clear way through either theoretical or experimental measures. In this accomplishment he was nearly the equal of Enrico Fermi, although working in a different area of physics. He had his most creative period at Bell Laboratories between 1940 and 1955. Having known him early on, I was never surprised at this aspect of his creativity.

Later he began studying differences in the characteristics of ethnic groups, but his work took a disturbing turn when he became convinced that there were measurable differences in intelligence between the races. While objective studies of physiological or other differences in such groups clearly have a place in science, it is quite a different matter to draw from that work generalized conclusions regarding the intelligence, temperament, or character of a group, and to use those conclusions as an argument for eugenics. Here, unfortunately, Shockley became mired in a morass of his own making because of his second characteristic: he was apparently unable to empathize with others and thereby understand that proposing eugenic measures in a diverse society is bound to be disruptive. Yet to the end of his life he advocated that such a course be followed. He was never able to admit that the methods of analysis which work so well in relatively clear-cut physical systems may yield highly controversial, or indeed completely false, results if applied fallaciously. Early on, I tried to reason with him, pointing out that at most one had strongly overlapping statistical distributions, but to no avail.

I saw an inkling of this second characteristic in the young Shockley, but did not take it seriously then. He was inclined to believe that society should be governed by a vaguely defined intellectually elite group, rather than by majority rule as in a democratic society. Unlike many other intellectuals, he was never led by this belief into any political system, either Marxist or Fascist. He was guided entirely by his own internal sense of logic—distorted though it was.

Early in the next year, 1933, William Hansen, my former mentor at Stanford, received an appointment at M.I.T. At the end of the spring term, Shockley wanted to drive home to Hollywood, Hansen wanted to attend an advanced summer physics lecture series at the University of Michigan, and I was returning to Stanford, so the three of us started west together. The term at Michigan had not yet begun when we arrived, but we were welcomed by Robert F. Bacher (Figure 5.21), who held a postdoctoral position and was widely known for the book he and Samuel A. Goudsmit (Figure 9.14) had published on atomic spectra. He took us in hand and gave us a fascinating tour of the department. At that time, he was in the midst of studying the hyperfine structure of atomic spectra derived from nuclear magnetic moments.

Bacher was destined to play several major roles in American science. Five years after our visit he was at Cornell University with Hans Bethe; together they prepared a series of excellent overviews of the status of nuclear physics which contained much original material. In another ten years he

Figure 5.21. Robert F. Bacher near the peak of his career. (AIP Emilio Segrè Visual Archives.)

was a prominent figure, first at the Radiation Laboratory at M.I.T. and then in the development of the bomb at Los Alamos. He then spent some time as a member of the Atomic Energy Commission, and his career was crowned by an appointment to the California Institute of Technology where he served on the faculty and as one of President Lee A. DuBridge's principal colleagues.

Our trip back to the West Coast was routine, with one notable exception. Shockley had switched his car registration from California to Massachusetts during the winter. The year before, we had been greeted warmly wherever we went on our way East with California plates, but we now encountered conspicuous hostility west of the Mississippi River. The people there apparently felt that irresponsible easterners were to blame for the Great Depression they were suffering through.

That summer, Shockley married Jean A. Bailey, a long-time friend of his family. Jean was an attractive, generous person who was dearly loved by all who came to know her. My wife and I spent many delightful hours with Bill and Jean and their family, both at their home in New Jersey and in the Adirondacks. Much to our sorrow, Jean died in the mid-1950s.

I happened to be on the Stanford campus when Hansen, having completed his appointment in Massachusetts in June of 1934, returned to Palo Alto to take a position on the faculty at the university. He wandered from room to room in the Stanford Physics Department, savoring fond memories everywhere. Unfortunately World War II was to take him away again for an extended period. The stresses associated with intense war work probably accelerated an underlying physiological weakness, possibly of genetic origin.

Many of my friends in industrial laboratories, not being burdened with teaching responsibilities, committee assignments, and the like, managed to find time for pastimes aside from their creative work in the laboratory. Shockley was no exception; throughout his life, he maintained an interesting set of hobbies. Before I knew him he had been adept at sleight-of-hand parlor tricks, and he kept his skills honed over the years. He amused himself during his student years at M.I.T. by keeping the staff there on edge with his subtle and not-so-subtle tricks. Later he was an avid climber beginning with hand-over-hand rock climbing and progressing to far more sophisticated rope climbing, including semi-professional assaults on some of the more difficult peaks around Mont Blanc.

He had another highly solitary hobby that, to me, symbolized a special side of his character. He enjoyed establishing ant colonies in large glass containers and training the ants to take circuitous routes in seeking food and returning to their storage base. This usually involved the construction of delicately balanced seesaws of straws which would tilt under the weight of an ant. The ant, near its home base, would climb on the lowered end of such a straw and, in moving past the fulcrum, would cause the straw to tilt so that the ant could reach the food supply. But as soon as the ant stepped off the seesaw, it would return to its original position so that the ant was forced to find an alternate path back. The return path was usually strewn with one or more similar seesaws. Shockley could spend hours watching his pets learning to overcome obstacles.

In the late 1980s, a dean of engineering at one of the large South African universities invited Shockley to visit South Africa to give a speech commemorating the invention of the transistor. Knowing of Shockley's controversial studies of ethnic groups, the dean emphasized that any mention of Shockley's views on such matters would be completely inappropriate as the South African government was struggling to find a way out of the morass of apartheid. Alas, when Shockley came to give his lecture he did just what he had been asked not to do, focussing not at all on the transistor but on his personal views of the relative virtues or deficiencies of various ethnic groups. It was a great embarrassment to one and all. Shockley spent a substantial part of his time in South Africa studying the trainability of the local ants.

This increasingly unpredictable aspect of Shockley's behavior as he grew older, made it very difficult to maintain any semblance of true friendship, particularly during the 1960s when I had major responsibilities as President of the National Academy of Sciences.

I spent my last two years at Princeton working on a variety of problems and examining prospects for the future.

Condon and I had a great deal of fun together during those years, attending meetings at which he was often a key speaker. On one occasion John Blewett and I accompanied him to Columbia University where he gave a colloquium talk. The weather turned bad as we were driving back very late in the evening, and Condon expressed concern about the possibility of our getting a flat tire—a problem much more common in those days than now. To relieve his anxiety, John and I said that we would not mind changing a tire in spite of the rain. He gloomily replied, "That is great news but we do not have a spare."

On another occasion, we made plans to drive to a meeting of the American Physical Society in Pittsburgh taking place between Christmas and New Year's. Our original plan was to drive there on December 26th. We were sitting around the Condon living room late on Christmas Day following a large dinner when Condon said, "Why don't we go now. The highway will be deserted." I promptly agreed and we started out on a night-long adventure. As we neared Pittsburgh, with the first streaks of dawn in the sky, he commented, "I guess this trip puts us in the same class as the Vikings."

Nearing completion of my graduate work at Princeton, I was offered a post as Harvard Fellow. Unfortunately the conditions attending the appointment were not very favorable for married students, so I elected to take a job as assistant professor at the University of Rochester instead as a result of a generous offer from Lee A. DuBridge, the new department Chairman there, since I indeed planned to be married.

Elizabeth Katherine Marshall (Figures 5.22 and 5.23) and I had met in the autumn of 1934 when I was asked to give a colloquium talk in the physics department at Bryn Mawr College. The Head of the department there was Dr. Jane Dewey, the brilliant daughter of the celebrated educational philosopher at Columbia University. Walter C. Michels, who had been a National Research Fellow at Princeton during my early years there, and A. Lindo Patterson, an x-ray diffraction physicist, were also on the staff. Michels had done graduate work at Caltech and enjoyed discussing his life there under Robert Millikan. Dr. Dewey eventually joined the research laboratory of the Aberdeen Proving Ground in Maryland, and Patterson and his wife Elizabeth, a biochemist, with whom we formed a close friendship, joined the newly established Fox Chase Cancer Research Center nearby. Patterson, who had studied under the two Braggs in England and Max von Laue in Germany, made basic contributions to the theory of x-ray diffraction as applied to determining crystal structures.

Michels gained considerable repute in the 1950s when the physics community, supported by the National Science Foundation, conducted a nation-wide review of the methods of teaching elementary physics. He became one of the leading figures in this undertaking, playing a major role in both formal and informal discussions.

Betty and I soon discovered that we had many interests in common, including music. She had spent most of her formative years in South China where her parents, George and Edmonia Marshall (Figures 6.8, 6.9, 6.10,

Figure 5.22. *(Left) Elizabeth Katharine Marshall in 1933 while a graduate student at Cornell University.* **Figure 5.23.** *(Right) Betty and I during our courting days in 1934. The photograph was taken in Bryn Mawr.*

and 6.11) had started out as missionary teachers. Her father proved to be an excellent administrator and was made Administrative Head of the western type of hospital on the Bund in Canton. He held that post for over a decade, and was there in the 1920s when the armies of Chiang Kai-shek and Mao Tse-tung battled one another in the streets of Canton at the start of the great civil war between the nationalists and the communists. The hospital still stands, and probably remains the best in Canton. George Marshall did not choose sides when it came to treating the wounded, and in trying to serve the needs of all, he initially encountered hostility from both factions. However, he eventually gained the respect of both contending groups because his efforts were obviously humanitarian.

Betty had attended the once famous Shanghai-American School in preparation for college and went to Wilson College in Chambersburg, Pennsylvania. This was followed by a year of graduate work at Cornell University which led to her being awarded an excellent fellowship at Bryn Mawr.

Betty's roommate at this time was Betty Armstrong, who was studying for a doctor's degree in geology but also had strong interests in mathematics and physics. She later married an electrical engineer, Ira Wood, a member of the Bell Telephone Laboratories. After her marriage, she devoted most of her attention to matters related to physics and engineering and, in the course of her professional work, wrote several books on the application of physics to everyday observations, such as the view from an airplane window.

Betty Marshall's two brothers (Figures 5.24 and 5.25), Lauriston C. and Robert N., were both physicists, and her first cousin, Lauriston Taylor (see Figure 5.28), was Chief of the x-ray physics section at the National Bureau of Standards in Washington. Taylor played a key role in medicine in the United States and internationally, working to develop standard doses for radiation therapy. Joining this happy and productive family added a great deal to an already interesting life.

The Taylor family had lived in Maplewood, New Jersey since the turn of the century. Betty's aunt, Nancy Sale Taylor, had grown up in Bedford, Virginia where the Sale family had suffered all the deprivations of the Civil War. She had married Charles Taylor (Figure 5.26), a metal assayer from Brooklyn whom she had met when the latter visited Virginia on vacation. Their elder son, Charles Holt Taylor (Figure 5.27), became a distinguished historian at Harvard University specializing in medieval French history. Lauriston Taylor (Figure 5.28) of the Bureau of Standards was their second son. Their youngest son (Figure 5.29) was a lithographer.

Because of his mother's southern background, young Charles Taylor was enamored in his boyhood of the Southern cause in the Civil War. He was permitted to replay famous battles of that war with his set of lead soldiers

Figure 5.24. *(Left) Betty's older brother, Lauriston Calvert Marshall, creative in several branches of science.* **Figure 5.25.** *(Right) Betty's younger brother, Robert Nelson Marshall. He began his productive career at the Bell Telephone Laboratories.*

Figure 5.26. *(Left) Betty's beloved uncle, Charles Holt Taylor, a metal assayer.* **Figure 5.27.** *(Right) Charles Holt Taylor, Jr., a brilliant medieval historian who spent most of his career at Harvard University. He prepared the official history of the D-day invasion of Normandy. (Courtesy of his son, Oliver Taylor.)*

but the outcome depended on him. If his behavior had been good, the South could win; otherwise, victory went to the North. Years later, after retiring from Harvard, he made a leisurely tour of the great battlefields of that costly war. When I saw him shortly after his return, his comment was, "It was really not very glorious, but instead sad and grim."

The Taylor home in Maplewood was a happy and busy residence with many friends and neighbors coming and going. Charles Taylor senior enjoyed playing the role of an eccentric practical joker and had cultivated a reputation as such in the community. He lived into his ninetieth year, and at his funeral, his friends and family spent much of their time regaling one another with anecdotes of his escapades. Fortunately, I passed muster with the Taylor family — Betty's nearest relatives.

Betty's brother Robert obtained a position at the Bell Telephone Laboratories upon graduating from Princeton and had the good fortune to retain it during the Depression years. He carried out the creative development of new audio microphones for both wide-angle and narrow-angle reception; his microphones were the quality standard at radio stations for nearly two decades. In his spare time, Robert was the leader of his local Cub Scout den; on one Saturday, they made an expedition to tour Princeton University. Betty and I had arranged to meet the group in Fine Hall, and just as Robert

Figure 5.28. *(Left) Lauriston S. Taylor. He was a major pathfinder in the standardization of x-ray dosage.* **Figure 5.29.** *(Right) The youngest of uncle Charlie's sons, Edward Taylor, a professional lithographer.*

and the uniformed Scouts trailed into the Hall, Albert Einstein came out of the building on his way home. He chuckled at the sight of the retinue and patted each boy on his head as he went by. The rest of the tour was clearly anti-climactic.

Since Betty's brother Robert had been in the Princeton class of 1930 and her brother Lauriston had spent two years there (1929–1931) in the department of physics as a National Research Council Fellow, Betty was greeted as a member of the family when she visited me on campus. In fact, both the Condons and the Robertsons took a warm interest in our developing romance and were delighted to have Betty as a houseguest on occasion. Condon was especially interested in my stories of the Taylor family and events in their home in Maplewood. When visiting the Bureau of Standards during a meeting there, he sought out Lauriston Taylor, who was then in his thirties, and greeted him with, "I gather that you are Uncle Charlie's little boy."

Chapter Six
University of Rochester

Betty and I were spending the summer months of 1935 in Schenectady, and while there, we drove to Rochester to explore living quarters. We found a modest but suitable apartment in a venerable building, which had once been a stately private home, on South Washington Street where her friends Margaret and Sidney Barnes were living. She had met the Barnes during her graduate year in the Physics Department at Cornell University. Sidney (Figure 6.1) had been involved in x-ray research with Floyd Richtmyer at Cornell and had come to the Physics Department at the University of Rochester with hopes of participating in a program in nuclear physics. Margaret Barnes, or Peg, was a lively, stimulating professional biologist. Our friendship with the Barnes was to last for a lifetime.

We also formed a close, lifelong friendship with our neighbors across the hall, the Leonard Lee Bacons (Figure 6.2). He was a young lawyer from a prominent Rochester family; his wife, Helen, was the daughter of a St. Louis physician, Dr. John Green, who visited occasionally with Mrs. Green and treated us to sumptuous dinners at local restaurants. Helen and Lee had a great sense of humor and we spent many delightful hours together.

Over the years, Betty and I crossed paths with the Bacons in a number of different situations and in various parts of the world, as Leonard and Helen pursued a series of exciting adventures. Bacon was with the European Forces in World War II, often in the vanguard of a front-line group and on a motorcycle, and afterwards went to work for the State Department in the Foreign Service. Three years in China culminated in a post as senior officer at the embassy in Nanking, where Leonard and Helen witnessed the city's fall to the communists in 1949. After several tense and dangerous months in which a long prison term for Bacon was threatened they, along with all other American officials, were expelled by the new regime.

Some ten years later, in 1959, we were together in Paris under the auspices of the North Atlantic Treaty Organization. Bacon had been posted to the NATO Defense College faculty while I had become the Science Advisor to the NATO headquarters. We spent a great deal of our time in Paris happily catching up on the intervening years.

On taking an early retirement from the State Department at the beginning of the 1960s, Bacon joined the staff of the National Academy of Sciences. He played a prominent role in the creation of the Universities Research Association (URA) which in turn was instrumental in the develop-

Figure 6.1. *Sidney Wilson Barnes (right) showing the control panel of his new accelerator at the University of Rochester to Ernest O. Lawrence. The photograph dates from about 1950. (Courtesy of the University of Rochester.)*

ment of the Fermi National Accelerator Laboratory in Illinois. The URA was not only successful in its original mission, but later assumed responsibility for the development of the proposed superconducting supercollider in Texas, which unfortunately became a matter of controversy in the House of Representatives, and presumably has been cancelled. Bacon became a much admired colleague of the high energy physics community. Once the four of us were settled in Washington, we were able to enjoy our friendship on a more stable basis.

The University of Rochester was founded in 1850 primarily to serve local students; when we arrived, the school was in the process of changing its status, as a large gift had been donated recently by George Eastman who had created the Eastman Kodak Company. A second campus was built on the Genesee River and the school began placing new emphasis on research in addition to its teaching programs. As was not uncommon in American academic institutions at that time, the university essentially had two faculties; the older group, which was heavily burdened with undergraduate teaching responsibilities, and the other, younger members which had a deep commitment to research as well as to teaching.

Doubtless some of the older staff felt a justifiable resentment towards the boisterous, relatively carefree younger group. The elder scientists were no less dedicated, but simply had fewer opportunities during their early careers. The truth is that in 1935 the scientific priorities of our country under-

Figure 6.2. *Leonard Lee Bacon, left, and his wife, Helen, right, during our Rochester days. The woman in the middle is one of Lee Bacon's cousins. Bacon, as a lawyer, was a mentor for the high energy physics community during and after the creation of the Universities Research Association.*

went a major shift. One encountered a similar division between older and younger faculty in many other institutions, as I subsequently found at the University of Pennsylvania, the Carnegie Institute of Technology, and even, but to a lesser degree, at the University of Illinois.

Lee A. DuBridge[1] (Figure 6.3) had been brought from Washington University in St. Louis in 1934 to Head the Rochester Physics Department with the expectation that it would become a prominent center of graduate research. He provided an enlightened environment for those of us who had the good fortune to be chosen by him to join the department. Moreover, the DuBridges brought with them to Rochester the warm midwestern hospitality characteristic of their former home. Doris DuBridge was a particularly warm-hearted hostess who quickly made newcomers feel welcome and comfortable; her kind hospitality was a wonderful antidote to the winters. Lying close to and on the lee side of Lake Ontario, the city tended to be overcast for extended periods, when the days were brief and gray. Warm friends and homes were just the refuge we needed.

Thus far, in addition to Barnes, he had selected Milton S. Plesset, a young

---

[1] Lee Alvin DuBridge died in January of 1994 as this book was going to press. I owe many debts to this much-admired leader of American science in our time, not only for providing me with my first position, but for special consideration on many other occasions.

Figure 6.3. Lee Alvin DuBridge at the peak of his brilliant career. (AIP Emilio Segrè Visual Archives.)

theoretical physicist who had received his Ph.D. at Yale University and then spent several very profitable years with Niels Bohr in Copenhagen.

An Institute of Optics was closely associated with the physics department and was a great stimulus to the working environment. It was headed by an English-trained Canadian, T. Russell Wilkins, best known for his social graces, although he did do some pioneering work on the use of thick photographic emulsions to detect particle radiation. There were, however, two outstanding professionals at the institute: Brian O'Brien and Rudolf Kingslake.

O'Brien had grown up in Denver and done graduate work in atomic spectroscopy at Yale University. He was widely conversant with physical and physiological optics and worked creatively and successfully in a number of fields. He eventually joined the American Optical Company after World War II after working on a multitude of special projects for both the company and several federal agencies, particularly the Department of Defense. Kingslake had been educated in England on the design of advanced optical systems. He left the University of Rochester in 1937 but remained in the city to join the Kodak company where he carried out brilliant work in optical research that was of major importance to the motion picture and television industries. He earned numerous awards both from industry and from the major professional societies of which he was a prominent member. His wife Hilda was the daughter of a distinguished scientist and was also richly talented. She has made significant contributions to historical research in the field of optics.

Having grown up near the Rocky Mountains, O'Brien frequently teased me good-naturedly about the Sierra Nevada mountain chain, with comments like, "It is unfortunate that California has nothing better than hills." At the same time, he developed a deep affection for the Adirondacks while in Rochester.

One particularly colorful member of the optics staff was J. Stuart Campbell. He had been well-regarded as an advanced graduate student at the California Institute of Technology during my time there. He was glad to find another Californian in chilly Rochester, and quickly became a member of the social circle that included the Barnes, the Bacons, and Betty and me. Campbell had grown up in Pasadena and loved to tell us of the life he led there. Sadly, he had been ill as a teenager with an ailment diagnosed as a form of Hodgkin's disease that was generally considered fatal at that time. Although he had been given heavy radiation therapy and apparently had a genuine remission, contrary to general expectations, he had not anticipated living very long, and much to our sorrow, he took his own life several years after we left Rochester. He had been a delightful companion to us, and we grieved that the emotional stress of his ill health had proved too great in the end.

The Head of the Department and Divisional Chairman of Biology was Benjamin H. Willier, a personable and talented individual. Two other distinguished biologists in the division were Curt Stern, a greatly admired geneticist and refugee from Hitler, and David R. Goddard, an amusing and loquacious young botanist. As it did just about everything else, World War II disrupted this group—Willier ended up at Johns Hopkins, Stern at Berkeley, and Goddard at the University of Pennsylvania.

Curt Stern, whom I came to know well, once confided in me that he did not have the temperament to work at the highly competitive frontier of biology. He preferred to find an interesting field somewhat off the beaten track where he could hope his efforts would lead to something remarkable but where, meanwhile, he could maintain control over the work.

One of the memorable events associated with the period at Rochester was attending a lecture by an English visitor, W. T. Astbury, who described his studies of the structure of long chain biochemical compounds by means of x-ray diffraction—the opening of a new era of biochemistry.

Rochester was fortunate in being remarkably well-shielded from the more disastrous effects of the Depression. The community was home to a number of successful companies such as Eastman Kodak, Bausch and Lomb, and Taylor Instrument, which were well-managed and did their best to maintain employment; Kodak, for example, was putting much effort into the development of color film at that time. There was one excellent physicist, Dr. Julian Webb, on the Kodak staff, carrying out research with equipment used for standardization and other purposes. He was a delight to be with, personally charming and with a fund of insights on quantum physics. Our local companies were also reasonably supportive of activities at the University. With their aid and with other public support, DuBridge succeeded in raising the funds to build a cyclotron, with Barnes overseeing its design and construction. This major achievement enabled our department to move into the mainstream of American nuclear physics research.

The city had a well-defined downtown area in those ancient days before the suburban shopping mall explosion. Our apartment was conveniently close to hotels, department stores, theaters, and the like. Dear to my heart was Odenbach's combination bakery and restaurant which preserved the best European traditions. A highly labor-intensive business, however, it vanished soon after World War II.

One of the attractive features of Rochester was its proximity to Cornell University, just ninety miles away in Ithaca. Cornell's traditionally strong physics and chemistry departments were further strengthened during this period by the addition of Hans A. Bethe (Figure 6.4), Robert F. Bacher, and Peter J. W. Debye. The departments at the two universities held joint colloquia and spring outings.

Some of the more research-oriented members of the Rochester faculty had formed a small monthly discussion club—the X Club—which I was asked to join soon after arriving on campus. Faculty wives were welcomed at these gatherings, and both Betty and I found it a rich and memorable experience. In addition to DuBridge and O'Brien, members at the time included Leonard Carmichael, a psychologist (Figure 6.5); George W. Corner, an anatomist and eventually a science historian; Wallace O. Fenn, a physiologist; and Goddard, Stern, and Willier. We remained linked to our remarkable fellow club members for many years to come.

In 1935, complete freedom for research and a modest teaching schedule

Figure 6.4. Professor Hans A. Bethe in his early Cornell days. (AIP Emilio Segrè Visual Archives.)

Figure 6.5. Leonard Carmichael during his days as head of the Smithsonian Institution. (National Academy of Sciences.)

gave me the opportunity to launch an ambitious project which had been on my mind since student days: to attempt to write a cohesive account of the various aspects of solid state physics, in order to give the field the kind of unity it deserved. This integration had been made possible by the development of quantum mechanics which offered the means of consolidation. The result, *The Modern Theory of Solids*, was published by McGraw-Hill five years later in 1940. It has since been reissued by Dover Press and presumably possesses at least archaeological value. I think it is safe to say that, in the writing of this book, Betty, who was soon deeply involved in the program, and I became familiar with every paper related to the field. The essential material was, of course, far smaller then. Perhaps the greatest value of the book was the attraction it provided for new, young investigators to undertake research in the area in the immediate post-war period.

It was, incidentally, during this same period that John Bardeen, first as a Harvard fellow and then as a member of the faculty at the University of Minnesota, decided that he would try to solve the mystery of superconductivity. His early attempts are merely of historical interest, but he continued to revolve the problem in his mind until he, Leon N. Cooper, and J. Robert Schrieffer were finally successful in the 1950s.

While working on my book, it soon became evident to me that the rate at which the field of solid state physics could be expected to evolve in the future would be determined to a considerable degree by the number of experimental investigators attracted to it, and who might work in close association with the theoretical developments. At that time there were perhaps twenty five institutions around the world doing experimental work in what might be called a systematic way on the fundamental properties of solids.

In the United States, for example, one thought of the work of Percy W. Bridgman at Harvard involving high pressure studies, and of several groups at M.I.T. Francis Bitter was starting his work there on high magnetic fields; Arthur von Hippel was involved in studies of the dielectric properties of solids; and Wayne A. Nottingham was concerned with thermionic emission from cathodes and related matters. Also at this time, Shirley Quimby at Columbia was exploring the elastic and inelastic properties of solids as they could be determined with the use of mechanical resonance methods employing piezo oscillators. At Rochester, Lee DuBridge and some of his students continued for a while the types of research on photoelectric emission from solids that he and Arthur L. Hughes had started at Washington University in St. Louis. And Clarence Zener, who had become established at City College in New York, developed a small but highly productive laboratory for studying the internal friction originating in various sources in solids under periodic stress.

The field was rather more active in Europe, with low temperature research in England, Holland, and Germany, and magnetic research at Pierre Weiss's laboratory in France. G. I. Taylor at Cambridge and others in Europe were carrying out basic work on plastic flow in crystals. Then, too, Robert W. Pohl's Institute at Goettingen was devoted to studies of the electrical and optical properties of carefully prepared specimens of the alkali halides under various circumstances. Also worthy of note was the work of Carl Wagner on diffusion in solids and related matters at Darmstadt. Several institutes in the Soviet Union, including that of I. Joffe in Leningrad, focussed on the electrical and mechanical properties of solids, while others there were engaged in basic metallurgy. Much good work was being done in Japan on the magnetic properties of metals, particularly in the laboratory of K. Honda.

Beyond these various efforts, the laboratories of electrical companies in both the United States and Europe were productively involved in solid state research, carrying out studies of solid state rectifiers, thermionic emission from hot filaments, and the development of luminescent agents for use in cathode ray tubes. In short, there existed what might be termed a global patchwork pattern of experimental studies, although nothing that could be compared to the unified effort that was being made in the field of high energy particle physics during the mid-1930s.

John H. Van Vleck (Figure 6.6), then at Harvard University, was at that time one of the great friends of young theoretical physicists in United States. He made a point of attending our brief lectures at the Physical Society meetings, discussing them with us, and in general making us feel like true professionals. He and his wife, Abigail, cultivated us during these immediate post-doctoral years and became our lifelong friends, meeting with us on many occasions.

Figure 6.6. John H. Van Vleck during his early period at Harvard. (AIP Emilio Segrè Visual Archives.)

As he believed very strongly that the field of physics was a highly unified one, Van Vleck was greatly distressed when the American Physical Society began to develop divisions associated with specialties. He called this the "Balkanization" of physics, and did his best to retard the process. Unfortunately the enormous growth of the society eventually did require some kind of organizational breakdown into categories.

As a child in Madison, Wisconsin, where his father was a professor of mathematics, Van Vleck became interested in the American railroad system and made a hobby of knowing all the principal lines and most of the schedules. He was, in fact, what was sometimes called "a railroad nut." When one of my close friends who frequently rode buses around the country for economy's sake became fatally ill, Van Vleck shook his head and said, "What do you expect? He rode buses."

As air travel became more and more commonplace, the quality of rail travel in America began to deteriorate. I once asked Van Vleck his opinion of the state of affairs. He said, "It has reached a point where each mode of travel is doing its best to get you to take another." Ultimately he was forced, reluctantly, to shift his loyalty to air travel.

Van Vleck had also become a motion picture addict during his boyhood; his passion struck an answering chord in me. He once said to me, sometime in the 1930s, "There are no really bad movies. The grade B pictures are interesting for a different reason." I never had an opportunity to compare notes with him on this subject in the decades following the 1950s, when standards changed so drastically.

One of the most memorable events of my two years at Rochester was a visit in 1936 by Professor Niels Bohr (Figure 6.7). As mentioned above, Milton Plesset had worked with him in Copenhagen and was now influential in ensuring that Bohr's American itinerary included a stop in Rochester. At the time, Bohr's major interest was the fact that the nucleus of the atom had more characteristics in common with a liquid drop than with a globule of gas of independent particles bound in a potential well. He felt it was vital to spread the word on this subject, and he lectured dramatically with his heavy Danish accent and much accompanying gesticulation. George Gamow, then in Washington, took the opportunity to make one of his much-appreciated quips, to the effect that Bohr was now advocating a mashed-potato model of the nucleus.

The festivities in Rochester honoring Bohr lasted for several days; and were then continued in Toronto on a far more expansive and formal scale in accordance with Canadian traditions of that time. The gatherings were joined by the highest university and city officials. We guests were treated like visiting royalty—provided we did not forget at appropriate occasions that the true seat of royalty was in England.

Not surprisingly, the appointment of DuBridge to the Department of Physics attracted a number of good students to Rochester, some of them would remain my lifelong friends. Our group of graduate students included Albert G. Hill, who eventually became director of the Charles S. Draper Laboratories in Cambridge, Massachusetts, and Douglas H. Ewing, who carried out his thesis research with me and later became technical director of laboratories at RCA. Robert J. Maurer, a native of Rochester, carried on

Figure 6.7. Niels Bohr in his study, probably taken in the 1930s. (Niels Bohr Institute and the AIP Emilio Segrè Visual Archives.)

his research directly under DuBridge; Robert and I spent much of our careers in close contact.

There were also three undergraduates with whom I was to remain linked. Most remarkable was Robert H. Dicke, who spent essentially all of his later career at Princeton University. As part of his very broad interests in physics, he played a major role in the search for the primordial radiation associated with the early stages of the creation of the universe. My two other friends were LeRoy W. Apker, who later joined the staff of the General Electric Laboratories, and Joseph B. Platt, who remained a close friend of DuBridge and eventually became president of the Claremont University Center in Los Angeles.

Since my parents had never met Betty, we set out to drive to the West Coast for a visit in early August of 1936. Taking a northern route that passed through Wisconsin, Minnesota, and the Black Hills, we made two important stops on the way. The first was in Madison, Wisconsin to visit the Condons; Ed was lecturing at the summer school there. Another summer lecturer was Paul P. Ewald, a distinguished authority in the field of x-ray diffraction who had just arrived in the United States having left Germany because of Hitler. Ewald's daughter was married to Hans Bethe, who was by then on the faculty of Cornell University. Ewald eventually settled in the United States, where he made significant contributions to his professional field through a combination of his leadership and wisdom; among his many other activities, he became the editor of an important new crystallographic journal.

Our second stop, after driving through Yellowstone Park and Jackson Hole, was to visit Uncle Louis (Figure 1.4) at his hundred-acre farm in Emmett, Idaho. We arrived just as he and a crew were completing the harvesting of his wheat fields—they brought in a truly bumper crop of which he was very proud. We came unannounced but he greeted us as though he had been expecting us for weeks.

Most of our conversation was about current farm and family matters, but we also heard some family history. Uncle Louis told us, with much merriment, of his adventures with my father in New York in the late 1890s when Louis was newly arrived. My father had already been established there for several years and was thoroughly familiar with the city. He promptly had his younger brother enrolled in language school and oriented toward the library and educational lecture circuit. Their dispositions and interests were very different, however, and Uncle Louis headed toward Idaho as soon as he was reasonably acclimated. His decision to leave was hastened when my father discovered that he had been consorting at a local bar with congenial colleagues instead of attending uplifting educational lectures. I could well imagine the tenor of my father's oratory on that occasion since I had experienced it a number of times myself. My uncle added with a grin: "As you can see, I have always been the black sheep of the family."

Uncle Louis unintentionally displayed his prowess with a rifle when he decided that we would have chicken for dinner. The chickens were wandering about a substantial penned-in area off the back stoop of his cottage; he took his 30 caliber rifle from the wall and stepped out on the stoop. Spotting an acceptable bird about one hundred and fifty feet away, he almost casually raised the rifle to his eye, and the chicken's head vanished—as merciful a death as one could hope for.

As cordial as he was, Uncle Louis proved to be slightly forgetful as a host. As Betty was settling our belongings in our bedroom, he arrived with a slightly sheepish grin and handed her a set of bed sheets, saying: "You probably would feel lost without these."

He married the widow of an old friend several years later in his late fifties, but died of a coronary problem during World War II, much younger than his siblings. His diet during his bachelor years, when he had been head of a lumber team, had not conformed to the ideals recommended by health-oriented dieticians; most of his meals had been prepared in a grease-laden frying pan.

One of the happy consequences of our visit was that we were able to bring our family in San Francisco the great news that all of the interconnecting roads between Boise and San Francisco had been paved, so that the drive was free of the old dirt washboard roads that had always been such a difficult feature of travel in the western states. The journey now involved no more than an easy two days of travel. Between the improvement in the roads and the development of ski resorts, Idaho was well on its way to losing its frontier status, and has since become one of the states in which Silicon Valley corporations establish secondary units.

When we arrived in San Francisco, Betty and I were greeted warmly by family and friends. In fact, we were so exhausted when it came time to leave that we traded in our car, arranged to pick up another in Detroit, and took the train back to Michigan.

Undoubtedly the most important event that occurred in San Francisco during that visit was the arrival of Betty's mother, Edmonia (Figures 6.8 and 6.10), from China. Betty's father, George Washington Marshall (Figures 6.9 and 6.11), had died early in 1934 and had been buried in Canton. Her mother had settled their affairs there, and not wishing to be a burden to the missionary system with which she had worked for forty one years, decided to return to the United States.

What a remarkable woman she was! She had been born in Bedford, Virginia in 1870, into a family that had several generations of lawyers in its background. To call Edmonia strong-willed would be a severe understatement—she decided early on that she should have the best education circumstances would permit. To that end, she repeatedly applied for admission to the University of Virginia, which was exclusively male at the time. Eventually the administration of the university relented to the extent that it agreed she could attend lectures sitting at the back of the rooms—she would

Figure 6.8. (Left) Betty's mother Edmonia as she looked when I first met her in 1936. Figure 6.9. (Right) Betty's father, George W. Marshall, during his period as the administrator of the Western Hospital in Canton during the 1920s.

receive no credit, and she was not to speak to the men. With that she went off to Nashville where she was accepted at Peabody Teachers College.

After graduating and starting a teaching career, Edmonia felt a strong religious calling. This resulted in converting from the Episcopal church of her family (Figures 6.12 and 6.13) to the Baptist church. Her immersion took place in her home town of Bedford and caused a considerable stir within her circle of family and friends. Then, at age twenty five, she took off for China to devote her life to missionary work there.

While in the mission language school in Canton, she met Betty's father, George Marshall, who had felt a similar calling but was a Presbyterian. They married a year or so later, after each had served in the field. Edmonia then found it expedient to join her husband's church so that they could work as partners—a decision that inspired her brother, Nelson Sale, back in Virginia, to compose the following ditty:

> Episcopalian born and bred
> Baptist by immersion
> Presbyterian all for love
> Still open to conversion.

China was a very unstable country when Edmonia and George arrived there, and they had a difficult time of it. There was no strong central government, and consequently, no real enforcement of civil laws. On several occasions the Marshall family, located at an outpost mission, had to flee to

Figure 6.10. (Left) Betty's mother in her early days of missionary work in China, displaying a native costume of the type she regularly wore in China. Figure 6.11. (Right) Betty's father in his early missionary days in China.

Canton to escape an attack by local bandits. Once, when George was head of a mission-supported boys school at the village of Shek Lung, midway between Canton and Hong Kong, bandits kidnapped several students and murdered them when their impossibly high ransom demand was not met. The school was permanently closed.

Through all of this, Betty's mother, who became very fluent in the Cantonese dialect, often travelled on her own, leaving her husband and children behind, in order to bring news and comfort to a mission family in some more remote outpost. She travelled by whatever means were available; in one letter she describes a two-day journey along the South China coast in a junk loaded with pigs. True faith was required.

As was mentioned in Chapter 5, Betty's father was such an able administrator that he was assigned the top administrative post at the western hospital on the Bund in Canton during a critical period in its political history.

During friendlier periods, the family spent summer months on one of the mountain tops on an island in Hong Kong harbor, or went on furlough to the United States to visit relatives. During World War I the Reverend Dr. Marshall occupied a special post at a U.S. Army camp on the Philippine islands.

Mrs. Marshall's arrival in San Francisco was cause for further celebration and conviviality. She was not a large woman and was already somewhat on

Figure 6.12. (*Left*) *Betty's grandmother, Edmonia Bell Sale, at the time of her marriage in 1863, during the Civil War. She was sixteen years of age at the time.* *Figure 6.13.* (*Right*) *Betty's grandfather, Lauriston Sale, in Confederate uniform, also at the time of his marriage in 1863.*

Figure 6.14. Professor H. Richard Crane. (AIP Emilio Segrè Visual Archives.)

the frail side at that time. With the exception of my father, most of my family, particularly those on the Bruhns' side, were large and had very healthy appetites. Mrs. Marshall looked on in wonder as they did away at dinner with an amount of food that would have supplied a comparable Chinese family for at least a week. She had adjustments to make on returning to the United States.

Betty, her mother, and I travelled east together on the train; on our journey we met an old friend and made new ones. Robert N. Varney, a good friend of Betty's brother Lauriston, had recently completed graduate work in physics at the University of California at Berkeley and was on his way to a new post at New York University. Also travelling on that train were Florence and H. Richard Crane (Figure 6.14. See also Figure 2.2). Richard was a native of Turlock, California and had carried out graduate work in physics at Caltech. A year earlier he had accepted a position on the faculty at the University of Michigan where he was to remain for the remainder of his career. These chance meetings were a great privilege for Betty and me, particularly the new one with the Cranes. He is an exceedingly rare individual, being both a highly accomplished physicist and an unusually sensitive and perceptive human being. As a physicist, he had the master's touch. It was a great pleasure to be with him whenever the circumstances arose.

Chapter Seven
General Electric Laboratory

In an effort to stimulate more directed fundamental experimental work, I joined the General Electric Research Laboratory on a full-time basis in 1937 with assurance of a budget sufficient to establish a good laboratory with an appropriate technician, namely Mr. Frank Quinlan who was highly qualified.

In moving from Rochester to Schenectady, Betty and I had to adapt to a somewhat more severe climate, since temperatures customarily went well below zero degrees Fahrenheit (–18°C) during a typical winter. However, there was also an advantage to be gained, as Schenectady enjoyed many more crystal clear days than Rochester – this was because the climate of the former was much less influenced by the presence of the Great Lakes. Skiing was just becoming a popular sport, and many of my friends spent their winter weekends on what today would be regarded as primitive ski trails and slopes. Woodstock, Vermont, and the Laurentian Mountains of Canada were particularly favored skiing areas for those in the Schenectady region. I occasionally joined these expeditions, mainly for the pleasure of the outing. I was, however, deeply involved in research and book writing, so that most of my physical exercise in winter involved jogging in the streets of Schenectady or the surrounding countryside.

Also of interest in the region were the caves in the nearby Helderberg Mountains, where many of the bats from New England wintered. It was on an expedition to the bat caves with William Shockley and James B. Fisk of the Bell Telephone Laboratories and my colleague Ralph P. Johnson that I first met Donald R. Griffin. Then a graduate student at Harvard, Griffin was deeply involved in studies of bat sonar – a field in which he was a prime innovator. In fact, through his books he later became widely known by close colleagues as the Batman. He eventually married a very attractive colleague, Dr. Jocelyn Crane, who had undertaken extensive research on the fiddler crab – naturally, it was said within the profession that Batman had married the Crab Girl. We spent many interesting years together later on at Rockefeller University.

The caves were densely occupied by hibernating bats which were easy to trap. Shockley took several home to his New York City apartment but found

93

that they did not survive long on a diet of houseflies, perhaps because of a nutritional deficit.

One of the special privileges of being in Schenectady was the chance to meet with a wonderful physicist at Union College, Vladimir Rojansky (Figure 7.1). I had first encountered him during my graduate student days when he spent a semester at Princeton. He had been a soldier in the white Russian army, and was in Harbin, China in 1918; there, a chance meeting with some Americans inspired him to go to the United States to seek an education. He arrived in Seattle and took his first job, as a fruit picker. A school teacher with whom he became acquainted recognized that he was an unusually gifted individual and aided him in enrolling in Whitman College. Along the way, she helped him eliminate from his speech some of the forceful but inelegant expressions he had acquired as a fruit picker. Later Rojansky carried out graduate work at the University of Minnesota under the guidance of John Tate, a celebrated physicist of the time.

Years of soldiering made Rojansky a gifted raconteur. He could spin out a story for an hour if he was so inclined, going into many byways but always holding your interest. Saul Dushman (about whom more later), who had apparently once been snubbed by some of the other faculty members at

Figure 7.1. Vladimir Rojansky, upper right, and his three physicist colleagues who were students together at Whitman College and the University of Minnesota. Clockwise from Rojansky, Walker Bleakney, Walter H. Brattain, and Everly J. Workman. Workman was a Caltech PhD. Brattain was one of the three co-inventors of the transistor. (AIP Emilio Segrè Visual Archives, Brattain Collection.)

Union College, commented to me, "Having a guy like Rojansky at that place is somewhat like throwing pearls before swine." Later, Rojansky joined Walker Bleakney's wartime research team at Princeton and eventually moved on to the Claremont Colleges in California after World War II.

Rojansky, as a good physicist, was intrigued by the discovery made in 1932, that the electron has a positively charged counterpart, the positron. He suggested that all fundamental particles of nature possessing a charge could have similar counterparts and wrote an interesting paper, proposing that there might be bodies in our universe composed of what he termed "contraterrene" matter composed of positive electrons and negative protons. Such matter may yet be made in measurable quantities in the laboratory in a so-called storage ring container circulating negative protons.

As early as 1935, while still a student, I had been invited to be a summer "guest" at the G. E. Laboratory. My "host" was Albert W. Hull (Figure 7.2), who was widely known for developing methods to determine the structures of crystalline materials using powders and x-rays, as well as for the invention of special alloys for metal-to-glass seals. W. D. Coolidge (Figure 7.3), the x-ray physicist, was then the Director of the Laboratory, having succeeded Willis Whitney, the lab's founder.

The Laboratory had hired very few engineers and almost no scientists between 1929 and 1935, as a result of the impact of the Depression. Ralph P. Johnson (Figure 7.4), who obtained his doctoral degree at M.I.T. working with Wayne Nottingham, and I were essentially the first new physicists added in over half a decade. A small group of organic chemists had been established under the leadership of Abraham L. Marshall in 1933 to take

Figure 7.2. Albert W. Hull. (Courtesy of the General Electric Research Laboratory and the AIP Niels Bohr Library.)

Figure 7.3. William D. Coolidge, the Director of the General Electric Laboratory. He carried his programs through from basic research to manufacture. (Courtesy of the General Electric Company and the AIP Niels Bohr Library.)

advantage of the rapidly growing opportunities in polymer chemistry. C. Guy Suits, the future Director of the Laboratory, had joined in 1930 as one of the last pre-Depression research physicist-engineers. He developed a highly sophisticated system of studies of the behavior of high pressure arcs under varying physical conditions.

My lab was included in the lighting research group under the leadership of Dr. Saul Dushman (Figure 7.5), an excellent physical chemist with a fine command of thermodynamics, who was also a remarkably wise and generous man. He was a warm-hearted counselor to anyone in the laboratory, regardless of rank, who came to him seeking advice, both personal and professional. He had made significant contributions to the theory of thermionic emission using the third law of thermodynamics. He was in his early fifties when I first joined the Laboratory and was busy writing a book on quantum mechanics for chemists. We spent many interesting hours together discussing quantum mechanics while at work, at his home, and at his delightful summer place on the upper reaches of the Hudson River, where it is a babbling trout stream. Dushman's wise quips and aphorisms, as well as what he euphemistically termed his "slang" expressions, remain deeply embedded in my mind.

Figure 7.4. (Left) *Ralph P. Johnson, an admired colleague and friend. (Courtesy of the General Electric Company and the American Institute of Physics.)* **Figure 7.5. (Right)** *Saul Dushman, my boss and a wonderfully caring individual. (Courtesy of the General Electric Company and the AIP Emilio Segrè Visual Archives.)*

Saul's wife, Anna, was both a queen and an angel. Her family had come from Lithuania, while his had originated in Rostov; he referred to her family as being of the aristocracy. Since Saul's everyday language had been acquired dealing with the workmen in the various shops of the company, it was well-seasoned, and Anna was continually reprimanding him to "stop talking like a roughneck."

On one occasion, Betty, the Dushmans, and I drove to Rochester together for a summer meeting of the American Physical Society. The DuBridges hosted a reception, to which we were invited. On the way to the party, Anna warned Saul in no uncertain terms to watch both his behavior and his language. He sat silently by until some topic of conversation was raised that was just too interesting for him not to join in; unfortunately, it was also too interesting for him to remember Anna's warnings. While leaving, he said apologetically to Mrs. DuBridge, "Please excuse me if I use a bit of slang now and then." She responded, "Do not worry, I had been warned." Poor Saul had little peace the rest of the day.

After World War II, Dushman turned his remarkable talents to additional writing and produced what became the definitive book of its time on high vacuum techniques. Daniel Alpert (Figure 11.24), one of the leading experts in the field, who led the way to much of the practical development of modern high vacuum technology, said that everything he learned with respect to principles came out of Dushman's treatise.

Saul suffered a stroke in his early seventies and spent two or three years as a house-bound invalid before passing away. On one of the numerous visits in the 1950s Betty and I made to see him, we found him seated before his black and white television set, watching some inconsequential program — his principal amusement. He said that it gave him a great thrill to realize that this marvel was finally operating and in so many homes since he, in his own way, had participated in its development through his research. He was certain that it would create a new world. He then told me of the numerous programs he watched with great pleasure, including all the standards of the day: Country Fair, The Howdy Doody Show, and the Lucille Ball program. Simply by way of making conversation, I mentioned that we had a friend who, during the summer, arose at 11:00 am, turned on the television, and was glued to the picture tube for most of the day until midnight. Not to my surprise, Saul gave me a baleful look and snapped, "Imagine an able-bodied man wasting his time on that junk."

In the summer of 1936, a group of us from the Laboratory, including Dushman and Irving Langmuir drove to Ithaca for a conference at Cornell devoted to thermionic emission. There were presentations by researchers from M.I.T., such as Wayne Nottingham, and from the Bell Telephone Laboratories, such as Joseph Becker. Langmuir entered enthusiastically into the proceedings since much of the discussion centered around matters in which he had long had a profound interest. One of the highlights of the meeting was the participation of Hans Bethe, who had just arrived from Europe to join the Cornell faculty and who would remain such for the rest of his career.

Attention in the lighting division of G.E. at that time was focussed on the development of two lamps which were much more efficient than the ordinary incandescent lamp: the sodium vapor lamp, which is still widely used in several forms, principally for street lighting, and the now common fluorescent lamp in which inorganic crystals which fluoresce in the visible region are stimulated with a low pressure mercury arc.

My work environment was enhanced by the presence of many stimulating colleagues. I particularly enjoyed working with Ralph Johnson, since we had many interests and friends in common. We published a series of review papers together on the band theory of solids.

Other members of the lighting group were Clifton G. Found, a very competent physicist of Canadian origin, who was in charge of the development of the sodium lamp, and Lewis R. Koller, who was generally conversant with the field of electronics, including thermionics, photo electronics, and luminescence.

In pursuing the practical side of my work, I spent a great deal of time with Newell T. Gordon, who was examining the qualifications of various fluorescent crystals and dyes that might be used for general illumination as well as for cathode-ray tubes. On the more fundamental side, I published a detailed theoretical interpretation of the results of experiments carried out in Robert Pohl's laboratory in Goettingen on the properties of fluorescent alkali halide crystals containing thallium. This work attracted sufficient attention that I was invited to speak at symposia in Europe, but the Laboratory budget was too lean to permit such a journey. I did, however, enter into animated correspondence with several European scientists whom I was not to meet until World War II, when I visited Europe for the first time.

In keeping with this, I had also followed with much interest research in the scientific journals, parallel to our own, being undertaken in European industrial laboratories, including that of Nikolaus Riehl then at the Auer Company. He was a pioneer in the development of luminescent materials for fluorescent lamps.

Another productive but more reserved member of the team was Gorton R. Fonda, who had been involved in the development of lighting equipment for many years. He subsequently wrote a useful book on practical aspects of luminescent materials. Surgery for what probably had been a mastoid infection had left this interesting and accomplished man with a slightly distorted face; it may have been discomfort with his appearance that caused him to withdraw from the type of hail-fellow-well-met spirit that was characteristic of most other members of the Laboratory.

During the 1930s, the General Electric Company had contract agreements with several European companies to exchange technical information and patent rights in some areas. These agreements made it possible, for example, for the company to manufacture and use tungsten carbide, which had first been developed in Europe, for high speed machine tools. As a result of these agreements, we had occasional visits from scientists from companies such as Philips in the Netherlands and the AEG in Germany. One of the visitors who remained fixed in my mind was Gustav L. Hertz, who was presumably a consultant to the AEG. He was the nephew of Heinrich R. Hertz, who had discovered the propagation of electromagnetic waves. A decade later, in the aftermath of World War II, Gustav Hertz was taken prisoner by the Soviet Union and headed a laboratory devoted to isotope separation. He was treated well there, but became what Nikolaus Riehl, mentioned above, who was in a similar situation later, termed "a bird in a golden cage." (See the book by Riehl mentioned in footnote 5 on page 168 giving details of his ten year period as a civilian scientist-prisoner in the Soviet Union. The book also contains a brief account of his work on fluorescent lamps.)

As the result of commercial ties and a mutual interest in fluorescent materials, I formed a friendship with a group at the RCA Laboratory in Camden, New Jersey, led by Vladimir K. Zworykin (Figure 7.6), that was conducting early work on television. Humboldt W. Leverenz of the RCA team was pro-

Figure 7.6. *Vladimir K. Zworykin, left, and Manfred von Ardenne. The latter was an early developer of the electron microscope whereas Zworkin was the first to commercialize it. The photograph was taken in East Germany in the early 1970s at an international conference on medical and biological engineering. Von Ardenne, incidentally, was captured by the Russians and served as a "scientist-prisoner" in Soviet Laboratories. (AIP Emilio Segrè Visual Archives.)*

ducing a wide variety of new luminescent crystalline materials for television screens. Our group at G.E. studied those materials for possible use in fluorescent lamps. In fact, one of his products, a form of zinc beryllium silicate activated with maganese, served very well until it was discovered after the war that beryllium compounds can be toxic for some individuals.

Leverenz had been a laboratory instructor in freshman chemistry at Stanford when I was a student there, so our relationship soon developed beyond the professional, and we became good friends. During a train trip to the Bay Area of California in 1941, when long-distance travel was still not very common, I telephoned his parents in San Jose and offered to stop by to pass on news about Humboldt, or "Lefty," as he was generally known. On being conducted into the living room of his parents' home, I found myself surrounded by burly westerners and subjected to a puzzling, almost hostile, cross-examination about myself and my friendship with Humboldt. Finally, Humboldt's father called off the questioning saying, "He is okay." It turned out that an impostor had recently wormed his way into the bosom of a family in San Jose by claiming to be a close friend of their son in the East. He proceeded to walk off with the family silverware. I was relieved to have passed muster.

Leverenz eventually married Edith Langmuir, the niece of Irving Langmuir, whom he had come to know through her brother, David, who also worked for RCA. The all-night party that commemorated their wedding has remained a high point in the memory of everyone who survived it. A heavy blizzard trapped the guests in the Leverenz house so it was almost a never-ending affair.

A lifetime friendship with Emanuel Piore (Figure 7.7) also grew out of the association with the staff at RCA.

Zworykin was a truly remarkable man, a Russian emigré who came to the United States to escape the Communist Revolution. He became committed to the development of television in the early 1930s and pursued his work with great zeal. He had invented the iconoscope camera for converting visual images into electrical signals and was busy with his staff developing the video viewing end of the system. In the mid-1930s one could see primitive but effective demonstrations of a black and white working system in the laboratory. I realized then that it would only be a matter of time until most homes had access to television in some form.

As a student in Russia, Zworykin had been interested in science, and had persuaded his father to allow him to go to the University in Petrograd to study physics. Unfortunately, he was jailed as a result of his participation in a revolutionary demonstration. His father arranged his release, but then insisted that he attend the Institute of Technology to study engineering.

Zworykin is unquestionably the Father of American television and was duly honored for his achievement by the television industry. On accepting one award, he commented with typical frankness, "The technology is marvelous but the quality of most programming is terrible." In addition to work-

Figure 7.7. Emanuel R. Piore. The photograph was taken in mid-career. (AIP Emilio Segrè Visual Archives, Physics Today *Collection.)*

ing on television, he also recognized the revolutionary potentialities of the electron microscope, first developed in rudimentary form by Manfred von Ardenne (Figure 7.6). Zworykin did much to make it an everyday tool for science and technology.

To his devoted staff, Zworykin was a benevolent martinet. In the right season, he frequently invited members of his team for a weekend at his summer home on the New Jersey shore. They would start back to Camden in his car at about 6:30 on Monday morning. When the dashboard clock of the car read 8:00 am, the start of the official day at the laboratory, the chief would order everyone in the car to stop chatting and begin thinking about the work waiting at the laboratory.

RCA was pushing ahead to market their system despite complications raised by the Federal Communications Commission; then World War II intervened. Actually, the delay proved beneficial since the basic technology was greatly improved in the interval.

My work at G.E. allowed me to roam at will through the entire Laboratory as well as several of the General Electric plants. In my travels, I made a number of good friends outside our own group: Simon Ramo, who eventually contributed to the creation of what is now TRW; Louis Navias, a sophisticated ceramist; Eugene G. Rochow, who went to Harvard and became a celebrated silicone chemist. Irving Langmuir and Katharine Blodgett were in the midst of their experiments on built-up films at that time; I can only hope that my frequent visits to catch up on the latest developments were not too much of a nuisance to them.

The Director of the Laboratory, William D. Coolidge, had revolutionized the technology for producing high intensities of nearly point-source x-rays with the introduction of sophisticated, rotating, water-cooled targets and well-focussed electron beams. Among his many gifts was one that was particularly remarkable. He would devote himself to basic laboratory research for an extended period of time; then, when he found something that might lead to a practical application, he would switch his attention completely to the development of whatever device it might be, and then to its manufacture, making major contributions at each stage. In this respect, he was a disciple of Thomas Edison—but with much more formal scientific training. In discussing stability, he once said to me, "Do not be concerned about the future of the laboratory. Our company would soon lose its markets if it discontinued research for as short a time as five years."

Coolidge was also the innovator in the production of ductile tungsten, a development which gave the company a great advantage in the incandescent lamp industry, which had previously depended upon carbon filaments. I spent a fascinating hour with him as he described the details of that research.

Early successes such as this, combined with patent agreements with foreign companies, made G.E. America's leader in lamp research and production. To avoid anti-trust problems, the company offered licenses of its patents to competitors. There was also at that time an agreement between G.E. and Westinghouse to divide the country's lamp market on a sixty-forty basis giving Westinghouse the smaller share. I believe this agreement was later struck down under national antitrust laws.

Pushing x-ray technology to ever-higher voltages was proving increasingly beneficial in medicine and other fields, and the research was continually advancing. During my stay at G.E. a close associate of Coolidge, Ernest E. Charlton, succeeded in developing a transformer system that would accelerate electrons to one million volts.

Sidney Barnes (Figure 6.1), my nuclear-physicist colleague at the University of Rochester, joined the Laboratory as a summer guest during the summer of 1937, and became fascinated with Charlton's system. He pointed out to me that if the voltage were to be reversed and the equipment used to bombard a beryllium target with deuterons, it would provide a uniquely powerful source of high energy neutrons—a source that would have been of great interest to the scientific community at that time when neutron research was much at the center of attention and sources were relatively weak.

I discussed this possibility with Dushman, who mentioned it to Coolidge. This led to a group meeting with Coolidge, Dushman, Hull, and Langmuir. I presented the case, describing in the process the current status of nuclear physics, with which the others were only vaguely familiar. While Hull was prepared to listen further, Langmuir was adamantly opposed to the idea on the basis of the principle that the nucleus was shielded by a repulsive, electrostatic barrier so high that nothing of practical interest could be expected to emerge from investments in the field. When I pointed out that the neutron was neutral and could enter the nucleus even if it were at thermal energies, he made it clear that he would never support any such line of research. The meeting ended on that.

The next time I discussed nuclear science with Coolidge and Hull was during the winter of 1944–45 when Arthur Compton, then the director of the Manhattan District Laboratories at the University of Chicago, asked me to give them a confidential account of the successes that had been achieved at Chicago, Oak Ridge, and Hanford, Washington. Coolidge was particularly interested in these developments since Frank Jewett (Figure 9.12), then the President of the National Academy of Sciences, had sought his advice early in the war when the practical significance of nuclear fission was being reviewed in the White House.

One of the special pleasures I experienced at G.E. was making the acquaintance of Willis R. Whitney (Figure 7.8), who had founded the Laboratory early in the century, when company laboratories were a novelty in the United States. He had been both student and faculty member in chemistry at M.I.T., or "Boston Tech," as he called it, before joining General Electric. Having built up the laboratory, he had difficulty dealing with the stagna-

Figure 7.8. Willis R. Whitney, the founder of the General Electric Laboratory. (AIP Niels Bohr Library, Physics Today Collection.)

tion of the early years of the Great Depression, and he retired from administration in 1932, leaving the Directorship to Coolidge, one of his admired colleagues. When I was at the lab, he had two passions: to find a preventive or a remedy for arteriosclerosis, and to see if he could purify water by freezing out impurities. In the latter investigation he was premature, for it took the development of the techniques of zone refining, which only came about after World War II, to demonstrate how such a goal might be achieved, at least in principle. Zone refining requires an essentially continuous process, whereas Whitney was focussing on a batch technique.

One of Whitney's strongest maxims was "Never be afraid to work against a theory." Since Dushman was deeply interested in theoretical as well as experimental developments, he was frequently the target of jocular comments from Whitney. Dushman, being unduly sensitive to criticism, took these jibes much more seriously than he should have and hesitated to introduce me to Whitney because of my theoretical leanings. He need not have been concerned. Whitney walked into my office one day soon after I arrived and said, "Say, you are new here. Let me show you this arteriosclerotic artery I have in a sack in my pocket!"

By the summer of 1937 business had improved sufficiently — perhaps because of a rise in military spending initiated by President Roosevelt — to allow the lab to take on a larger group of summer visitors. One guest was Leonard I. Schiff (Figure 7.9) who had just completed graduate work at M.I.T. He was not only brilliant but had many diverse interests, including a special talent for music. He, Betty, and I became good friends and remained close until his premature death.

Robert Hofstadter worked with me in my laboratory as a summer guest

Figure 7.9. Leonard I. Schiff during his Stanford days. "He knew everything." (Courtesy of Stanford University and the AIP Niels Bohr Library.)

in 1938. Years later, with amusement and pride, he informed me that the fluorescent crystal in the detector which he had used in his Nobel Prize winning experiment at Stanford, in which he demonstrated that the proton has internal structure, had been made for him by Frank Quinlan, my technical associate, during his stay in our laboratory.

Another visitor to the laboratory that summer was Louis N. Ridenour (Figure 8.2), who had just joined the Physics Department of the University of Pennsylvania to help develop a nuclear physics program there. He had obtained his degree in high-voltage physics at Caltech in association with Charles Lauritsen and had spent two years as a postdoctoral fellow at Princeton. He was an individual with a remarkably wide range of talents and a brilliant insight into many fields. Our paths were to cross in significant ways on a number of occasions until his death of a stroke in 1959 at the age of forty eight.

Unfortunately, the autumn of 1938 brought on another wave of economic depression and the Laboratory was compelled to close down a half-day each week and curtail other activities. It regained momentum in its own right after World War II.

Among the major events at the lab in 1937 was the arrival of Wayne Nottingham on sabbatical leave from M.I.T. He was placed under Dushman's care and provided with everything he asked for. He soon made his presence felt in the entire lab, as he never failed to offer open criticism of aspects of the facilities that he thought were behind the times. The glassblowers had developed a vacuum pump of which they were very proud. When Nottingham gave them plans for one of his own designs, they offered him one of their prized ones; he refused it. Later he telephoned to say that he would take two of theirs. "I assume we should cancel your other order?" inquired the head of the glassblowers shop. "No," said Nottingham, "I want to use them as crude backup for my own." Betty and I made many outdoor expeditions with Wayne and his wife, Vivian.

One of the principal technical and business consultants to the Laboratory was Dr. Zay Jeffries, a metallurgist who had been involved in numerous activities at both G.E. and elsewhere and had been appointed Technical Director of the Lamp Department in 1936.

In 1938, when the fluorescent lamp was sufficiently well-developed to be sold commercially, Dr. Coolidge held a conference chaired by Jeffries to discuss what such a venture might mean to the company. Jeffries was optimistic; he believed that we could safely anticipate a market for the lamps that might, in the course of time, be worth as much as five million dollars per year. What he could not have foreseen was the coming explosive expansion of factories that lay just ahead when World War II broke out. The fluorescent lamp quickly became adopted for the lighting of factories and offices, and has remained a mainstay ever since. Its efficiency in generating a mixed spectrum of visible light was over ten times higher than that of its only significant rival, the traditional incandescent lamp.

In 1937, Condon, who had been at a fixed rank and salary since he had joined the Princeton faculty, decided to look for another post and accepted an appointment as Associate Director of Research at the Westinghouse Research Laboratories in East Pittsburgh. Soon after this move, he passed through Schenectady on his way to a meeting in Boston, and I had the opportunity to introduce him to Coolidge, Dushman, and Hull.

The management of the G.E. labs had graciously granted me the privilege of accepting invitations to lecture at the summer sessions of various universities, and I spent several weeks at the University of Pittsburgh in 1937 at the invitation of Elmer Hutchisson, the Head of the Physics Department, meeting Nevill Mott (Figure 7.10) and George Sachs, the German-refugee metallurgist then in Cleveland. A similar visit to the University of Michigan in 1938 brought me into close contact with George Uhlenbeck and Samuel Goudsmit, well-established immigrants from Holland and full-time members of the faculty, as well as the Dutch physicist, Hans A. Kramers (Figure 10.5), a summer lecturer.

The visit to Pittsburgh gave me the opportunity to visit the Condons, who were now settled there. In keeping with the traditions established at Princeton, they acquired a large wooden frame house in the Wilkinsburg area, a short distance from the Westinghouse Research Laboratory. Their home was not only sufficiently roomy to accommodate the family but also became something in the nature of an informal center for countless visitors to the laboratory and to the Pittsburgh area in general.

Condon was always innovative and had established a program of laboratory fellowships for new doctoral candidates. Several of my good friends started their careers with these appointments, including Sidney Siegel (Fig-

Figure 7.10. Professor Nevill F. Mott as a young scientist. (Courtesy of Sir Nevill Mott and the AIP Niels Bohr Library. Photograph by Lotte Meitner Graf.)

ure 10.3) and Thomas A. Read, who had been thesis students of Shirley Quimby at Columbia University, and Daniel Alpert (Figure 11.10), who had carried out graduate work at Stanford with William Hansen. James S. Koehler, a colleague at the University of Pennsylvania (about whom more later) also held a fellowship briefly before joining the Physics Department of the Carnegie Institute of Technology during World War II.

In 1937 Clarence Zener (Figure 5.11), whom I had known at Princeton, accepted a position at what was then called the City College of New York, an institution founded in 1847 which then offered excellent, low-cost education to New Yorkers. Traditionally it could boast a fine faculty, and many of its graduates went on to illustrious careers. Zener had become interested in internal factors influencing the damping of solid materials when they are set into oscillation. Although the college did not normally offer experimental facilities to its faculty, Zener found an unused broom closet which he took over, and he was soon turning out interesting data. He carried on his experiments both at City College and at other institutions, and finely honed his experimental techniques. In fact, in later work, he used these methods to determine the rate of diffusion of interstitial carbon in iron over many decades. Early in World War II he relocated to Watertown Arsenal in Massachusetts, where he continued his measurements, and also became involved in applied problems of ordnance.

I visited Zener several times at City College to follow the course of his work and had the privilege of witnessing some argumentative undergraduate students in heated discussions about the situation that was developing in Europe. Most had great hopes for the future of communism and the Soviet Union, but there seemed to be a fairly clear-cut division between those who fastened their hopes on Stalin and those who hoped that Trotsky would find his way back to leadership. This level of radicalism interested me a great deal since in my own student days not so long since I would have enthusiastically joined in the debate.

Radicalism among students was by no means a new thing. Bright young minds are quick to note the many imperfections in the society in which they live. Moreover, admiration for the Soviet Union was easy to understand at that time since it appeared to stand in firm, militant opposition to Hitler in the mid-1930s. There was probably another significant factor influencing the City College students. Many of their parents were active or passive socialists who had come from Russia, where the middle class had little to say in the political system, or were from the countries in central Europe where the failure of the revolutions of 1848 had resulted in a reactionary backlash. To some of these parents, a form of socialism seemed to be an appropriate antidote to deprivation and not out of keeping with the traditions of the United States. My own father, in fact, exhibited strong traces of such thinking as a result of his struggles as a young man, and this, of course, influenced me to a degree. Many of the students at City College heading toward professional careers could see little gain in taking the reins of society away from "plutocrats," who they believed really ran the country, and turning them over to the middle and working classes by democratic means, as recommended by their socialist parents. They favored rather a more elitist system run from above by a self-elected cadre of which they might well be part. To them the Soviet system seemed to offer a solution to the more obvious defects in the world about them.

Most such students eventually shook off such views as the ugly facts regarding the repressive features of the Soviet Union came to light—the brutal, hypocritical purges of the second half of the 1930s and, even more striking, the Hitler-Stalin pact of 1939 which cleared the way for the start of World War II. Unfortunately, a good many students never became free of their naive notions and have had substantial influence on our national policies, meanwhile enjoying the full benefits of the great freedoms the United States offers. They worked in various ways within organizations and through the media to promote the Soviet system or its equivalent, denying the great massacres in the Ukraine and other horrors while doing their best to glorify individuals such as Chairman Mao and Fidel Castro.

My own deliverance from any admiration for the communist system came about in my graduate years when the *Daily Worker*, the official publication of the Communist party, was regularly available at our student residence hall at Princeton. I quickly came to understand the hypocrisy and corruption of the leaders in the movement.

In an article which appeared in the *New York Times Magazine* in the 1970s under the title "Why I Never Go Back To City College," a distinguished alumnus described his own participation as a City College student in debates on the merits of various forms of communism. Frequently he and his fellow students continued their discussions in his home after class while his mother plied the group with chocolate and cookies. On one occasion, after she had closed the door on the departing visitors she exclaimed, "You students, smart, smart, smart—stupid!"

Most of my full-time colleagues at the General Electric Laboratory were investing in homes in the suburbs of Schenectady. There was, however, much excellent rental real estate near the center of the reasonably lively town. Betty and I decided to rent, and so we were able to acquire a cottage on Lake George, which became an anchor in our lives. Intimate familiarity with the Adirondack region brought the same joy that I remembered from happy days in the California mountains.

Our initial plan had been to purchase a plot of land on the shore of the lake and start with a tent platform or an equivalent simple structure. In the course of our search, however, our agent showed us a simple, tastefully designed summer cottage in a bay on the eastern side of the lake. We fell in love with it at once, but did not have the cash in hand to purchase it. Fortunately the owner took a liking to us. He was the celebrated Professor Milo Hellman (Figure 7.11), an orthodontist-anthropologist at Columbia University, with an affiliated appointment at the American Museum of Natural History, who had first recognized the hominoid characteristics of the fossil pre-human now termed Sivapithecus. He allowed us to spread out the payments in a way we were able to manage. Ralph Johnson initially acquired an interest in the place and my father provided a guarantee for a small bank loan which we assumed along the way. We have retained the cottage ever since. It became a year round refuge, not only for Betty and me and our

Figure 7.11. Dr. Milo Hellman, orthodontist-scientist who played a major role in establishing the hominoid nature of dryopithicus (Sivapithecus?). (Courtesy of his daughter Mrs. John L. Bull.)

friends, but for our son Jack, his wife Elise and their three children, Eric, Carey, and Jennifer.

My years with the General Electric Company were very happy and productive ones. I particularly enjoyed the special freedoms that were given me. Unfortunately, the Depression caused a certain stagnation at the lab that stood in sharp contrast to the situation at many research-oriented universities which were moving ahead spiritedly in spite of their financial problems. I could not ignore alternative opportunities and eventually decided to accept one. Conditions eventually improved at the industrial laboratories, but the effect was not fully felt until after 1945. Looking back, it was fortunate that I left the G.E. lab when I did. It did not develop a unified program during World War II but instead splintered into small groups, each of which went its own way, working on problems as they came along. I would not have had the great flexibility or support that the academic environment provided later on. The laboratory did not resume an upward evolutionary path until after the war.

In the autumn of 1938, while still in Schenectady, I received an invitation to attend a meeting devoted to atomic and nuclear physics at the George Washington University in the District of Columbia in mid-January. The organizers were George Gamow (Figures 10.7 and 10.8) and Edward Teller (Figure 7.12).

The meeting (Figure 7.13) was a major historical event. This was not because of its planned agenda, history being rarely that tidy. Excellent and stimulating physicists and chemists were present and fine presentations and discussions were carried on, but history did not appear until Niels Bohr, who had just arrived from Denmark, came directly to the meeting and announced the discovery of nuclear fission by Hahn and Strassmann in Berlin. Three conference participants, Merle A. Tuve, Lawrence R. Hafstad, and my former Princeton classmate, Richard B. Roberts, from the nearby Carnegie Institution, immediately returned to their laboratories to check the accuracy of the announcement and returned dramatically the next morning to provide confirmation through ionization measurements. The dawn of a new age was at hand. The American physics fraternity, which hitherto had been almost invisible, was eventually to be raised to demi-god status by the work which followed from this and other discoveries in World War II. Moreover, the intrigues and altercations within the profession, which also grew, became more and more visible to the rest of society.

By the time of that famous meeting at Washington, Teller[1] had been in the United States for nearly four years and was widely known in both the physics and chemistry communities for brilliant insights into many scien-

---

[1] A biography of Edward Teller is contained in the book by S.A. Blumberg and G. Owens, *Energy and Conflict* (G.P. Putnam's Sons, New York, 1976).

Figure 7.12. Edward Teller during his days at the University of California at Berkeley. (Courtesy of the Lawrence Radiation Laboratory and the AIP Emilio Segrè Visual Archives.)

tific problems covering a wide range of atomic and nuclear science. He and Gamow had added stimulation to the already well-developed scientific life in Washington through their academic-style colloquia.

Teller moved to Columbia University soon after the meeting but then left for the Metallurgical Laboratory of the Manhattan District in Chicago and finally, in 1943, went to Los Alamos where the group that would work on the nuclear bomb was assembling. He soon determined that sufficient attention was being devoted by others to problems associated with nuclear fission, both for steady state reactions and explosives. He could therefore turn his own attention to the problem of releasing large quantities of energy by fusing the light elements, particularly deuterium, which is far more plentiful than uranium and its isotopes and is relatively easy to obtain in pure form. In this way, his work at Los Alamos during World War II made him something of a maverick or loner, working off the mainstream of the laboratory but on a crucial problem.

By the time of my early extended visits to Los Alamos in 1947 and 1948 (see Chapters 10 and 11), Teller had gone to the University of Chicago to return to basic research. I was privileged to use his vacant office at Los Alamos for my work, and I had access to his classified files, which he had left behind. I devoted a number of spare hours to them. It was clear that he had made a great deal of progress toward understanding the special, very high temperature, physical environment in which deuterium and tritium could be expected to undergo rapid fusion. Achieving such conditions in reality is, of course, another matter, and the possibility seemed remote at the time. In view of what has happened since, however, there is every reason to believe that good scientists in the Soviet Union, including Andrei Sakharov, were also beginning to think about the problem in practical terms at that time. I have always felt that, guided by the genius of Sakharov and others and the pressures exerted by both Stalin and Khrushchev, the Soviet Union would have developed a hydrogen bomb without question even if we had not. Both the challenge and the opportunity were much too great for a country focussed on expanding its world power to ignore.[2]

Teller returned to Los Alamos in 1949 after President Truman decided, against the recommendation of Robert Oppenheimer's General Advisory Committee of the Atomic Energy Commission, to go ahead with research on fusion weapons, or hydrogen weapons as they are popularly called. The rest is well-known history. The first U.S. bomb was tested successfully in 1952; a primitive Soviet hydrogen bomb was exploded the next year, and a

---

Figure 7.13. A group photograph of the meeting of physicists at the George Washington University in the District of Columbia in January of 1939 during which a newly arrived Niels Bohr announced that O. Hahn and F. Strassmann had discovered nuclear fission in uranium. The meeting was organized by George Gamow and Edward Teller. The code of names is given below. The original focus of the meeting was on atomic and molecular theory but the topic of discussion changed abruptly with Bohr's announcement. M. A. Tuve, L. Hafstad, and R. B. Roberts confirmed the existence of the large energy bursts associated with fission in an overnight experiment at the nearby Carnegie Institution laboratories carried out during the meeting. A year or two later, most individuals in the photograph were deeply involved in military-oriented research in one way or another. The photograph records, in a sense, the transit of physics in the United States from a true age of innocence, in which the field was scarcely known to the general public, to one in which it became linked to social issues in prominent ways for better or for worse. The individuals in the group are, by row, from left to right, as follows. First row: O. Stern, E. Fermi, J. A. Fleming, N. Bohr, F. London, H. C. Urey. Second row: F. Brickwedde, G. Breit, F. B. Silsbee, I. I. Rabi, G. E. Uhlenbeck, G. Gamow, E. Teller, M. G. Mayer, F. Bitter, H. A. Bethe, H. Grayson-Smith, J. H. Van Vleck, R. B. Jacobs, C. Starr, M. H. Hebb, C. F. Squire. The last two rows, in sequence of heads: J. H. Kuper, A. I. Mahan, R. D. Myers, R. B. Roberts, C. L. Critchfield, Baroff (U.S. Patent Office), A. Bohr, C. F. Meyer, K. F. Herzfeld, R. C. Lord, D. R. Inglis, O. R. Wulf, P. K. S. Wang, T. H. Johnson, F. L. Mohler, R. B. Scott, E. H. Vestine, L. Rosenfeld, F. Seitz, G. H. Diecke, J. E. Mayer, J. H. Hibben, M. A. Tuve, H. M. O'Bryan, L. R. Hafstad, K. P. Cohen, H. J. Hoge, A. L. Sklar, F. D. Rossini. (Courtesy of George Washington University.)

---

[2] An outline of the early history of the Soviet program on fusion weapons can be found in the book *Target America* by Steven J. Zaloga (Presidio Press, Novato, CA, 1993). The initiative was provided by I.V. Kurchatov soon after the uranium program got under way following the end of World War II. Sakharov became the most innovative member of the special team working under the leadership of Igor Tamm.

Figure 7.13. See legend on page 112.

full-scale USSR bomb was tested in an air-dropped system in 1955. Developments in the two countries were almost parallel.

In taking the very important step that he did in 1949, Teller made enemies of some individuals with whom he would otherwise have been friends, and he became friends with others who were allies in his causes, but with whom he would not normally have been companionable. There is an ancient Chinese curse, "May you live in interesting times." That is basically the burden that Teller has had to bear.

I have been involved with Teller in many activities in the years since 1949, mainly but by no means entirely having to do with national defense, and I have conceived a great admiration for his courage, and genius, and for his very straightforward but highly reasoned approach to exceedingly difficult problems. His willingness to cut corners and his not-uncommon practice of abandoning tact have caused him to make mistakes and to make more enemies than may have been necessary, but one has never had occasion to be uncertain about either his goals or his inner convictions.

In the 1970s, I might have had some doubts about Teller's ultimate place in history since there were many reasonable people who felt that the Soviet Union might end up providing a positive good to humanity in the future. While that opinion is still held in some intellectual circles, the increasing disclosures regarding the practices of the Soviet leadership seem to me to make it clear that Teller's viewpoints will, for the most part, be vindicated. The basic danger his reputation faces rests on the fact that popular and textbook historians tend to develop stereotypes of prominent figures that bear only a superficial resemblance to real people. In this connection it is of major significance that, during a visit to the United States, Andrei Sakharov praised Teller for his dedication to preserving the strength of the free world during a very dangerous period.

Chapter Eight
Philadelphia and Pittsburgh— World War II

In 1939 Gaylord Harnwell (Figure 8.1), who had been on the faculty at Princeton and who had recently become Head of the physics department at the University of Pennsylvania, made me an offer of an associate professorship I could not refuse, as it would allow me to add members to the staff and work with an important team of experimental and theoretical physicists in solid state physics. Louis Ridenour (Figure 8.2) also had a hand in this decision.

Harnwell was the only child of a well-to-do family and had attended Haverford College as an undergraduate. His parents thought that he might profit from a graduate education at Cambridge University in England, but he realized, after a year or so, that the methods of education practiced there were not well suited for one coming from the kind of undergraduate life he had lived in the United States. He elected to continue graduate work at Princeton where the range of courses was much broader and matched his background more closely.

Those who joined me in Philadelphia were Andrew W. Lawson, a student of Shirley Quimby's from Columbia University with very diversified talents, and a grandson of the distinguished geologist of the same name; Park H. Miller, an experimental physicist from the California Institute of Technology; Robert J. Maurer (Figure 8.3), a student of DuBridge's who had spent a period with Arthur von Hippel at M.I.T.; James S. Koehler (Figure 8.4), a new graduate in theoretical physics from the University of Michigan whom I had met there during the summer of 1938; and Hillard B. Huntington, also a theorist, who had been a student of Wigner's for part of the time I had been at Princeton. Our working life in Philadelphia was idyllic in most respects—until the rumblings of World War II reached us.

The Physics Department at the University of Pennsylvania was quartered in two adjacent buildings which had been constructed in the middle of the 19th century and were situated across from the main library. One, called the Randal Morgan Laboratory, had at one time been an orphanage. Most of the older faculty were primarily physics teachers and were pleasant, cul-

Figure 8.1. (Left) Gaylord P. Harnwell, taken in the 1950s when he was president of the University of Pennsylvania. (AIP Niels Bohr Library.) Figure 8.2. (Right) Louis N. Ridenour, probably taken in the late 1940s. He had a deep instinctive appreciation of the importance of emerging technology. His brief three years at the University of Illinois left an enduring legacy. (AIP Niels Bohr Library, Physics Today Collection.)

tured colleagues. Up to that point in its history, the department's primary purpose had been to provide special service courses to other departments and colleges such as chemistry, engineering, and medicine.

The President of the university at that time was Thomas S. Gates, who had been a prominent Philadelphia lawyer. Many members of the university faculty were from eastern Pennsylvania. The Medical School had a distinguished history going well back into the last century.

One older member of the physics faculty, Dicran Kabakjian, was particularly interesting. He was of Armenian origin and had become an expert in the chemistry and physics of the natural radioactive elements. In addition to teaching, he ran a private business out of his suburban home manufacturing radioactive needles to be used as inserts in tumors for cancer therapy. In the 1980s, a new owner of the Kabakjian house learned of this old business and decided to check the level of radioactivity in the house. It turned out to be so high relative to current standards that the house was condemned. There is, however, no evidence that anyone who had lived there was affected in any significant way.

Another member of the faculty, Professor Thomas D. Cope, was colorful in a different way. He was a font of information not only regarding the

Figure 8.3. *Robert J. Maurer, right, Claude Dugas, left, and me at the top of Notre Dame Cathedral in Paris during August of 1949. The photograph was taken by Maurer's brother, the Reverend Armand A. Maurer.*

history of the department but also about Benjamin Franklin's relationship to the university. Cope had a small collection of ancient research equipment which he claimed had belonged to Franklin, saying: "It's in our tradition."

I encountered Detlev W. Bronk (Figure 13.1) for the first time at the University of Pennsylvania. He was the Head of the Johnson Foundation, an institution devoted to research in the neurosciences. It had been founded with a donation from a senior official in the Radio Corporation of America who was enthusiastic about science. Bronk, who had originally been trained in the physical and engineering sciences, had become interested in biology when he worked with A. V. Hill and E. D. Adrian in England, and became a successful physiologist. He greatly enjoyed meeting new people, and welcomed me to the campus soon after my arrival.

I was asked by the Dean of Engineering if I would be willing to teach a night school course for working engineers from the local community. I chose the topic "physics of metals" and soon had in the class a group of twenty five or so active metallurgists from various industries. We ended up running it as a seminar which continued until war work demanded all of our time.

Inspired by the class, I joined the American Society for Metals and got a good feeling for the profession by attending the local lectures and business meetings of the society. I soon learned that the metallurgists used their regu-

***Figure 8.4.** Harriet and James S. Koehler, lifetime friends and colleagues.*

lar gatherings at the club bar to trade important non-competitive technical information across company lines. Collectively, they considered themselves a guild and their loyalty to one another was comparable to that which they gave their employers.

Betty and I were already familiar with Philadelphia and enjoyed its many pleasures, including the symphony orchestra. We were also happy to be close to so many friends from our Bryn Mawr and Princeton days. Initially we lived in apartments on Spruce Street near the campus. However, on the assumption that we would be at the University for many years, we acquired a newly built house in Wynnewood with easy access to transportation into the city. Our first two years in Philadelphia were very pleasant ones; we felt at home there and, in addition to the music and theater, enjoyed Saturday afternoon walks along the Wissahickon River in Fairmount Park in the spring and autumn. Harnwell and his vivacious wife, Mollie, took a great personal interest in the members of the Department and we spent many happy evenings at their home.

Harnwell, who had been accepted immediately by the local community, succeeded in raising sufficient funds to move ahead with a program in nuclear physics. He and Ridenour designed an electrostatic accelerator to be located in a vertical, cylindrical pressure tank in an open space behind the physics buildings. Construction was underway when I arrived early in 1939. In the meantime, my colleagues in solid state physics became actively involved in experimental and theoretical research of their own choosing.

Harnwell and Ridenour also successfully applied for a federal grant which allowed the Department to hire a number of well-qualified unemployed workers, such as machinists, cabinet makers, and draftsmen—a program of the Federal Works Progress Administration. Philadelphia traditionally had been a highly industrialized area characterized by what were known as specialty "small parts" industries, as opposed to the heavy industries associated at that time with the Pittsburgh area and parts of the Midwest. There had been much unemployment in the region and we were able to obtain the services of a remarkable group of skilled workers. The same program provided a grant to support indigent but talented students. During that period, Betty and I were working very hard to complete the survey volume on solid state physics which we had begun in Rochester and continued in Schenectady. The student typists and the capable draftsman hired on the government grant were enormously helpful in speeding the work along.

Harnwell was also especially helpful in taking the book to publication. Early on I had contracted with a well-known publishing company which had decided to enter the field of scientific publishing. As the manuscript began to take shape, however, their enthusiasm waned and their editor requested that many significant sections be condensed or eliminated entirely. On learning of the situation and my deeply felt frustrations Harnwell, who had had several books published by McGraw-Hill, spoke to the management there and arranged to have my manuscript accepted in its entirety; it was published in the special physics series previously edited by Richtmyer and now edited by Lee DuBridge. I have ever since cherished a special lifelong gratitude to Harnwell and DuBridge for their generous act. The book went to press in the autumn of 1939 and appeared in 1940. Had it been delayed just then, it probably would not have come out until after World War II.

Harnwell, incidentally, had promised his wife that he would not spend late nights at the laboratory; as a result he devoted his evenings at home to writing physics textbooks. Many students had Mollie to thank for those excellent books, which satisfied a great need at the time they were published.

The Physics Department needed additional staff in the academic year 1939–40, and I recommended that Harnwell hire the theoretical physicist, Leonard Schiff (Figure 7.9). We had met and become friends with Schiff in Schenectady in 1937 when he was a summer guest of the General Electric Laboratory, as mentioned in the previous chapter. He had spent the intervening two years with Robert Oppenheimer at the University of California at Berkeley as a postdoctoral fellow. He made an instant hit in Philadelphia and quickly became a much-admired member of the staff. Ridenour commented one day: "Schiff knows everything." The years at Berkeley had greatly expanded his horizons.

It would, incidentally, be hard to overstate the influence that Oppenheimer had on the stimulation of theoretical physics in the United States. On returning from postdoctoral research in Europe in 1929, he accepted a joint appointment in the physics departments at Berkeley and at Caltech. He soon became a magnet for an increasing number of graduate and postdoctoral scholars committed to theoretical physics. These young colleagues inter- acted in a remarkably productive way not only with Oppenheimer, but also with one another, ultimately forming what was essentially a national school.

It is beyond question that Oppenheimer was one of the most creative of his generation of American-born theoretical physicists, probably even more so than the published record indicates. He also brought out the best quali- ties of those who had the good fortune to work with him, as was illustrated by their productivity in later life. Not surprisingly, he inspired in his young colleagues a devotion bordering on religious worship both during and after their immediate association with him.

In addition to my stimulating research and graduate teaching, I had the enjoyable task of shepherding a group of very capable advanced under- graduates through the basic courses. These were highly intelligent, earnest, hard-working students, all of whom later had successful careers. One of them, Robert Weinstock, did graduate work at Stanford and eventually be- came a Professor at Oberlin College. Another, Julia Teitelbaum, became ac- tive in the women's professional movement after the war and wrote force- ful letters to me on the subject in later years. One of our new graduate stu- dents, Herbert Fusfeld, eventually developed an organization which brought prominent scientists and international businessmen together for informa- tive discussions in a highly affective way.

Robert Hofstadter joined the Department for two years after completing graduate work at Princeton but did not seem to find the atmosphere conge- nial and left to take a position at City College in New York. He became a member of the junior teaching staff at Princeton after the war, but in 1950 he went to Stanford where he became a distinguished member of the depart- ment, eventually winning a Nobel Prize for his work on the scattering of high energy electrons by protons which demonstrated that the particle had composite structure.

My parents came to Philadelphia in May of 1939 on their way to making an extended tour of Europe. My father had just retired; he wanted to meet his relatives and to see parts of Europe which had been inaccessible to him when he left in 1895. When they arrived in Philadelphia the international situation appeared ominous—Hitler had just invaded Czechoslovakia after stating unequivocally that he would not do so. John von Neumann had warned all of his friends that a great war was about to start. I attempted to alert my parents to the situation, to no avail. My father erroneously thought that all parties were in agreement, that it would be madness to start another large-scale war. My parents then proceeded to journey through England, France, Germany, Switzerland, Austria, and Italy. We met in New York City on their return in August. The world situation looked even more ominous but my father, having on this occasion lost his remarkable prescience, retained his optimism. The war started before they reached California on their way back from New York.

When my father saw the large pressure tank that would house the accelerator being erected behind the Physics Department, he said he thought it was foolhardy to build what could potentially be a gigantic bomb so close to buildings occupied by so many people. He proved prescient in this case, although fortunately no one was injured. The insulating columns of the electrostatic generator had been deemed fireproof, but this proved not to be the case under the actual working conditions, and the system caught fire. Fortunately the pressure valve at the top of the tank functioned as intended; the very high pressure generated by combustion was relieved so that there was no explosion. Unfortunately both Harnwell and Ridenour were away at the time and I had the unpleasant duty of telephoning the former to tell him what had happened.

In February of 1941 I took a trip to the West Coast by train to see my family. It was my first visit to the Bay Area in five years and I found many changes. Among other things, the bridges across the Bay and Straits were now completed. In visiting Berkeley I was asked to give an impromptu colloquium talk, and I presented some of the views on the theory of dislocations in crystals that Thomas Read and I had developed together in Pittsburgh. Oppenheimer was in charge of the colloquium and invited me to his home with a small group for the evening. Naturally, we all discussed the frightening developments in Europe and it was clear that he was trying to make up his mind about what to do next. Thoughts of actually constructing a nuclear bomb still lay two years ahead, although everyone was already concerned about what the discovery of fission might imply.

I was invited to lecture at the university summer school at Michigan in 1941 — this would be the last "ordinary" undertaking before larger events took control of our lives. Two of my fellow participants were Wolfgang Pauli (Figure 8.5) and Victor F. Weisskopf. Many of the students on campus that summer were from Latin America; our government was doing its best to cultivate strong relations with the Latin Americans, in the hope that they would stay neutral in connection with the war in Europe.

Pauli had recently been married to an intelligent and attractive Swiss professional woman who was clearly the ideal companion for him. Among other things, she asked if I would teach him how to smoke a pipe since she thought that that might help ease the various tensions under which he labored. Actually, that was a dreadful period for him in several ways; not the least of his worries was the diversion of the physics community from basic research. He was heard to mutter: "These patriots!"

Someone once said that if Pauli treats you well it means he does not consider you worth insulting. During that summer he treated everyone very well but I hopefully ascribed this to the good influence of his wife. Once,

Figure 8.5. Wolfgang Pauli, the great scientist in his youth. (AIP Emilio Segrè Visual Archives.)

while attending an informal party at the home of Professor Otto Laporte, I was asked to pour beer from a two-quart bottle. These were the days before air conditioning, and in the humid room the chilled bottle became very slippery. It slid from my hands, rolling and spinning on the floor and spraying the lower part of the room with beer. I started to apologize profusely but Pauli tapped me on the shoulder and said, "Do not apologize. The party was quite dull until you did that."

The outbreak of war in Europe had relatively little effect on Betty and me at first but, as will be seen, it made increasing inroads in our life as time went on. By 1942, I found it necessary to give up virtually all teaching and basic research; I was spending more and more time away on special missions related to war-time research and development. Ridenour had gone to the radar laboratory at M.I.T. in 1940 and Harnwell was called away in 1942 to direct a Navy laboratory in San Diego devoted to radio and acoustics. This was, incidentally, the time when Eckert and Mauchly were beginning their work on the first electronic computer, the ENIAC, in the Engineering School of the University.

After reviewing the situation in the Physics Department, Harnwell decided that in his absence it would be wisest to appoint Leonard Schiff the Acting Head since Schiff was much less involved in external programs at that time. This was undoubtedly the right decision under the circumstances: the Department had very little in the way of auxiliary staff to keep things running. The situation, however, did raise questions in the back of my mind about my own future at the University. There was always a chance that neither Harnwell nor Ridenour would return. It was expected that the former would become the president of a major university in the not-too-distant future. And as the nuclear program Ridenour had established had gone up in smoke, there was little to hold him in our department. Moreover, neither Schiff nor I were fully tenured or well-established at the University. He could easily get a much better offer from another institution, changing the entire situation in the Department. In fact, he was called to Los Alamos in 1945 to help in the final stages of the work there that would lead to the so-called Trinity test of the bomb. Fortunately for the University of Pennsylvania, he returned as soon as the war was over.

In mid-1942 the position of Department Head at the Carnegie Institute of Technology in Pittsburgh became available and Condon proposed me to fill it. I knew the Institute only superficially through having attended meetings of the American Association for the Advancement of Science and the American Physical Society in the city. While on an exploratory visit, I learned that the Department, being in an engineering college, had an able and seasoned support staff. One of the key members of the physics staff, Emerson M. Pugh, had obtained his Ph.D. at Caltech. Others, such as Charles Prine, William Mitchener, and Charles Williamson, were fine, experienced teachers with a wide range of intellectual interests.

The President of the institution was Robert E. Doherty, an engineer who had helped the General Electric Company develop an in-house educational program, and who had then served on the Yale faculty. The Dean of the college (or Director, as he was then called) was Webster N. Jones, a Harvard-trained chemist who had played a role in the development of the rubber industry. Doherty proved to be an excellent, logically minded administrator and community leader. Jones was a humane, sympathetic administrator who found some decisions painful to make and relied on Doherty's judgment in many important areas.

The Department of Physics also had associated with it a small research group under the leadership of Otto Stern (Figure 8.6), a brilliantly creative scientist who had come to the United States as a refugee from Germany in 1933 thanks to a substantial grant from the Carnegie Corporation. He was accompanied by his colleague, Immanuel Estermann, another excellent scientist, whom I had already met in the early 1930s when he spent a sabbatical at Berkeley. Also attached to this group was a tenured faculty member, Oliver C. Simpson, who was a fine physicist and chemist in his own right.

Beyond this, I already knew and greatly respected all the members of the outstanding Department of Metallurgy which Robert F. Mehl, the Head, had assembled. One, Charles S. Barrett, was a leading x-ray structure analyst. Another, Gerhard J. Derge, a specialist in metallurgical chemistry, had been a graduate student in chemistry at Princeton during my time there.

It soon became evident that my outside interests were seen as both normal and desirable to the administration of the institute. Such commitments

Figure 8.6. Otto Stern, probably taken during the height of his days at the Institute in Hamburg. (AIP Niels Bohr Library. Photograph by Francis Simon.)

not only indicated that I was involved in significant national programs, but that I could be of help opening doors for others in the institution.

Since the new opportunities offered at the Institute were too attractive to refuse under the circumstances, Betty and I ultimately decided to accept the position and arrived in bustling, dynamic, smoky Pittsburgh at the end of 1942. Betty found a fine flat on Ellsworth Avenue near the campus. It was large enough that her mother could stay with us as she had in Wynnewood. There was also room for an intermediate-sized grand piano which she enjoyed for the remainder of her life.

We also had room for Betty's Uncle Charlie. Her Aunt Nancy, who lived in Maplewood, New Jersey, had died quite suddenly in 1936. After her death, Uncle Charlie began spending his winters with various relatives and was often with us in Pittsburgh after we moved there in 1942. He enjoyed being with Betty's mother, his sister-in-law, and loved the city with all its industry. He knew how to make friends easily, and returned home late each afternoon with a new tale of some exciting adventure. On one occasion he talked his way into the main switching tower for the freight trains that rolled through Pittsburgh. Under instructions, he was allowed to manipulate the switching controls for what he termed a "jolly good time."

Leaving Philadelphia was not an easy move for us; we were deeply attached to the community there and were sad to leave our beautiful home. We especially hated to part from the Harnwells since we had become very good friends and had much admiration and respect for them. We had supposed that we would spend the rest of our career at the University of Pennsylvania; however, we soon made additional lifelong friends in Pittsburgh, among whom were J.C. Warner (Figure 8.7) and his family. Warner was the Head of Chemistry and would succeed Doherty as President when the latter retired in 1949. Another new friend was Richard H. Teare, the Head of Electrical Engineering, whose brother William I had known at the General Electric Research Laboratory.

While I became engaged in many new responsibilities, Betty accepted an appointment to teach physics and mathematics at what was then called the Pennsylvania College for Women, and is now Chatham University. In the process she expanded our circle of friends in the Pittsburgh community.

During this period, several of my colleagues at the University of Pennsylvania left to take up special wartime duties. Both Andrew Lawson and Hillard Huntington joined the Radiation Laboratory at Cambridge. Koehler became a Westinghouse Fellow, spending first some and then all of his time at the Carnegie Institute of Technology, so that we again became colleagues. Robert Maurer also agreed to join us in Pittsburgh.

One of the most remarkable train journeys one could take during World War II was that between Washington, D. C. and Pittsburgh on the Baltimore and Ohio railroad. The train left Washington at about five thirty in the afternoon and wandered through Maryland to Martinsburg and Cumberland and then along the Monongahela River to Pittsburgh, arriving somewhat before midnight. About twenty five miles outside of Pittsburgh, the train entered a veritable Dante's Inferno. Steel-making furnaces lined the river all the way into the city, roaring and smoking until one really felt as though one were in the heart of Vulcan's forges.

Returning from Washington to Pittsburgh on this train early in 1944, I

Figure 8.7. Dr. John C. Warner, taken during his days as President of what was then the Carnegie Institute of Technology. (National Academy of Sciences.)

encountered George B. Kistiakowsky with whom I had worked closely in Pittsburgh (about which more later). He had finished his work in Pittsburgh and was on his way to Chicago and Los Alamos, where he would remain for the rest of the war developing explosive systems for nuclear weapons. The war in Europe seemed to be at a definite turning point and we spent some time discussing what might follow. I asked him if he thought that some form of political corruption might eventually overtake the Soviet Union, assuming it emerged victorious. He pondered my question and said, "I believe the key individuals in the system are honest at present, but no one can vouch for the future." It is clear that he had very mixed feelings about the Soviet Union. He still had relatives there, including a brother, so that he wished his homeland well, but he also realized that he was now a complete stranger to its way of life.

Soon after I arrived in Pittsburgh, the administration of the Institute was asked to establish engineering programs for young recruits in both the army and the navy. Most of those selected for such programs had either previously been in college or had intended to go to college. We realized right away that the school could accommodate only one of the two educational programs, and after looking into the two plans, I felt quite strongly that the navy program was better planned and, being more selective, would bring in better-prepared students. The administration, however, decided to accept the proposal from the army, citing the fact that it had had an army program in World War I as the basis for its decision. The campus was soon flooded with young soldiers in uniform and the teaching staff was heavily loaded with classes. Unfortunately, the army had not let it be known that it was actually using the university campuses as reservoirs for troops who would be mobilized for the invasion in Europe. Once the plans for that invasion had solidified early in 1944, the recruits were taken away and the campus was effectively drained. In checking at institutions which had chosen the navy program, I found that no such disruption occurred.

Otto Stern, mentioned earlier, was one of the great physicists of our time. He came of a prosperous German family and developed an early interest in physics. One of the most stimulating experiences of his younger days was a two-year association with Albert Einstein, who was a decade older, first at the Charles University in Prague and then in Zurich (1912–1914). They spent many hours together, alone or with illustrious visitors, discussing the mysteries that were unfolding in the field of quantum physics, and the two men became close, lifelong friends.

Their speculations regarding quantum theory led Stern to expand the technique of molecular beams whereby the properties of a well-collimated stream of identical atoms or molecules are studied as they pass through appropriately designed apparatus. This technique was first used by a French physicist, Louis Dunoyer; however, Stern and his colleagues improved and expanded the system substantially and obtained many new and interesting results with it. Their most celebrated experiment, carried out at the University of Frankfurt in the early 1920s, was the one commonly referred to as the Stern-Gerlach experiment. In it, a beam of silver atoms, which possess the magnetic moment associated with a single valence electron, is passed through an inhomogeneous magnetic field which deflects the atoms as a result of the magnetic forces acting upon the magnetic dipole moments. The concepts of classical physics would have led one to suppose that the deflection, being associated with random orientations of the dipoles in the plane normal to the direction of motion of the beam, would generate a continuous streak of silver atoms in the direction of the gradient of the magnetic field. Instead, only two, well-separated beams emerged. An understanding of this latest atomic puzzle had to await the development of wave mechanics.

Stern was made head of an institute for molecular beam research at the University of Hamburg in 1923 and there developed a sophisticated research center frequented by many of the outstanding physicists and chemists of the day. He lived a genial bachelor existence in the city, frequenting the theater and opera, and enjoying the best restaurants, the finest cigars, and a myriad of other pleasures. It clearly was a high point of his life during which he became somewhat fixed in his ways.

I. I. Rabi spent 1928 and 1929 at Stern's Institute as an International Education Board Fellow and came back to the United States with a deep comprehension of the issues at the frontier of physics. He also brought back to Columbia University the art and science of molecular beam technology. He and his colleagues used this knowledge effectively to increase the range of experiments that could be made with it. At the end of World War II both Stern and Rabi were awarded belated Nobel Prizes for their work. The latter wrote a moving account of their relationship which appeared in *Physics Today* following Stern's death in 1969.

Stern came to the United States in 1933 at the age of forty five, quitting Germany as soon as Hitler came to power. Among the many indignities he suffered at the hands of the new government was the confiscation of a portrait of his friend Einstein. He once commented to me, "I saw the rise of the Nazi party with great concern. Some of the members were decent and others were clever. However, none of the decent ones were clever and none of the clever ones were decent."

Dr. Thomas S. Baker, the President of the Carnegie Institute of Technology, welcomed Stern cordially and provided him with the best facilities the Institute offered at the time, as well as a research grant from the Carnegie Corporation. Stern and his Hamburg colleague, Immanuel Estermann, soon had a very productive laboratory up and running. Their young American colleague, Oliver C. Simpson, an ingenious and thorough experimenter, was a great help to them in this undertaking.

Unfortunately for Stern and his group, Dr. Baker became ill and had to leave office in 1935. Baker's successor, Dr. Robert Doherty, was an excellent engineer in the classical sense, but had little appreciation of Stern's great work and reputation. He promptly appropriated some of the reserve funds which had been meant to support Stern's work, and turned them to more general university purposes. One of my tasks soon after arriving was to try to impress upon the administration the fact that we had one of the great scientists of the century on our campus. In this I was partially successful. Needless to say, the Nobel Prize eventually awarded to Stern carried its own message, but by that time he had lost most of his interest in the institution.

Milton Plesset, who had been a colleague at the University of Rochester and was a native of Pittsburgh, had told me that, during his occasional visits home, he had enjoyed a series of stimulating, leisurely conversations with Stern at the latter's apartment on various aspects of modern physics. As a result, I was encouraged to follow Plesset's pattern and call on Stern at his apartment once or twice a month, in addition to our meetings at the laboratory. In the process, I enjoyed the privilege of reviewing with him many elements of his very rich professional experience.

Stern retired from the institute in 1946 and moved to Berkeley, California, where he bought a small house overlooking the bay. Unfortunately one of his best friends in Berkeley, Professor G. N. Lewis, the physical chemist, died just as Stern was making the transition. Having lost his friend, he carried on in Berkeley in a fairly solitary way, almost ignored by colleagues on campus, who were preoccupied with their own creative work as dynamic leaders in high energy physics and nuclear chemistry. When he attended the physics and chemistry colloquia at the University his presence went almost unnoticed. A notable exception was Professor Emilio Segrè, who had worked with Stern in Hamburg; it was Segrè who prepared the special memoir for Stern which appears by tradition in the records of the National Academy of Sciences. I visited Stern whenever I could during trips to the West Coast. He had once expressed the belief that there might be a basic relationship between Planck's and Boltzmann's constants. He spent much time in Berkeley mulling over the issue without, however, any great expectation of success.

Stern's bitterness towards the Nazis was such that he refused all invitations to visit Germany after the war. He also refused any monetary compensation for the violation of his tenure and the confiscation of his property. He once said to me, "I could not possibly go back for a visit. The taxi driver of the car I was riding in might well have been a member of the staff in one of the extermination camps. It would prey on my mind."

Estermann, Stern's former colleague from Hamburg, stayed on at the Carnegie Institute of Technology until 1952 when he joined the London branch of the Office of Naval Research. He ended his career in Israel, teaching at the Israel Institute of Technology. His parents had been part of the original Herzl group of settlers in what was once Palestine.

Oliver Simpson took a leave of absence during the war to work in the Manhattan District laboratory at the University of Chicago with Professor T.E. Phipps, a physical chemist with whom he had done thesis research at the University of Illinois. At the end of the war Simpson was offered and accepted the directorship of a major portion of the Chemistry Division of the Argonne National Laboratory, and eventually was made head of the lab's solid state division.

In the late winter of 1942–43, I took a business trip by train to the West Coast and arranged to travel with Edward Condon. Condon had established a cooperative relationship between the Westinghouse Laboratories and the University of California at Berkeley, where E. O. Lawrence and his team were hard at work on plans for separating isotopes by electromagnetic means. Only later, however, did I become privy to that information. It so happened that Oppenheimer was on the same train; he had just been charged by General Leslie Groves with the responsibility of developing what would become the Los Alamos Laboratory. The recruitment of suitable staff was undoubtedly much on his mind. In the course of that train ride, he invited Condon to join him, and when I returned to Pittsburgh ten days later, Condon was preparing to move to New Mexico.

As it turned out, Condon and General Groves, the head of the Manhattan District which directed the entire nuclear energy program, did not get along at all well. The factors involved were complex, but perhaps the most important was that Condon did not feel that Los Alamos was a good site for a major scientific laboratory because of its location. He was particularly concerned about the isolation he and his family would experience because of the need for very strict security, and the fact that the water supply was marginal. The upshot of this friction with Groves was that Condon reappeared in Pittsburgh a few months after leaving. General Groves was of no help to Condon years later when, during the McCarthy era, he was facing great trials as director of the Bureau of Standards.

Condon once told me that he had developed an abhorrence for military uniforms during World War I when he was treated roughly by a training officer during an ROTC drill. On that occasion, he said he became so angry that he threw down his training rifle, walked away from the drill, and never returned. Small events cast long shadows.

Being a creature of fairly regular habits whenever possible, I frequently arrived at the entrance to the campus of the Carnegie Institute of Technology, having walked there from my apartment, at the same time as the janitor in our physics building, who arrived by streetcar. He was an industrious, elderly Czech, whom I knew only by the name of "Mr. Steve." I gathered that he had immigrated to Pittsburgh as a young man in order to work in one of the factories, probably a steel mill, and was now too old for that heavy work. We usually discussed the course of the war as we strolled together for a few minutes. Mr. Steve was intelligent and perceptive and was very much concerned about the fate of his homeland. He treated me quite formally at first but eventually felt more at ease. One day he said, "Professor, I worry about Europe after this terrible war. It will be like an old man, down and out. I worry that it may never come back again." I was forced to concur with his sentiments. On another occasion he said, "Some of my friends say communism will save Europe but I do not think so. It seems to me that the hammer and sickle that they use on their flag is about right. They catch you around the neck with the sickle and hit you on the head with the hammer. I hope my people have better luck." It did indeed take a great deal of time and effort to prevent Europe from being permanently "down and out." Perhaps the most important factors, granting the help of the Marshall Fund and a long period of peace that was by no means an accident, were good internal leadership and the memory of practices which had produced success and prosperity in the past. There are times when well-established traditions can provide valuable guides if the people have the privilege of following them.

Harnwell returned to Philadelphia in 1946 after completing his work in San Diego and made me an attractive offer to resume a position there at a tenured level. By that time, however, I was too deeply immersed in my work at Pittsburgh and at the nuclear laboratory at Oak Ridge (as will be described in Chapter 10) to consider going back. Harnwell continued to rise in stature and influence at the University of Pennsylvania and was made President in 1953. He not only showed great vision as president in promoting the growth of the institution itself, but also took maximum advantage of the support offered by the community to produce a remarkable transformation that was mutually beneficial to both the university and the neighborhood surrounding it.

Ridenour returned to Philadelphia briefly but then became Dean of the Graduate School at the University of Illinois in 1947. His wartime career at the Radiation Laboratory at M.I.T. had been a most remarkable one. As a result of his great talents and brilliant insights, he rapidly became one of the most effective leaders in the laboratory. I had the pleasure of witnessing his rise in the course of many visits. As the applications of radar increased, and the strength of our forces in the United Kingdom grew, it became customary to send laboratory representatives there to provide technical advice. During his trips abroad in this special service, Ridenour, not surprisingly, not only served the laboratory well as a technical advisor but also succeeded in cultivating the members of the upper levels of command socially. He made friends easily, both for himself and for his institution.

Some presumably exuberant escapade, to which I was never privy, eventually led to a request from the command in England that Ridenour return to the United States and remain there for the duration. When I next saw him at the laboratory, he was back at his bench in a small room, just where he had started on joining the laboratory. He was assembling his own equipment with enthusiasm and carrying on as though this was a perfectly normal state of affairs. As the war was ending, he took on the assignment of organizing and editing the twenty seven volume Radiation Laboratory Series on radar. This was a superb accomplishment; the series was the major means by which the enormous amount of knowledge gained in the field during the war years was disseminated. Our paths were to cross again at the University of Illinois.

Leonard Schiff meanwhile left the University of Pennsylvania for Stanford University in 1947 and became Head of the Physics Department there a year later. I visited him at Stanford a number of times. He followed Paul H. Kirkpatrick, an x-ray physicist, who in turn had followed David Webster. As anticipated, Schiff demonstrated brilliant leadership in his new post and brought the Department to new heights of achievement prior to his premature death at the age of fifty five in 1971. The memoir for Schiff written by Felix Bloch for the National Academy of Sciences is a moving account of his career.

Chapter Nine

Applied Research
1939–1945

Any hope I might have had of returning to a completely sheltered academic life when joining the University of Pennsylvania was a vain one. Not only would it have taken much more willpower than I possessed to turn my back on scientifically interesting aspects of applied work, but I would also have to have had the good fortune to live in a world without wars. The more immediate requests, however, came from private industry and were a consequence of my published research at the General Electric Laboratory.

Soon after I arrived in Philadelphia, I received a letter from A. M. Erskine of the DuPont Laboratory in Newark, New Jersey asking if I would be willing to visit there for a day to discuss some problems regarding the tendency of several of their inorganic colored pigments to darken in the presence of sunlight. The Newark laboratory, located on Vanderpool Street, was attached to a plant that produced so-called dry colors—colored pigments which were used in many products ranging from road markers to automobile paints to printing inks. The Plant Director was Edward R. Allen, who had been on the staff of Cornell before joining DuPont.

This invitation launched me on a thirty-five-year association with various Departments of the DuPont Company—an association that provided me with benefits in terms of intellectual stimulation and personal friendships that far exceeded the monetary compensations. The uncooperative pigment which had prompted the visit was chrome yellow, a form of lead chromate. With the aid of a research fellowship at the University provided by the company, we were able to demonstrate that the lead chromate was inherently unstable with respect to the loss of oxygen by decomposition, and that this instability caused its color to shift from bright yellow to a dull and uninteresting green. Exposure to sunlight enhanced the change.

There were some fascinating developments on colored pigments underway at this time; one came about with the introduction of metal-organic compounds, which proved to be very stable and allowed for the production of a great variety of interesting colors, including blues, greens, and maroons. The copper pthalocyanines were among the first in the family but many others soon emerged. The discovery of their usefulness had been made in England, but DuPont was exploiting their possibilities to the limit with an able team first led by Dr. Donald S. Killian.

Our success with chrome yellow and other similar problems inspired J. Elliot Booge, who was head of a company plant in Newport, Delaware which dealt with white pigments, to request a consultation with me. White lead or lead carbonate pigments, which had been in general use for centuries, could produce so-called painter's colic, a deadly disease arising from lead poisoning; their use had finally been banned in the early 1930s.

My maternal grandmother is buried in Mount Olivet cemetery just south of San Francisco. During a visit there with my father when I was young, he took me aside to show me a large plot of ground that had been reserved for members of the house painters' union. The headstones showed that most had died in their thirties and early forties, doubtless leaving behind widows with young children. This was a part of the tragic consequences of the use of lead-based paints.

Responding to the dangers of lead, industry turned to other, safer compounds, the most promising of which was titanium dioxide. This comes in two crystalline forms, both of which have a relatively high index of refraction. The group at Newport posed two major problems: first, what is the best way to produce suitable grades of the slightly preferable of the two forms, namely rutile; and second, is there a serious commercial competitor to titanium dioxide? The first problem was solved by the development of a process in which the oxide was generated by the combustion of titanium tetrachloride; this method is still widely in use. While the burning of titanium chloride is inherently simple, the all-important goal of obtaining a satisfactory pigment-grade material involved much research which I both followed and contributed to with great interest.

The search for possible competing alternatives to titanium oxide went on for several years, in the course of which we combed the literature for possible colorless compounds having a high index of refraction. In examining these potential rivals, we needed to determine first, which compound was the easiest to acquire or produce and second, which would ultimately render the highest quality pigment that had all of the properties we required. The upshot of this work was that the special advantage of titanium oxide was confirmed. The strongest contender, stochiometric silicon carbide (SiC) could, in principle, be a competitor; however, no method was found to produce from it pigment-grade forms having the precise one-to-one ratio of silicon to carbon needed for ideal whiteness at a sufficiently low cost.

As might be expected, I made a number of very close friends through my association with the DuPont chemists. Among them was Dr. C. Marcus Olson, who made forays with me into many technical problems, as well as into the shops of many Persian rug dealers.

Soon after I started working with the group in New Jersey, I was asked to visit the Laboratory of the Corning Glass Works in Corning, New York. The Director of the Laboratory was Jesse T. Littleton, who had done extensive studies on the mechanical strength of various glasses. However, the individual who had actually made the request was C. Hawley Cartwright, a brilliant but physically fragile physicist who had carried out graduate work at Caltech and had spent a postdoctoral period in Europe. Among his other accomplishments, Cartwright had perfected a means of putting non-reflecting coatings on optical lenses, and had rendered a great service to the military by making the process available to them for use on their old service binoculars.

One of the significant peripheral rewards of my visits to Corning was the privilege of meeting Eugene C. Sullivan, the inventor of Pyrex glass. He was then just turning seventy years of age and was a sprightly, widely informed individual.

It was well-known that some glasses would fluoresce in the visible region of the spectrum if stimulated by ultraviolet light. The issue that took me to Corning concerned the possibility of using such glasses in the form of an inner coating as a substitute for the crystalline inorganic materials being used in fluorescent lamps. One expected a substantial reduction in cost of lamps if this proved feasible.

Extensive studies carried out subsequently over several years showed that none of the glasses developed for this purpose had the requisite conversion efficiency. Their best value was only one-third or so as large as the values for the best crystalline materials. The fact that this work was going on during wartime made it impossible to complement our investigations with the type of auxiliary basic studies that I would have enjoyed participating in and that normally would have been possible at the university with the help of students. Such research would have given us a deeper understanding of the source of the low efficiency of our test glasses. It is relatively easy to produce thin, optically homogeneous films of glasses which could have been used in studies similar to those commonly carried out with single crystals.

Investigations of the optical properties of glasses took on a new life starting in the 1960s when it was recognized that it is possible to produce glasses in fiber form which have very low optical absorption coefficients over substantial parts of the spectrum; this property makes them very efficient, exceedingly broad-band communication channels. Another emerging interest for researchers in the 1960s was the potential for using special glasses as sources of laser light.

Edward Condon, then at Westinghouse Research Laboratories, asked if I would visit during the summer of 1939 to deliver a few lectures on solid state physics and to hold discussions with a group of research fellows he had appointed to the laboratories. These individuals, who had already received their doctor's degrees, were given two-year postdoctoral appointments with the understanding that they could carry on open publishable research of their own choosing, much as they would have at a university, but with the added advantage of having the facilities of the laboratory at their disposal. The invitation to me raised a somewhat delicate ethical question since I had recently been an employee of the General Electric Laboratories; yet I could not ignore the fact that Condon was an old and valued friend who was asking for my help. Condon and I resolved the issue by agreeing I would not become involved in any activity at Westinghouse that was related to the privileged matters on which I had worked in Schenectady.

As mentioned previously, two of the Westinghouse fellows, Thomas A. Read and Sidney Siegel, were already well-known to me from their days as graduate students when they worked at Columbia University under the guidance of Shirley Quimby. Read was studying the internal energy loss of single crystals of metals which were set into vibration using quartz piezo oscillators in the manner developed by Quimby. In the course of his studies, Read had obtained incontrovertible evidence that some of the loss was associated with the reversible motion of a form of lattice defect termed a dislocation, which had been the subject of previous speculative studies by several individuals such as G. I. Taylor in England, E. Orowan in Germany, and C. M. Burgers in Holland. Read's new evidence prompted the two of us to prepare a series of review articles in which we attempted to describe, in a more or less systematic way, the roles that dislocations could play in affecting the various properties of crystalline materials. This topic came to life again in a very dramatic way after World War II, when F. C. Frank in England demonstrated that dislocations can serve as highly efficient catalysts for the growth of crystals from solution or vapor. His work resolved the mystery of why crystals can grow readily under some conditions once they are nucleated.

Condon repeated his invitation to me for the summer of 1940, but thereafter, I became far too deeply involved in war research to continue the association.

Westinghouse had acquired the right to produce copper oxide power rectifiers of a type that had been invented by Lars O. Grondahl at the Union Switch and Signal Company in the 1920s. Consequently, the Laboratory maintained an active research staff to study the properties of the rectifiers, both to improve them and to test their quality. The leader of the laboratory was Earl D. Wilson, who had worked closely with V. Zworykin when the latter had been on the Westinghouse staff prior to joining RCA. Associated with Wilson was Carl C. Hein, an excellent chemist with whom I became close friends. My discussions with this group prompted the company to give a fellowship grant to the University of Pennsylvania to study the fundamental electrical properties of cuprous oxide. The graduate student who held the fellowship, Stephen J. Angello, subsequently joined the Westinghouse Laboratories and spent the greater part of his career with the company.

Through the Westinghouse lectures, I became acquainted with two other remarkable individuals, Benedict Cassen and Joseph Slepian (Figure 9.1).

Figure 9.1. Joseph Slepian. (AIP Emilio Segrè Visual Archives.)

The former, who had completed graduate work at Caltech, had a wide range of interests in basic and applied physics and had been involved in the development of x-ray equipment for medical use before Condon offered him a position that allowed him to pursue any subject that interested him. We spent many hours together in scientific and technical discussions.

By 1941 it had become evident that rockets were again going to play a role in warfare in one form or another. They had been used mainly for incendiary purposes in various conflicts prior to the Civil War, but had been replaced by more accurate field guns, particularly rifled guns. The vast expansion of industrial productive capacity in the meantime however, had made it possible to reconsider the use of rockets containing explosives in massive barrages, where quantity would compensate for accuracy. I recall Cassen making a prophetic comment at this time: "The evolution of rocketry in this war may eventually make truly intercontinental war possible." Cassen eventually returned to his earlier interest, which was biomedical physics. We had the pleasure of meeting again at the Atoms for Peace meetings initiated by President Eisenhower in the 1950s.

Joseph Slepian, a brilliant engineer, had been involved in a number of significant research developments, including the perfection of rectifying mercury arcs of high power. On one occasion, following a lecture I had given on band structure in solids, he raised a question concerning the possibility of using known materials to develop what is now called a light-emitting diode. His comments prompted a very lively discussion of the subject; unfortunately, the technological developments required to make LEDs possible lay well in the future.

In the early autumn of 1939, soon after World War II had broken out in Europe, I was approached by Herschell Smith and William J. Kroeger of the Research Laboratory of Frankford Arsenal in Philadelphia, a historic arsenal that dated back almost to the eighteenth century. It had been relatively inactive since the end of World War I, but was now expanding its research and production, particularly of small arms, with some focus on armor-piercing bullets. Smith and Kroeger asked if I would be willing to visit Frankford and discuss a number of problems they were facing with their start-up. Since the arsenal could be reached conveniently both by subway system and by the local Pennyslvania Railroad, I agreed to do whatever I could. As part of this expansion, an excellent group of physicists including Colin M. Hudson and Thomas Read, with whom I had worked at Westinghouse, joined the group.

The officer in charge at Frankford was Major Leslie E. Fletcher, an ordnance specialist. One of his chief advisors was a metallurgical consultant, Samuel Tour. Fletcher was subsequently sent to North Africa as ordnance officer to instruct the troops in the use of a new weapon, the bazooka.

While Major Fletcher undoubtedly wanted me as a consultant to the Research Laboratory, he nevertheless enjoyed needling me in a slightly annoying way about my academic status when I first began working with the Arsenal. On one occasion when I suggested that he get a blackboard in his office as an aid to discussions, he said, "You profs would be tongue-tied without a piece of chalk in your hand." Fortunately his jibes became less frequent as our work proceeded. Then one day, when I had been particularly helpful to him, he said, "You're not a bad guy for an academic, Fred, but what would you think of a prof who would ask a question on a final examination on a topic that was not discussed either in the textbook or when he lectured?" Apparently, Fletcher had suffered a catastrophe of this kind during a post-graduate year at M.I.T. while in officer training.

The problems we dealt with at the arsenal were varied in nature. Some were associated with understanding the mechanism of armor penetration, others with the design of bullets, and still others with a form of corrosion termed "season cracking" that affected brass cartridge cases. As our studies progressed other consultants, such as Hans A. Bethe of Cornell and George Sachs of the Case Institute of Technology, were brought in to review special topics. Bethe, for example, wrote an excellent paper on the mechanism of penetration in homogeneous armor. Sachs, who had carried on research in Germany on residual stresses in deformed metals, was especially productive in resolving the problems associated with corrosion.

The arsenal was also visited occasionally by Cyril S. Smith (Figure 9.2) who later did a great deal of work on metallic plutonium at Los Alamos. At that time he was employed by the American Brass Company, whose primary interest was establishing quality standards.

Yet another regular visitor was William E. Deming, a physicist-statisti-

Figure 9.2. Cyril S. Smith, probably taken in the mid-1940s. (Courtesy of Argonne National Laboratory and the AIP Emilio Segrè Visual Archives.)

cian, then employed by the Bureau of the Census, who instilled in us a working knowledge of the techniques of statistical analysis in order to achieve quality control in production. Much of the methods he brought with him had been developed earlier in the century by a group of statisticians at the Western Electric Company led by Walter Shewhart, under whom Deming apprenticed.

Deming subsequently achieved international fame when he introduced the same procedures into Japan after World War II and became a national hero there. One of the members of the Arsenal staff, C. West Churchman, soon became the in-house expert in this field.

An excellent group of metallurgists also joined the staff to work on the composition of projectile steel and the development of steel jackets for cartridges to replace the more expensive brass jackets.

Another full-time staff member at the Arsenal was one of Betty's former classmates in physics at Bryn Mawr, Barbara Raines, who helped develop a program devoted to the production of standardized quartz piezoelectric oscillators which I followed with great interest. Remembering the difficulties her group encountered in developing stable electrodes, I marvel today at the wide proliferation of highly accurate, inexpensive quartz watches.

When I visited Princeton on one occasion, Harry Smyth asked me to join a rather informal study committee which was examining the effect upon various structures of over-pressures resulting from explosives in the air and ground. H. P. Robertson, Walker Bleakney, and Abraham Taub, a former Princeton classmate, were among those looking into the matter. The committee eventually evolved into Division 2 of the National Defense Research Committee. It was first headed by John E. Burchard of M.I.T. and then by E. Bright Wilson of Harvard. I am indebted to Professor M.P. White for extended access to the book *Rockets, Guns, and Targets* dealing with Burchard's activities in World War II.

Bleakney formed a well-populated experimental group to carry on this work at Princeton. In the process he developed what came to be called the Bleakney shock tube, which made it possible to generate shock waves of varying intensity under controlled conditions with relatively simple apparatus. He was joined by several other physicists including Vladimir Rojansky, Merit P. White, who had been at the Illinois Institute of Technology, and Curtis Lampson, a former student at Princeton. I remained active with the committee until 1943 when I became associated with the Manhattan District. Albert Einstein occasionally attended our review meetings in Princeton.

H. P. Robertson moved to England early in the war to pursue special intelligence work. I was to meet him again in Europe in 1945, just as the war there was ending. Smyth soon became deeply involved in the work of the Manhattan District.

The activities of Division 2 served to make it something of a coordination center, or clearing-house, for the national program in classical ordnance. My work at Frankford Arsenal, as well as my activities at the Naval Proving Ground at Dahlgren, about which more later, were regularly reviewed in our meetings. The Department of Ordnance of the Army refused at first to establish any formal link with the Division. This attitude changed abruptly, however, as a result of a Presidential order, probably drafted by Vannevar Bush or James B. Conant, who presumably learned of the problem through Burchard. Thereafter, the members of the committee had relatively free access to army ordnance laboratories throughout the country, including the lab at the Aberdeen Proving Ground in Maryland where a number of excellent scientists congregated. The director of that laboratory was Robert H. Kent, a Harvard-trained physicist. We also established good working relations with the Naval Research Laboratory.

After the fall of France in the summer of 1940, I was approached by Louis T. E. Thompson, the Chief Scientist at the Naval Proving Ground at Dahlgren on the Potomac River south of the District of Columbia, and Leonard B. Loeb (Figure 9.3), a professor of atomic physics at the University of California with whom Betty's brother, Lauriston Marshall, had studied as a graduate student. Thompson had worked with Arthur G. Webster, a famous classical physicist, on ballistic problems in World War I and had since been engaged in related activities, joining the Proving Ground in 1923. Loeb was in the Naval Reserve and was planning to join Thompson in Dahlgren. He had also arranged for some of his former students, such as Norris E. Bradbury (Figure 9.4) of Stanford University, Robert N. Varney of Washington Uni-

Figure 9.3. Professor Leonard B. Loeb as Naval Captain during World War II. (AIP Emilio Segrè Visual Archives.)

versity in St. Louis, and Allan V. Hershey to join him there in establishing a ballistic laboratory. After some discussion, I agreed to serve as a quarter-time consultant since I was unwilling to give up entirely my other activities.

The Naval Proving Ground, in contrast to Frankford Arsenal, was devoted to developing major shipboard armament, including both light and heavy armor, and to anti-aircraft and large-scale armor-piercing missiles. Other academics such as Ralph A. Sawyer of the University of Michigan and Francis W. Dresch, whom I had known from student days at Stanford, were also on the staff.

A notable feature of the Proving Ground was the presence on temporary duty of navy line officers who were due to take up active posts on combat vessels and who were preparing themselves by working with the weaponry at the base. One of these was a brilliant young commander, W. S. Parsons, who came from a family of naval officers. He eventually served as weaponeer on the Enola Gay, which dropped the nuclear bomb on Hiroshima, working as a participant in the Manhattan District on that mission.

In 1940 it was unusual to witness the testing of a single plate of armor at the Proving Ground in the course of a week. Two years later, the area reverberated almost continuously with the roar of large guns and the clatter of smaller ones as the nation geared up towards its full production capacity.

I drove myself back and forth from Philadelphia during the early years of my association with the Naval Proving Ground at Dahlgren, since having my own transportation gave me a great deal of flexibility. This arrangement had, of course, to be discontinued once gasoline rationing took effect. I began traveling instead by train to Washington, which also entailed the use of a somewhat antiquated bus service that followed the road on which

Figure 9.4. *Norris E. Bradbury, taken in the early 1960s in his office at Los Alamos. (Courtesy of Los Alamos National Laboratory and the AIP Emilio Segrè Visual Archives.)*

John Wilkes Booth fled from Washington. I sometimes wondered if I wasn't riding on a bus he might have used. The bus crossed the Potomac on a high bridge very close to Dahlgren and made a stop there. The ride to Dahlgren was fairly reliable but the return bus, which started from some place south such as Norfolk, was frequently behind schedule.

Often, while I was waiting at the gate for the bus back to Washington, someone who had driven down from Washington on business would offer me a ride back. On several occasions, the someone was a group of labor organizers who had been seeing to some business at the Proving Ground and were on their way home by way of Washington. As a rule, they carried on with their conversation as though I were not in the car. I soon gathered, to my amusement, that they had been bootleggers in the days of prohibition and had changed careers when the Volstead Act was repealed and the right to unionize became federal law. It also became evident that a new and powerful social element was emerging in the United States. These men were quite different from the most radical labor agitators of an earlier time who were prepared to destroy the structure of society, and were well-known in my native San Francisco. The new leaders were more interested in manipulating the system for their own personal benefit.

The laboratory group with which I became associated at the Proving Ground developed a scaled-down test system using three-inch projectiles and comparable armor plate in order to study the relative effectiveness of various projectile designs. While armor was relatively easy to obtain, procuring special orders of projectiles was almost impossible if one went through formal channels. I managed to find a shortcut by having my metallurgical colleagues at Frankford Arsenal order appropriately designed projectiles from their extensive machine shop—a relatively routine procedure. John Desmond, a Harvard-trained metallurgist at the arsenal, was an accomplice in this work. My frequent arrivals at Dahlgren carrying several special projectiles made me very popular with the laboratory staff.

Louis Thompson took me aside at one point to show me a report he had written in the early 1930s concerning the possibility of developing a self-propelled aerial bomb that could penetrate the relatively light deck armor used in naval warships. He was reprimanded by the naval command for making such a proposal and was told to cease discussion of it immediately; the topic was classified and placed completely off-limits. He showed me, and presumably others, the report only after news reached us of the sinking of a British aircraft carrier in the Mediterranean with a similar bomb invented and used by the Germans.

The environment at Dahlgren changed abruptly in late 1941 when a new commanding officer took over the Proving Ground. He soon ran the station as though he were commanding a battleship in a potential combat zone. He restricted the freedom of travel of all personnel, both civilian and uniformed, and reexamined all laboratory consultantships. Mine was canceled. Leonard Loeb managed to have himself transferred to a post in the San Francisco Bay area; Dr. Thompson left to join the Norden Company; and Norris Bradbury was eventually (1944) transferred to Los Alamos, doubtless with the help of Parsons and the scientists there. The station became, in effect, a restricted area for the remaining scientists, whom I did not see again until after the war. Bradbury became the director of Los Alamos in 1945 upon the departure of Robert Oppenheimer, and held the post until 1970.

In 1941, Lee Dubridge called me from the Radiation Laboratory, where he had been Director since its start the previous autumn, and asked if I would visit there to discuss a significant problem with which our physics group at the University of Pennsylvania could possibly be helpful. In order to transpose the envelope of the high frequency returning radar signals to a lower frequency range for the convenience of various detection purposes, it was necessary to mix them with a frequency produced by a local oscillator in a non-linear device from which a lower, so-called beat frequency, could be extracted. After much trial and error, colleagues in the United Kingdom had found, possibly in a German technical journal, a method that used a silicon point contact rectifier analogous to the cat-whisker rectifiers used in the primitive crystal radio sets of my boyhood. The advantages of silicon were its semiconducting properties and its comparative chemical and physical stability. It had been common practice up to that point, in both the United Kingdom and the United States, to employ crude, impure metallurgical silicon, of a kind routinely used to de-oxidize molten steel, to which small amounts of aluminium were added as a "doping" agent. Units made in this way were highly variable and far from stable, and we were asked to see if we could improve them.

Our experimental group, led by Andrew Lawson and Park Miller, organized equipment to conduct the study and rapidly confirmed that metallurgical-grade silicon rendered units of low quality. They were assisted by one of our capable advanced graduate students Marshall D. Earle who eventually left to become a naval technical officer. I discussed the problem with Elliot Booge at the DuPont center in Newport, Delaware. He set a team to work which was soon able to produce relatively pure elemental silicon by the reduction of silicon chloride with zinc in a gas phase reaction — a system which fitted in closely with the research program, described earlier, in which we studied alternative white pigments. When the system was working well, the group achieved a purity level of 99.999%. Our colleagues at the university soon found that this material, when appropriately melted with additional agents to enhance the semiconducting properties in a controlled way, yielded rectifying units that were far more satisfactory.

Silicon looks so nearly metallic at first sight that one can hardly believe it is anything but a metal. Once, an experimental reactor in the DuPont laboratory produced an ultra-thin, flake-like form of silicon; I was amused to see that it actually had a transparent, horny appearance not unlike tinted flakes of mica.

As the use of radar increased, many other laboratories affiliated with universities, industry, and government agencies joined the program, under the guidance of the very able staff at the Radiation Laboratory, led by Henry C. Torrey, to increase our knowledge of the field and the quality of the materials produced; germanium, for example, a sister element to silicon, also received a great deal of attention in the labs. This work is described in detail in Volume 15 of the Radiation Laboratory series. The DuPont material was a real boon to radar technology, not only during the war, but well into the following decade.

Harry C. Kelly, who earlier had completed graduate work at M.I.T., helped expedite early work on the diodes as a member of the Radiation Laboratory. He joined General MacArthur's staff in Japan at the end of the war and strove valiantly, and with considerable success, to preserve and nurture basic academic scientific research there in the difficult years immediately following the war. Initially the military personnel of the occupation forces were highly suspicious of any basic research in the physical sciences being carried out in Japan, particularly atomic or nuclear research. Kelly was effective first in halting the destruction of equipment in university laboratories that was underway, and then in facilitating the primary stages of reconstruction. He was considered virtually a holy man by the Japanese scientific community, and, as might be expected, he was soon dubbed "Hari Kari" by his admiring American friends, echoing the Japanese pronunciation of his name. On his death, the Japanese government requested a portion of his ashes to form the basis of a commemorative shrine.

The intense research on the electrical properties of silicon and germanium during the war, carried out by a relatively large group of investigators, completely transformed attitudes toward the pure crystalline forms of these elements. They were no longer regarded merely as exotic materials but became flexibly useful components of circuit elements which could be manipulated to show various properties by the addition of small amounts of other elements. In this sense, the war-time research on diodes laid the groundwork for the invention of the transistor shortly after the war and ultimately the development of the integrated circuit. A new age of electronics had dawned.

Betty's brother Lauriston was fully occupied as a division director of the Radiation Laboratory at M.I.T. during part of the war. He had become a recognized expert on vacuum tube electronics, and he expressed to me what might be called professional annoyance at the fact that the radar engineers were being forced to depend upon semiconducting diodes as essential elements in their devices. On several occasions, he said to me, "If I get a free weekend I would like to invent a vacuum tube that would make it possible to eliminate those off-beat things." I was amused at his attitude, since I did appreciate the very special role the diodes were playing. While the invention of the first transistor did not surprise me, the invention of the integrated circuit was quite beyond my imagination at that time. As it happened, it took fully ten years and a great deal of federal funding to bring the first primitive concept of the integrated circuit to any truly practical application; the new era really did not begin to flower until about 1970. When I discussed this breakthrough with John Bardeen in the 1980s, he acknowledged that it had been beyond his vision as well; he had seen the transistor principally as a companion to, or a more efficient substitute for, the vacuum tube.

The advance of transistor technology ultimately demanded material that was much purer than that produced by DuPont during the war, as purer material would interfere less with the migration of so-called minority carriers. By the mid-1950s, other companies had attained much greater levels of purity, and soon captured the business.

Betty's other brother Robert occupied a special place in his mother's heart. She particularly enjoyed his humorously impertinent comments when something aroused his interest. He wrote to her regularly, and early in World War II, he told her that he had shifted his activities from commercial work to a particularly important area of military research. Mother's conversations with Betty and me made it clear to us that he was indeed a major force in the war effort and that we should be grateful that he was playing such an exceedingly important part. Much to my delight, Robert appeared at one of our regular review meetings on crystal diodes to describe his own contributions, which involved the production of units for a specialized radar system. When I returned home, I commented casually, "Oh mother, I ran into Bob." She seemed startled and said in effect, "How did *you* happen to see him?" "Oh," I responded, still casual, "we are working on the same devices." I do not know if my stock went up or his went down.

Lee DuBridge's request for assistance at the Radiation Lab had enabled us to help create a very productive program on silicon rectifiers at the University of Pennsylvania which continued on there under the leadership of Leonard Schiff.

Soon after joining the Carnegie Institute of Technology, I went back to the Radiation Lab to see if there were other problems there that would be appropriate for an academic staff to address. We were invited to think about participating in a new research program requested by the Navy which was under serious consideration. The problem was that the officers on the bridges of naval vessels had difficulty reading radar traces on conventional cathode ray screens in the daytime because of the bright light on the bridge. The difficulty was particularly severe in the South Pacific. The Navy wanted to know if a suitable alternative could be found which would make the cathode screen more like a printed page, with a dark trace on a white background.

Scientists in the United Kingdom were already experimenting with screens which showed dark on white. Unfortunately, the screens tended to retain portions of their images from one radar sweep to the next so that they became cluttered with overlapping images. A group of us at the Institute agreed to establish a program to study the problem, in cooperation with Wayne Nottingham at M.I.T. and a group at the G. E. Lab under the direction of my old boss, Saul Dushman.

Very soon, we focussed our attention on the use of evaporated layers of potassium chloride on the interior of the cathode ray tube. These layers darkened in a color band closely matching the peak sensitivity curve of the eye, namely the yellow-green part of the spectrum. Maurer, Stern, and Estermann, assisted by a group of younger colleagues, took on the problem. Unhappily, while we learned a great deal about the creation of evaporated layers and the properties of color centers in the alkali halides and even learned much about the so-called burned-in traces, we never achieved a really suitable system. As a result, our group dropped the program at the end of the war. The navy, however, continued to pursue the matter for a number of years using industrial contractors. We learned after the war that the British and Germans had attempted to follow the same path, and had been equally unsuccessful.

I note that even today the fluorescent radar screens on the bridges of ships are surrounded by flexible rubber light shields to enable those on deck to read them in bright daylight.

In 1942, the National Defense Research Council established a laboratory at Bruceton, Pennsylvania near Pittsburgh to explore the feasibility of developing small, specially designed explosive charges that could penetrate tank armor. The charges contained a concave, recessed, cone-shaped metal insert, or liner, which became compressed during the wave-like ignition of the surrounding explosive. The collapse of the cone from the tip forward produced a linear jet of metal along the axis of the cone that travelled at hypersonic speed. While this type of focussing effect had been observed qualitatively in the previous century by many observers (it was sometimes known as the Munroe effect), new research demonstrated that it could be the basis of a very effective anti-tank weapon. It was the crucial component of the so-called bazooka, a hand-held, rocket-propelled weapon that enabled an infantryman to attack an armored vehicle.

The new laboratory was placed under the direction of George B. Kistiakowsky (Figure 9.5), the highly talented Russian-born chemist, with whom I later shared a train travelling from Washington to Pittsburgh (Chapter 8). He was joined by Franklin A. Long from Cornell and Duncan P. MacDougall, already at the Bureau of Mines in Pittsburgh. The group rap-

Figure 9.5. *George B. Kistiakowsky at the peak of his career. (AIP Emilio Segrè Visual Archives.)*

idly and expeditiously determined the best combination of explosive and cone geometry to maximize the weapon's effectiveness. In the process, Kistiakowsky became an expert in the field of chemical explosives and their design for special uses. This knowledge proved invaluable at Los Alamos when he eventually shifted his activities there.

Incidentally, the German army captured some of the first specimens of the bazooka to arrive in North Africa and was soon using a very powerful duplicate—the *Panzerfaust*.

When the Bruceton laboratory was being formed, James B. Conant, the head of the National Defense Research Committee, arranged with Webster N. Jones, his former classmate at Harvard, to have the Carnegie Institute of Technology handle the federal contract for the laboratory as a public service. I once commented to Kistiakowsky that a business discussion with Jones tended to start in the middle and work its way to both ends. He disagreed, saying that in his extensive experience with Jones the pattern of any given discussion had to be represented by a many-branched tree.

In 1943, after the work at Bruceton had achieved its most dramatic successes, the production of devices was well established, and I had joined the staff of the Carnegie Institute, Kistiakowsky visited Jones and me to ask if we could set up a group to investigate counter-measures to the bazooka which would provide protection against the metal jet. My colleague, Emerson Pugh, expressed a great interest in the problem, and with his leadership we initiated what proved to be significantly effective defensive systems. The research involved setting up a small explosives test laboratory in a ravine at the edge of the campus. Initially there were some complaints from our neighbors because of the noise, but they diminished after we erected sound barriers. The work continued in close cooperation with Bruceton until the end of the war.

Passing through New York City in 1943, I had the good fortune to meet my friend, Ralph Johnson (Figure 7.4), who, through some connections I had made earlier in the war via Betty's cousin, Lauriston Taylor, had been appointed a civilian scientist with the Air Force in Europe. Johnson proceeded to fill me in on one of his activities in operational research.

It was well-known that the German anti-aircraft batteries were most effective when the American bombers were returning from their raids since that gave the Germans time to track them and get their anti-aircraft weapons and crews in position. As a counter-measure, the group with which Johnson had been working decided to initiate what they called "shuttle bombing." The strategy was a simple one: Our bombers instead of returning over Germany immediately after a raid, would keep going and land in the Soviet Union. They would remain there perhaps a day or two, long enough to reload their planes, and then make another bombing run on their return to England. The Soviet military appeared reluctant, but finally agreed to the plan. However, when the first group of allied bombers reached the Soviet Union, they found the landing fields completely unprepared for them. The arms that should have been provided for their defense had been used instead for other purposes. Within a short time, German bombers flew over the field and destroyed the allied planes. It became clear that the Soviet Union was not anxious to have the Western allies on its territory in any form. By chance I noted an account of this abortive experiment, soon after its attempt, in a tiny insertion in *The New York Times*. It was never given wide publicity.

Early in the autumn of 1943, Eugene Wigner asked John Bardeen and me to visit the Metallurgical Laboratory of the Manhattan District at the University of Chicago to discuss the possibility of our joining his theoretical group there. The Director of the laboratory was Arthur H. Compton (Figure 9.6). Wigner and Enrico Fermi (Figure 9.7) were heading up two of the main divisions of the Laboratory which had overall responsibility for the design of the plutonium-producing reactors at Hanford, Washington, as well as for the operation of a relatively small laboratory at Argonne Forest near Chicago. A third, chemical, division was under the direction of James Franck (Figure 9.8). Glenn T. Seaborg had a large, quasi-independent group of experimental chemists hard at work in the laboratory at this time. They were

Figure 9.6. (Left) Arthur H. Compton. (AIP Emilio Segrè Visual Archives.) Figure 9.7. (Right) Enrico Fermi, probably taken during his Los Alamos days. (AIP Emilio Segrè Visual Archives. Photograph by E.D. Wallis.)

deeply immersed in the complex problems of retrieving plutonium from material produced in the reactors, as well as other facets of radiochemistry. The university also operated a larger laboratory at Oak Ridge, Tennessee, where experimental amounts of radioactive materials were produced in a one-megawatt graphite reactor.

In December of 1942, Fermi had clearly demonstrated the feasibility of building a controlled self-sustaining nuclear reactor. Wigner and a small group consisting of Edward C. Creutz (Figure 9.9), Alvin M. Weinberg[1] (Figure 9.10), Gale Young, and Sidney M. Dancoff, assisted by some younger staff, such as Francis L. Friedman and David Gurinsky, had anticipated Fermi's success well before the reactor was operated and had been exploring the most practical way to achieve large-scale production of the fissionable element plutonium generated in such reactors.

Wigner enjoyed great good fortune in choosing Creutz, Weinberg, and Young as colleagues. The three proved to possess extraordinary talents that perfectly suited his needs. Young had been a mathematics instructor at a Midwestern college, following a stint at the University of Chicago, and

Figure 9.8. (Left) James Franck, taken in the early 1950s. (AIP Emilio Segrè Visual Archives, Mayer Collection.) Figure 9.9. (Right) Edward C. Creutz, probably taken in the 1960s. (AIP Emilio Segrè Visual Archives. Paul Oxley Studio.)

---

[1] Weinberg has written a book, *The First Nuclear Era*, giving a detailed account of his and related experiences connected with this and subsequent work. It is to be published by the American Institute of Physics Press.

Figure 9.10. *Alvin M. Weinberg consulting with Eugene Wigner. Probably taken at Oak Ridge in the 1950s. (Courtesy of Alvin M. Weinberg.)*

Weinberg had just completed his doctoral degree there in theoretical physics. They were both diligent and highly effective partners in Wigner's theoretical work. He said to me once, perhaps pointedly, "They never make a single mistake." After the war, both men went with him to Oak Ridge where Weinberg ultimately became the Director of Clinton Laboratories (see Chapter 10). Young would later join with John R. Menke (Figure 10.3) to form a company that provided consulting services for the planning and construction of nuclear reactors.

Creutz had begun his career as an experimental high energy physicist at the University of Wisconsin, had accepted an appointment to work with the Princeton cyclotron, and then, at Wigner's suggestion, had agreed to join the Manhattan District. He had originally expected to work in an area closely connected with neutron physics, but soon found that there were very important technical problems of a highly practical kind whose solutions required the type of experimental skills he possessed. In working on these, he became expert in many fields of technology, from metallurgy to helium welding to chemical engineering, and became an indispensable facilitator in the rapid development of reactor technology.

When Wigner brought the famous bottle of Chianti with him to salute the group during the final test of Fermi's reactor in the West Stands of the University of Chicago, he was not merely confident that the reactor would work. He and his close colleagues were already well on the road to deciding what the next stage of technological development could be as they moved towards large-scale, plutonium-producing systems.

In the meantime, the DuPont Company had been brought in by General Leslie R. Groves, head of the Manhattan District, to carry out the final design and construction of the reactors at Hanford, Washington, that would produce plutonium. He apparently felt more secure in taking this route rather than having the Metallurgical Laboratory take direct responsibility.

During the previous year, Wigner had decided that water-cooled graphite reactors operating in the hundred-megawatt range represented the most favorable solution at the time. Although this proposal was made fairly early in Wigner's study, the DuPont staff examined a number of alternatives; they were very deliberate in their explorations, as they were learning the basic principles involved in the process, but ultimately settled upon Wigner's solution.

Because of what he regarded as this unconscionable delay of many months, Wigner was in an irate mood by the time Bardeen and I arrived. He felt that much valuable time had been lost educating the DuPont team. Fortunately, both Fermi and John A. Wheeler, a brilliant theoretical physicist from Princeton, had joined with the DuPont group, under the direction of Crawford H. Greenewalt, to help expedite their education in basic scientific matters. Wigner and his team reviewed the working drawings for the reactors in minute detail and much to the benefit of the program.

Early on, there had been some question as to whether graphite could be produced in a sufficiently pure form to be usable in nuclear reactors, but the manufacturers soon demonstrated that they were able to achieve reasonable results. Herbert MacPherson (see Chapter 10) played an important role in achieving this result. The production of pure uranium, however, was another matter. This problem might have been referred to the pigments department of DuPont with which I had worked earlier; in the event it actually was taken up very effectively by a group under Frank H. Spedding in the chemistry department at the State University of Iowa at Ames. Spedding had been an expert in the field of the rare earths and he and his staff rapidly developed successful techniques for furnishing satisfactory material in quantity.

In any event, work was in full swing when we arrived in Chicago for our visit. Wigner had become very concerned about the possible effects of neutron bombardment on the integrity of the interior structures of the reactor and he wanted to examine the issue in as much detail as possible, using whatever principles we could from theory and experiment. A good experimental team had been established in the Chemical Division under the leadership of Milton Burton, who had a lively group of young colleagues. They employed both the Oak Ridge reactor and a cyclotron located at Indiana University to gather basic information on the nature and the rate of damage to graphite that could be anticipated.

Leo Szilard (Figure 9.11), who was also a member of the Chicago Laboratory, and who was following the various aspects of the work as something of a roving advisor, pointed out that the disorder produced in the graphite by neutron bombardment represented stored energy that could, over time, increase to a point where the graphite would become unstable and, in returning to the ordered state, produce a large amount of heat in a relatively short time, much as occurs in the freezing of a liquid. Such a heat pulse could compromise or even destroy the effectiveness of the reactor.

Bardeen, who was involved in important research on torpedoes in Washington and had a young family at home there, decided not to accept the invitation, but I spent most of the next eighteen months in Chicago. I was joined by Robert Maurer, who carried out a very sensitive series of measurements on the energy stored in graphite that had been irradiated either at Oak Ridge or in the cyclotron in Indiana. Maurer also cooperated with Creutz in exploring the extent to which bubbles could form in liquids at low pressures when subjected to ionizing radiation. This work revealed the effects which were re-discovered by Donald A. Glaser and eventually used in the bubble chambers employed by high energy physicists.

Our work demonstrated that the concerns expressed by Wigner and Szilard were all justified and that special procedures had to be developed to prevent an excess buildup of the neutron-generated disorder.

In the course of the systematic analysis of the changes that might be anticipated in the various components of the reactor, I came to realize that the uranium metal slugs, which received very heavy bombardment from fission fragments, deserved special attention. Quite apart from the potential

Figure 9.11. Leo Szilard playing out his role as The General. (AIP Emilio Segrè Visual Archives.)

lattice disorder was the fact that about twenty percent of the fission fragments were noble gases which could diffuse under bombardment and create gas pockets in the slugs. A program was established at the Carnegie Institute under James Koehler to determine the parameters governing self-diffusion in uranium, and some of its other physical properties. Such work, accompanied by calculations, led us to conclude that the slugs would become greatly distorted and blistered during operation, as indeed proved to be the case.

Once the Hanford reactor had been up and running for a while, one of my DuPont friends confided in me that he and others on the DuPont staff had originally considered us all crazy alarmists, but were now impressed with the accuracy of our original analyses.

It may be added that after the war a graphite reactor in England experienced a thermal runaway of the type Szilard had predicted. Moreover, when the group at Los Alamos constructed an experimental reactor, the uranium slugs were inserted into channels with very close tolerances to ensure good heat transfer. The reactor ran for only a day or so before the distortion of the fuel elements which our Chicago work had anticipated caused it to fail. The Los Alamos staff did not have access to our work because of the compartmentalization of information deemed necessary for national security.

Another important service was rendered by Katherine Way. She had joined Wigner's group in the early stages of its expansion and began systematically collecting the available information on the capture cross section of the various elements for neutrons as a function of neutron velocity. The so-called Kay Way Tables were not only useful for the information they contained but also guided experimenters in carrying out measurements that would fill in important gaps.

In the course of their research on the properties of the fission products, the chemists at the Laboratory produced many unusual compounds. F. William Zachariasen, the remarkably talented x-ray diffraction analyst on the faculty of the University of Chicago, greatly aided this work by carrying out x-ray diffraction studies. Working with him was one of his former students, Dr. Rose L. Mooney. I was continually amazed at the remarkable insight into crystal structures Zachariasen possessed; he could quite often describe many details about the nature of a compound simply by glancing at the diffraction photograph. His Scandinavian temper was as sharp as his perceptiveness, and often exploded when he discovered that some chemist had given him a false report on the chemical composition of a sample provided, a not uncommon event.

Soon after Fermi had accomplished the building and operating of the first successful graphite uranium reactor, he and a colleague, Walter H. Zinn, proceeded with their design for a one-megawatt heavy-water reactor. The new reactor was to be used for experimental purposes, and would be constructed at the laboratory in the nearby Argonne Forest. It went into operation in 1944 and proved useful for neutron research as well as for the production of isotopes. Zinn had been an expert in the field of x-ray diffraction and carried out a series of experiments on neutron diffraction. I had the pleasure of working with him on the translation of standard x-ray diffraction theory for purposes of working with neutrons.

During this period, Fermi gave several typically lucid lectures, in which he expressed his own views of the future use of nuclear energy. Since it was believed at that time that the potential supply of uranium would be small, he thought it would remain a government monopoly and that its use would probably be restricted either for weapons or for the propulsion of naval craft, probably submarines.[2]

To help him administer the laboratory, Arthur Compton brought in some of his colleagues from the Physics Department, as well as his former graduate students. The most effective of these was Samuel K. Allison, an x-ray physicist who was a tenured member of the Physics Department. He could be counted upon to deal promptly and in a most friendly and efficient manner, with most of the problems that came up. Dr. Norman Hilberry, a former Compton student, was also quite effective, serving as Compton's more immediate advisor. Taken as a whole, the atmosphere was congenial, the major source of stress being Wigner's deep concerns about the pace at which the DuPont Company was proceeding with its share of the program, with the consequence that the Germans would develop a bomb first.

Two other individuals one saw frequently were Professor Arthur J. Dempster, a distinguished mass spectroscopist and member of the physics department who was involved in work on isotopes for the laboratory, and Professor William D. Harkins, a physical chemist who had proposed the existence of the neutron back during World War I when the relationships between atomic mass and atomic charge were first being clarified. Harkins was apt to dwell rather frequently on the topic of his prescience. However, it is quite possible that it was indeed he who first drew attention to the possible existence of such an initially elusive particle.

Betty and I became close friends of Professor and Mrs. Robert S. Mulliken in the course of our frequent visits to Chicago during this period. We rented their apartment for part of the summer of 1944, which was an enjoyable alternative to the inexpensive hotel rooms I normally occupied. Mrs. Mary Helen Mulliken was of Austrian ancestry and had an interesting collection of musical memorabilia going back to the time of Ludwig van Beethoven.

Many members of the lab, including the Wigners, the Weinbergs, and the

---

[2] Immediately following World War II, Fermi returned to basic high energy particle research at the University of Chicago, but also became interested in investigations being carried on by a number of scientists searching for signals from outer space that might be generated by other intelligent beings. On attending a summarizing meeting devoted to this topic, he noted the failure of the search and queried with humor and in his typically accented drawl: "Where is everybody?" — a very profound question.

Creutzes, lived quite close to the University and provided friendly gathering places for discussions about the state of the world.

J. C. Warner and his family moved to Chicago from Pittsburgh in the autumn of 1943 when he became involved in a special series of studies in cooperation with the Monsanto Company in St. Louis, which was working for the lab as a subcontractor. Because of our close friendship, the Warners soon became part of the extended Fermi-Wigner family when the project staff got together for social events.

One day, while Dr. Warner was travelling, Mrs. Warner heard a scuffle going on upstairs. After it had quieted down, there was a pause, and then a knock on her door. She opened it apprehensively to a security agent who showed her his credentials and said, "We have just had a bit of trouble picking up a couple of Soviet agents who have been living upstairs for the past few months. They have been using listening equipment to pick up anything that has been going on in your apartment. Everything is under control." Or was, once she recovered from her shock.

One of the most gratifying consequences of my involvement with the group at Chicago was that I was able to arrange for Otto Stern to join the lab as a consultant. This work enabled him to have a number of reunions with his old friend, James Franck. I cherish the memory of their animated discussions during which they attempted to predict not only the future of science but the fate of mankind. During one such meeting, Franck told me that when, during his advanced student days before World War I, he informed his parents that he planned to become a scientist, they responded almost as though he had said he wished to take holy orders. They were completely supportive.

Two of our brightest physics students at the Carnegie Institute, Marvin L. Goldberger and Harold C. Schweinler, were drafted into the Army soon after graduation in 1942. Goldberger had worked closely with Stern and Estermann on a series of experiments connected with the dark trace tubes. On joining the Laboratory in Chicago, I succeeded in having the two transferred from army training to the laboratory as "scientists in uniform." Goldberger cooperated with me in my work on neutron diffraction, and we published a joint paper in 1945. Following the war, he stayed on at the University of Chicago to obtain a doctor's degree with Enrico Fermi, and of course went on to become a celebrated theoretical physicist. Schweinler eventually became a professor of physics at the University of Tennessee.

Every few months General Leslie Groves (see Figure 9.19), the military head of the Manhattan District, assembled the Chicago laboratory staff in a lecture room to reiterate the importance of our work and the need to observe the security rules. While some members of the staff tended to denigrate him somewhat, I found him a fascinating individual. He reminded me very much, in fact, of some of the earnest, intelligent, practical, technical people I had known in industrial engineering practice. Groves was not as boisterously confident during the war years as he appeared immediately following the successful demonstrations of the bombs, and as history seems to remember him.

It was said at the time that the Manhattan District program had been turned over to the Army Engineering Corps for two good reasons. First, the project would involve a great deal of construction work, as indeed proved to be the case. Groves had been in charge of the construction of the Pentagon building in Washington which had been completed just before he was given this assignment, so there was no doubt of his capabilities. Second, it was felt that the engineers would not attempt to out-maneuver or second-guess the scientists, as the members of the Army Ordnance Corps might have. In this they were undoubtedly correct, if my own experience with some ordnance officers was any indication.

Groves, in contrast, catered extravagantly to the needs of the scientists, saying to us once jokingly, "I sometimes wonder why I was picked for this job but guess that it is because I am not afraid to spend money. Let me tell you that you people are doing a great job to help keep up my reputation." And it was indeed true that in serving the needs of the program, the corps extended itself to the very limits of its remarkable abilities. Viewed from my own perspective, there was no significant waste. The main worry related to major, unanticipated failures in the program—which never occurred.

In the years following the war, I was invited to the occasional reunions of the Alsos team, led by Samuel Goudsmit and Boris Pash, that had studied the German wartime nuclear programs. General Groves, to whom the team had nominally reported during its active years, also attended these gatherings. It became clear that, while the scientists among us had easily found new and exciting challenges to which we could devote our energies after the war, most of the active military leaders had tended to experience significant letdowns. Those who, like Groves, were older, were especially apt to live in the past and to take advantage of such occasions to relive old glories.

I last saw Groves in the 1960s under circumstances which I remember with more than a trace of sadness. I had taken an aisle seat on the New York-to-Washington shuttle plane with a briefcase full of work to take care of, and noted only dimly that an older man with a cane was seated by the window. When we reached Washington I arose to disembark and, glancing back, saw that my neighbor was a much-aged Groves who apparently had suffered a stroke. We nodded to one another as I realized with embarrassment and regret that I had missed an opportunity to talk with him once again. He probably knew me then only as "someone" from the project.

There is a slightly off-beat story about General Groves, perhaps apocryphal, that deserves some form of preservation. Soon after Groves was made head of the Manhattan District, he took up offices in the so-called War Department building on 21st Street between C Street and Virginia Avenue—a building that is now a wing of the State Department. He was wondering how to handle all the scientists he was about to inherit. Someone told him that at the nearby National Academy of Sciences was an individual, Dr. Frank B. Jewett (Figure 9.12), serving as President of the Academy, who had run a great industrial research organization, namely the Bell Telephone Laboratories. Jewett was very experienced in dealing with scientists, and it was suggested that Groves consult with him.

Since Groves wanted to maintain the upper hand in the meeting, he called Jewett quite formally and asked if Jewett would be willing to visit him at his offices one block up the street. Jewett, undoubtedly aware of Groves' new assignment, agreed.

Groves started out by saying, "I have to deal with these scientists and I think I know how to do it. I will make up my mind about what should be done and then I'll tell them to do it. If they do not respond, I'll tell them again. If that does not work, I'll tell them a third time, hoping that eventually they will get the idea. What do you think of that?"

Jewett nodded and said, "That reminds me of the new surgery for tonsillitis."

"New surgery for tonsillitis?" queried Groves.

"Yes," said Jewett, "You remove the tonsils not through the mouth, but through the rectum." They became fast friends, and the General never did try out his special technique for dealing with scientists.

Among my most notable memories of that time in Chicago were the hours spent discussing scientific and world affairs with Leo Szilard,[3] frequently in

Figure 9.12. Frank B. Jewett, probably during his period as President of the Bell Telephone Laboratories. (National Academy of Sciences.)

---

[3] An account of Szilard's life and career appears in the book by William Lanouette *Genius in the Shadow* (Charles Scribner's Sons, New York, 1992).

the evenings when we would dine together, either alone or with a small group at the Quadrangle Club. Szilard not only had a thirst for the knowledge and opinions of others but also was a font of more-or-less developed ideas of his own which he had pondered over many decades. We had known one another since my student days with Wigner.

One of his habits was to stop by my office mid-afternoon and say, "Seitz, let us have tea somewhere." We would find a small restaurant where he would proceed to empty the sugar bowl into a cup of tea while he discussed the course of the war, or anything else that was on his mind. Looking back, I can find very few occasions when he expressed negative opinions on matters. He was always looking for positive solutions to difficult problems or situations. Like the legendary Odysseus, he was never at a loss.

Szilard is properly given credit for expressing the view early in the 1920s that entropy is a measure of information, amplifying the views of Ludwig Boltzmann. Recently I took the trouble to read the original version of his basic paper on this subject. Previously I had known of it through the textbook literature. The paper deals with the difficulties that Maxwell's demon would encounter if it attempted to control the flow of gas molecules through an orifice between two containers. It is indeed a classic essay although, for reasons of his own, Szilard embellished it with much extraneous material, perhaps to impress the leading faculty of the time—the *Geheimrats*.

Szilard had faced hard times during his student days in Berlin. He once told me that back then, when he was hungry, he found that he got a kind of satisfaction staring through restaurant windows watching other people eat. Perhaps because of these earlier privations, he was very generous to members of the service staff in the laboratory at Chicago. If he learned of anyone who was in special need, for example someone with a sick relative, he frequently made what he called a "loan" to the individual, while it was clear to everyone that he kept no records and never expected to be repaid.

On one occasion when a group of us were having dinner at the Quadrangle Club, Szilard came hurrying up to our table. Somewhat breathlessly, he explained that he had just had a call from New York, and that he had to travel there that very night. He asked if we could help him out with a collective loan to pay his train fare. We of course agreed immediately, asking only to be told how much he was short of the sum he required. With that he started rummaging through his various pockets, and had soon produced some two or three times as much money as he could possibly need, mainly in crumpled bills. He re-assembled his money, apologized good-naturedly for the intrusion, and hurried out.

One of Otto Stern's phobias about his train rides between Pittsburgh and Chicago was the thought that he might arrive at his berth and find that, due to a booking error, it was already occupied by a woman with an infant, leaving him helpless to complain. To forestall such a catastrophe, he was always at the downtown railroad station a good hour before the gates opened, so that he could charge in ahead of the crowd to claim his berth.

Szilard greatly disdained this practice. His own habit on his trips to New York City was to board the train at the 63rd Street station in southern Chicago where it stopped only for a few moments. He usually arrived at the station with scant seconds to spare, frequently just as the train was pulling out.

During our many evenings together in Chicago, Szilard and I sometimes discussed politics and political systems. A staunch believer in elitism, Szilard did not admire the democratic system as he felt it distributed the authority to make important decisions far too broadly among individuals who had only a superficial understanding of major issues. At the same time, he appreciated Lord Acton's maxim: "Power corrupts; absolute power corrupts absolutely." Moreover, he had no illusions about the vile nature of the current communist or fascist systems which, unlike many intellectuals, he saw as kindred evils. He simply believed that a better system could be invented.

After the war, perhaps because he felt completely cut off from the establishment in Washington, he did cultivate links with the Soviet Union, and did what little he could to urge our government to provide Marshall Fund aid to that country. Perhaps if he had been younger he would have been tempted to become even more deeply involved with the Soviet Union. Given his forthright ways, the association probably would have been disastrous for him. He was never devious, but an ethical and humane person. His greatest weakness lay in his almost absolute confidence in his own opinions—a trait that served him exceedingly well at times, but was also very limiting. The close links Szilard formed with the Soviet Union caused strains in his long, friendly relationship with Wigner who, like von Neumann, had little admiration for dictatorships.

My last significant activity during the war came about as something of a surprise. By the spring of 1945, work at the Chicago Laboratory had quieted a great deal as the plants at Hanford were running relatively smoothly once the crisis associated with the, not entirely unanticipated, discovery that one of the fission products was a strong neutron "poison" had passed. Some of the staff, such as Fermi and his immediate circle, and a few others such as Edward Creutz, went to Los Alamos, where matters were reaching a climactic stage. The staff that remained, including Wigner's group and the chemists, were either following developments under the supervision of the Chicago laboratory or were planing future research projects.

In March I received a request to visit the office of Secretary of Defense Henry F. Stimson at the Pentagon in Washington. The Secretary and his staff had decided to establish a small office at our military headquarters in Europe to collect information on technical advances made by the Germans during the war that might be of special interest either to our military or to industry. The concept had originated with Edward M. Bowles, a professor at M.I.T., now in uniform, who had been serving as science advisor to the secretary.

The new organization was given the name Field Intelligence Agency, Technical (FIAT). I was told that the head of the organization would be H. P. Robertson (Figure 9.13) who, as mentioned earlier, had been in England since early in the war. The staff would be a mixture of U.S. and U.K. personnel, and I was asked if I would be willing to serve as a civilian in uniform with the honorary rank of colonel. After talking it over with Betty and with my colleagues at the Carnegie Institute, I agreed to join the group for the summer while awaiting an outcome of the war in the Pacific, where a crucial invasion of Japan was imminent. I wanted, however, to be free to resume academic responsibilities if and when the war ended.

While making final travel arrangements in Washington in mid-April, I ran into William Shockley, who was also serving the Secretary's office as an advisor, offering his expert views regarding the effectiveness of the bombing in Japan. The bombing by U.S. planes had become intense as our forces neared the islands. Shockley had concluded that the damage to factories, public buildings, and homes was severe and that the civilian economy was all but destroyed. The length of time the Japanese could be expected to go on before they were forced to surrender would be determined by their willingness to persist in a hopeless, suicidal struggle. Experiences like Japanese Kamikaze raids and the fierce battles on the islands had shown us that their resistance to defeat could be ferocious, and that any invading army could expect to suffer heavy losses at their hands.

I flew across the Atlantic in a C54 transport loaded with military personnel and joined the FIAT staff, which convened at Versailles just as the war in Europe was ending. Paris, which had been liberated the previous

Figure 9.13. Howard P. Robertson, known as "Bob," probably taken in the 1940s. (AIP Niels Bohr Library, Physics Today Collection.)

year, was in a new holiday mood and was swarming with American soldiers on leave. On arriving in Paris, I was somewhat amused to note that a large number, perhaps forty percent, of the flags flying in the city were communist. The communist party had been busy paving the way for its future political activities.

In the days immediately following the end of the war in Europe, the bomber pilots took their ground crews for exciting celebratory rides over Paris and environs, flying in squadron formation and at roof-top level. One of their regular flight paths ran along the Champs Élysée. Of course, the noise of the bombers was deafening, and as the aircraft went overhead, civilians in the street would look up and wave happily to them. Several weeks later I saw similar buzzings by American bombers in several cities in Germany; but here, they inspired terror rather than elation in the hearts of the civilians in the streets. Some instinctively dashed into any place that might serve as a shelter from the expected firestorm. I could do no more than feel a sense of relief that the real horror was at an end in Europe.

Our first office was established on the palace grounds at Versailles in the Trianon Hotel, a once very plush place that had been made part of the headquarters of the German army. It has since been restored to its original grandeur. Initially I was given quarters in a private home on the opposite side of the palace grounds from the Trianon, so that I had the great pleasure of crossing the gardens and park each morning, listening to the cuckoo birds in the woods. Once the American troops began to leave Paris either on their way home or to the Pacific Theater, I was given more luxurious quarters in a hotel in St. Cloud.

The children of Paris still showed signs of the malnutrition suffered during the occupation. This, however, did not interfere with their play or humor. They loved to mimic my fractured formal French when I spoke to them. Americans were no longer a novelty.

Robertson had already arrived, accompanied by an English technical officer, Major Gill, who had been in academic work and had then become involved in technical intelligence activities in England. He was a pleasant and intelligent man with whom I spent a great deal of time. Margery Onthank, an American officer in the Women's Auxiliary Corps and the wife of an army colonel, joined our staff along with some other individuals from British Intelligence. Major Gill had several friends, whom we met frequently at meals, who were what might be called hard-core British Intelligence. Unlike Gill and the other Intelligence staff in our office, these individuals were regularly uncongenial and their accounts of their work indicated that they had much more unpleasant, predatory dispositions.

Robertson was physically and mentally exhausted from years of hard work in England. His last position had been as head of the intelligence team gathering information on the V2 rockets—an intense, and frustrating job. When I arrived at Versailles he appeared to be at a loose end, as if he were trying to decide what he should do once the war ended. He was clearly not in the mood to be crossed so we all made sure to watch our step. On one occasion he insisted that I join him in the General's Mess to which he had access but I did not. As he ushered me in, he commented, "This will show them what I think of them."

The state of destruction, dislocation, and general turmoil one witnessed in Europe, and not least in Germany, demonstrated again what appears to be a general principle: Dictators have little interest in the well-being of the populations over which they have assumed control. They have personal ambitions, goals, and hatreds and manipulate the system so as to achieve their personal desires. One would have to leaf through the pages of history with great care and patience to find anything resembling what might be called a benign dictatorship.

Almost as soon as I arrived, our office began to receive translation copies of reports based on secret recordings of the private conversations of top German officials of various ranks. Some of these individuals were defiant in the face of defeat, while others expressed contrition and the hope that the civilian population would not be treated too harshly.

Any prominent German scientist who was thought to have worked in any way with atomic or nuclear physics was taken into custody. A not unreasonable list of those to be detained had been drawn up I know not how, perhaps back in the United States. In visiting one of the hotels in Paris where the scientists were being held for interrogation, I chanced to meet Walter Gerlach who had worked with Otto Stern on the famous Stern-Gerlach experiment of the 1920s described previously. Gerlach, I learned later, had been head of a team attempting to construct a nuclear reactor. When I mentioned my current association with Stern, Gerlach said that the experiment with Stern had been carried out under the most difficult conditions imaginable, because they had essentially no money and very little in the way of equipment in the period just after World War I. Everything had to be manufactured in the university shop, essentially from scratch with whatever scraps of material they could find. He added that he thought it remarkable that so many well-known American scientists were involved in a productive relationship with the military. "With us," he said, "everything was crazy." "(*Alles war Unsinn*)." Most of the German scientists were taken to England for further interrogation and were later brought back to Germany and released. They were still in England when the nuclear ("atomic") bombs were dropped on Japan. Otto Hahn has described his own experiences during those days in an autobiography. In the meantime (1992), the English translations of the conversations of prominent German scientists interned at Farm Hall, an estate in England, recorded by clandestine means, have been made public.

Since my military credentials allowed me to travel as I wished, I decided in mid-May to spend a few days in England and went to visit Nevill Mott in Bristol. The English people were just beginning to appreciate the fact that their war was over at long last. One and all were basking in the peaceful spring weather. The soap-box orators in Hyde Park were claiming that Churchill and the Conservative Party would be replaced by Socialists in the coming election. I thought this preposterous in view of Churchill's wartime leadership. They were right, however.

Late in May, our office moved up to Frankfurt, where I was quartered in a private home taken over by our army. Most of the city was in ruins but the headquarters of the great I. G. Farben Corporation was intact, and was taken over by General Eisenhower and his staff. There were two jokes of the day: one that the building had been spared by our bombers so that Eisenhower could use it; the other that it was untouched because it had been the only building the bombers had actually tried to hit. We shared offices there with the general and his staff for several weeks, and then moved into a small office building near the main railroad station.

My travels in May and June brought me into frequent contact with American soldiers who had recently been involved in heavy combat and who were now on their way to re-deployment. Having survived the European conflict, they would now be shifted to the Pacific Theater for the invasion of Japan. I was struck by the way in which their, often grisly, sense of humor had enabled them to adjust to and cope with the horrors they had been through, and those they could yet expect. Humor is indeed a survival mechanism; sometimes it is the only means by which we can preserve our sanity when dealing with circumstances that otherwise must drive us mad.

The level of destruction of most of the German cities as a result of Allied bombing was impressive, although not surprising since we had seen photographs of the damage in newspapers and newsreels. Many buildings which appeared to be only slightly damaged turned out to be very dangerous since, having been weakened by the bombings, they could collapse spontaneously at any time. What was surprising to me was the degree to which a people can adjust to such terrible difficulties when there is no alternative, taking advantage of whatever shelter and other facilities are available. Somewhat to my surprise, I found somewhere in any bombed-out town I visited, when I cared to look, a group of civil servant architects drawing up plans for reconstruction. This reflected an ancient tradition of city planning in the days when cities were forts.

It took no great imagination to realize that under the best conditions it would be a quarter-century or so before Germany returned to anything approaching normal life. I also concluded that it would require the passing of several generations—possibly a century or even two—for the country to recover psychologically from the traumas of that terrible war, especially the knowledge of the crimes of its leaders. Nothing that has happened in the intervening half-century has caused me to alter this opinion. Granted that the people wanted to rebuild, it was also clear that some outside help would be needed to get them started. This came eventually in the form of Marshall Plan aid, along with the introduction of a new currency which started out valued at four marks per dollar. During the next four years or so, until a solid currency was introduced, trade was carried on largely by means of barter, and, remarkably, packs of American cigarettes became an accepted form of currency. Almost inexplicably, packs with blue labels, on which the U.S. federal tax had been paid, had about twice the trading value of the tax-free packs with yellow labels that had been purchased in a military canteen. As far as I could tell, a yellow-label pack was equal to a day's pay for an unskilled worker. The United States military, however, used a special form of scrip that served as currency within military stores.

The German government had conscripted many foreign workers from neighboring countries into service to help run their factories—these were the so-called slave workers. While most of the people undoubtedly felt a deep resentment at being impressed into such work against their will, others formed attachments to the German civilians with whom they labored and in some cases performed heroic rescues during bombing raids. They apparently came to feel that they and the "common" German people were all in the grip of evil forces beyond their control.

When the war finally ended, the displaced persons were at liberty to do as they pleased. It was common in May and June of 1945 to see long lines of foot travelers on the roads heading toward their homes bearing some representation of their national flag to indicate that they were not Germans. Those whose homelands were now occupied by the Soviet Army faced a difficult decision, since they did not know what their fate would be at the hands of that army and looked for alternatives. Some sought employment with the Allied Forces, while others, less responsible, took to looting vacant factories and office buildings. A strange belief that developed among this group was that a single microscope lens had a very great value, comparable to that of a precious gem. In visiting academic and industrial laboratories, I frequently found valuable microscopes that had been taken apart piece by piece in order to retrieve the lenses. The metal parts from which the lenses were removed were usually scattered about like so much junk. Many of those from Eastern countries who returned to Soviet-controlled areas were mercilessly shot on arrival.

Any form of trade between American personnel and the Germans was technically forbidden, but the bartering started almost immediately once the Army of Occupation was settled in. German families would trade whatever valuables they had managed to save in exchange for food or other necessities. During the invasion and early stages of occupation there was considerable looting, or what the soldiers emphatically called "liberating," anything that might have some special value to them. In extreme cases this included items such as Mercedes sport cars and race horses. Robertson frowned on this practice saying that liberating property was a form of theft and that there was nothing the Germans or anyone else had that he wanted badly enough to exchange it for his sense of personal integrity. I would like to think, however, that he might have made an exception for a bottle of premium brandy.

While in Frankfurt I occasionally went to the movie theater reserved for the service personnel. The ushers were all middle-aged German women. One of them heard my broken German and thought that I might be able to answer an important question for her. She asked, "Do you think you could get them to show `Gone with the Wind' here?" That film had been released in 1940, after the war had already started and was by now something of a legend to German movie-goers.

Robertson and I spent some of our evenings in the residence of the Don Cossacks Russian male choir, a group of refugee Russian Cossacks who had migrated to the United States after World War I and were organized into an ensemble by their leader, Serge Jaroff. They had been well-known on American college campuses and were now entertaining Allied troops. On one occasion I asked Jaroff about a feature of many Russian songs which start off dirge-like but suddenly change pace to lightning speed. Did this shift in tempo have a name? "No," he said, "It's just Russian. Either too slow or too fast, nothing sensible."

Our office never really succeeded in achieving a well-defined purpose during that summer. A flood of Americans from various organizations, both governmental and private, passed through in search of useful information; and a number of German scientists visited us to offer their services or inventions. I made an extended trip to several excellent laboratories in Germany and Austria devoted to advanced research in powder metallurgy, which was widely used in Germany. I was also asked to visit a plant in Hanau nearby to see if it should be permitted to manufacture saccharin as requested by the local Germany authorities. An examination showed that it was a relatively ancient pigment and dye factory, much like the one at the DuPont plant in Newark, New Jersey. I advised approval of the production of the artificial sweetener.

Some of the tasks had their humorous side. A young, very earnest U.S. Air Force officer insisted on taking me to the nearby Taunus mountains to inspect what he claimed was the bombed ruin of a source of "death rays" that could stop the engine of a flying aircraft. I failed to convince him that it was merely a bombed-out early warning radar.

Having an interest in how the universities had fared under National Socialism, I visited as many as I could and found an almost universal pattern. Very few of the older faculty members who were well established by 1933 had joined the party. However, it was clear that no new candidate could expect to join the staff unless he or she were a member. Had the National Socialist regime endured, the universities would eventually have become completely politicized. Administrators and faculty members made me feel welcome during my visits, in spite of the upheaval they were going through.

Under Hitler's government there were essentially no well-organized cadres of academic scientists working in cooperation with the military, as was standard practice in the United States and the United Kingdom. Max Planck, the doyen of German academic science, had sought an audience with Hitler soon after the latter had been appointed Chancellor in 1933 in an attempt to persuade him to cancel his plans to dismiss Jewish scientists. Hitler's response was so vicious that Planck was left a shaken man. The lesson derived from this exhibition of Hitler's special irrationality and vehemence was not lost on other senior scientists. The sensible ones limited their activities strictly to university affairs.

Immediately after World War II, several prominent scientists, particularly Samuel Goudsmit (Figure 9.14), the Chief Scientist of the Alsos mission, said that Werner Heisenberg (Figure 9.15), the major theoretical physicist of his generation who had remained in Germany after Hitler's rise to power, had actually worked on the design of a nuclear bomb for the military. He had, however, so poorly understood the fundamental principles that his research was completely misdirected, and he had failed in his quest. This view came to be widely accepted in the United States.

Heisenberg, in turn, claimed that, in fact, he had had a very clear concept of what was needed to make a bomb. Because of his aversion towards the Nazis, he persuaded the German military leaders that, even if a bomb were possible, its development would seriously drain the resources of the war effort.

Heisenberg was one of the German scientists at Farm Hall who were incarcerated for a number of months after the war in Europe ended. Information gleaned from the translations of the clandestine recordings of their private conversations makes the situation somewhat less enigmatic and seems to confirm Heisenberg's version of events. Within a very short time

Figure 9.14. Samuel Goudsmit, probably taken during his days at the University of Michigan in the 1930s. (AIP Emilio Segrè Visual Archives, H.R. Crane – H.M. Randall Collection.)

Figure 9.15. Werner Heisenberg, probably taken in the early 1930s. (AIP Emilio Segrè Visual Archives, Lande Collection.)

after learning of the dropping of a nuclear bomb on Hiroshima, Heisenberg gave his German colleagues in the detention center a lucid account of the possible means to construct the bomb, and the industrial support necessary to its manufacture. The recordings make it clear that he had a quantitative understanding of the problems involved in the design of a bomb. It also seems clear that he had not been willing to do anything that would encourage Hitler and the military to proceed with the development of such a nuclear device. As a scientist, Heisenberg was a giant whose stature was equal to that of the other titans of his generation, such as Bethe, Dirac, Fermi, Oppenheimer, Peierls, and Wigner. As such, his version of the situation has credibility. He undoubtedly held foolish, naive, or mistaken social or political opinions at times, but he was not alone in that respect.[4] This does not mean, however, that his behavior during the war will cease being a matter of controversy — some very bitter.

---

[4] Perhaps the most thorough attempts to deal with this complex issue are contained in the books *Uncertainty* by David Cassidy (W.H. Freeman, New York, 1991) and *Heisenberg's War* by Thomas Powers (Knopf, New York, 1993). Powers had the advantage of access to the Farm Hall transcripts. In discussing matters with scientists who remained in Germany, one is impressed with the level of ignorance that inevitably prevailed among them as a result of being cut off from the outside world by the war and from their government by mistrust. J. Hans Jensen, a member of the group, once said to me: "We knew anything we heard was probably false, but did not know what was true."

With respect to the German uranium program, Nikolaus Riehl (see footnote 5), a level-headed and observant individual, states that while it is more or less true that there was little enthusiasm on the part of the German scientific community to provide Hitler with a nuclear bomb, the main reason for the failure to exploit any significant phase of nuclear energy was the lack of any intellectual depth on the part of Hitler and his influential associates. The concepts related to nuclear energy were beyond their ken. Noisy rockets and aircraft were matters they could appreciate and support without hesitation. This assessment fits in with the brutally cavalier way, mentioned above, in which Hitler treated Max Planck in the interview in which Planck pled for the retention of the Jewish academic scientists.

Soon after the war, the Soviet Union began to staff some of its laboratories with German scientists who were either official captives[5] or "guests." Heisenberg, then in Goettingen, was invited to visit Moscow to lecture but turned down the invitation with a comment, paraphrased from one of Aesop's fables, to the effect that "many footprints lead to Moscow but none seem to return."

I made several visits to Heidelberg to visit the University. After a little research I located two of my father's sisters, who lived in an apartment along the river. The elder sister had been in charge of a women's secondary school until 1934 when she was retired and replaced by a party member; she was in any case close to retirement age. Her younger sister had been a school teacher in nearby Mannheim and had moved to Heidelberg when Mannheim was heavily bombed. They complained little, although it was

---

[5] Dr. Nikolaus Riehl, an excellent physical chemist, had carried out his doctoral research under Otto Hahn and Lise Meitner (1927), and then became involved in basic research in the Auer Company dealing with specialty products. He was taken as a civilian prisoner by the Soviet authorities in 1945 and spent ten years in the Soviet Union. He was promptly drafted into a position of scientific and technical leadership for the production of reactor grade uranium at a development and manufacturing center some forty five miles east of Moscow. The center reported to Lieutenant General A.P. Zavenyagin in Moscow, but the overall program was under the suzerainty of Levrenty Beria, who visited Riehl twice for reports of progress in the course of events. The pressure placed on those in the organization was enormous and the working conditions were necessarily primitive. The group benefitted to a degree from the Smyth Report and espionage. The atmosphere relaxed after the successful test of the first Soviet bomb in 1949. Soon thereafter, Riehl, much to his surprise, was awarded a Stalin Prize, First Class, with very rich perks, including fine homes and a fortune in gold. He and his family were, however, detained as closely confined special prisoners until 1955. Riehl, of mixed German-Russian parentage, had been born and lived in Leningrad until age seventeen, and was entirely at home in the Russian language. An account of his experiences appear in a small book, in German, having the title in English: *Ten Years in a Golden Cage*. (Rieder Verlag, Stuttgart, 1988). The book is no longer in print but I received a photocopy through the generosity of Riehl's family and Professor H.C. Wolf of Stuttgart University and have translated it into English with the hope of having the result published. The Archives of the American Institute of Physics contain interviews with Riehl made in 1984 and 1985. He presents details of German reactor research between 1939 and after he returned from the Soviet Union. He also comments on Heisenberg.

clear that life had been very difficult. They also commented on some of the kindness they, as elderly women, had experienced in encountering American soldiers. Following their directions, I visited the region nearby where my father had spent his early years and where a third, married sister lived.

My elder aunt, Anna the Rector, had formerly lived on Philosophenweg, the beautiful road that winds up along the hills on the right bank of the Neckar River and overlooks the castle and the medieval bridge. She had the foresight to move to a more modest apartment house in the less glamorous part of the town well before the American army came and took over all of the homes on the Philosophenweg, leaving the tenants to shift for themselves. I myself was billeted in one of the elegant homes there occupied by American officers.

From my aunts, I learned that two cousins, who had been in the Signal Corps of the German army, were both now somewhere in American prison camps. The younger one, Ernst Gaus, had been taken through Heidelberg on his way to detention, and had managed to drop off a note which a young girl delivered to Aunt Anna. Anna had played an important role in the education of Ernst and his cousin Fritz, seeing them through law school at the University of Heidelberg. I was not to meet my cousins until 1949. Ernst, who is exceedingly adaptable and who had spent several years studying English, soon found himself in a managerial position in an American post exchange while a prisoner of war.

The Alsos Group, which was studying the German attempts to exploit nuclear fission, used our offices as one of its main headquarters. The two leaders of the group, Colonel Boris Pash (Figure 9.16) and Samuel Goudsmit, were in and out regularly in connection with their various investigations. I had learned of the existence of the Alsos group through the Chicago project grapevine. From discussions with the group in Frankfurt, I gathered that the Germans had not achieved a workable nuclear reactor and were nowhere near to developing a bomb. It was also clear that the Alsos team had just spent several strenuous months with and ahead of the advancing Allied army collecting vital information. Their greatest difficulty lay in finding the right places to look for informed individuals as a result of the great on-going confusion.

Colonel Pash was a former FBI agent well-suited for the dynamic role he played. Goudsmit was a much admired physicist of Dutch origin who had joined the University of Michigan in 1927 after he and George Uhlenbeck had discovered the degree of motion of the electron which is known as "spin." We were to see a great deal of one another after the war as a result of mutual interest in the American Institute of Physics and the American Physical Society.

Goudsmit enjoyed telling stories, particularly of his earlier days. While a young faculty member, he had read several papers written in the 1930s by the brilliant English astronomer, Arthus S. Eddington, in which he attempted

Figure 9.16. Colonel Boris T. Pash in a typical moment of action somewhere in Germany. (AIP Niels Bohr Library, Goudsmit Collection.)

to predict the future course of physics. These predictions were widely ridiculed within the profession in spite of Eddington's solid reputation as an astronomer. Goudsmit worried that he might suffer the same fate as Eddington, and confided his fears to Professor Hans Kramers, one of his mentors, who kindly reassured him: "Do not worry. That happens only to rare geniuses. You will find as you grow older that you merely become more and more stupid."

On the recommendation of Kramers and Paul Ehrenfest, Goudsmit received an appointment as postdoctoral fellow at Niels Bohr's Institute in Copenhagen. His excited family got together and it was agreed that an uncle who ran a haberdashery store should provide him with a fine new set of clothes to replace his shabby student attire. It was summer at the time, and the uncle forgot to include a new hat with the ensemble. As a result, Goudsmit arrived in Copenhagen for the autumn term elegantly dressed but wearing his old battered hat. When he met the great Wolfgang Pauli, the latter said to him, "Judging from your clothes, I would say you were not a physicist, but your hat gives you away."

Robertson and I accompanied Pash and Goudsmit to Goettingen in the British zone of occupation to meet with our counterparts in an office the United Kingdom was using to review the state of German science. A movement was underway there to make the University in Goettingen the centerpiece of cultural support to the German academic community. During this visit I had the privilege of spending an hour with Max Planck (Figure 9.17) and his wife, whom Goudsmit had found stranded and virtually destitute in a small village as they fled from Berlin. I also met Richard Becker, who had given Wigner his first position in Berlin, and Robert W. Pohl (Figure

Figure 9.17. *Max Planck, right, with Niels Bohr, probably taken in the 1930s. (AIP Emilio Segrè Visual Archives, Margrethe Bohr Collection.)*

9.18), whose work on color centers I had been following since my student days.

It is remarkable that Goudsmit was able to maintain any perspective in his work, for he learned during his activities that his parents had died in a concentration camp. He was still able to recognize that there were individuals inside Germany who despised the Third Reich as much as he did; these people he treated sympathetically, somewhat to Colonel Pash's displeasure. It was only much later that the immensity of his own loss and the brutally vile nature of everything associated with the death camps came home to him and made him an understandably bitter man, not in his everyday relationship with colleagues, but toward Heisenberg.

While in Goettingen I went out of my way to visit Professor Pohl's laboratory. It was off-limits to German faculty and professional staff. I did, however, help Dr. Erich Mollwo, one of Pohl's associates, service a valuable set of wet-cell batteries.

Colonel Boris Pash, the military leader of Alsos, had developed an obsession. It was well-known that the Germans had pilfered the French radium standard and it was believed to be hidden somewhere near Berlin. Since it was expected that the Russian army would extend its occupation well to the west of Berlin, he wanted to find the standard before the Russians moved in, so that it might not be lost forever. This quest took him on long trips during which he had frequent contact with the Russians. Being of Russian

Figure 9.18. Professor Robert W. Pohl. (AIP Emilio Segrè Visual Archives, Brattain Collection.)

descent, he could converse adequately with them and, in fact, enjoy some convivial gatherings. On such occasions he said to his Russian hosts: "Remember, we shall drink from the same bottle!!"

One morning he arrived in our office elated. He had driven all night from some town near Berlin; he had discovered what he thought was the radium standard. Not appreciating the hazards involved, he had taken the specimen out of its lead case, wrapped it in paper and stuffed it into his rear pocket, the easier to carry it. I happened to be present in Robertson's office when Pash arrived; he reached into his pocket, tossed the packet on Robertson's desk and said: "Here is the French radium standard!" Robertson and I raced out of the room as fast as our legs would carry us, much to Pash's bewilderment as he stayed behind. Several hours later I acquired a lead coffin, suitable for radioactive materials, from Professor Boris Rajewsky, a radiologist at the University of Frankfurt. We turned the specimen over to the U.S. Army to have it taken care of properly. Pash apparently did not escape unharmed from this escapade and was ultimately discharged with a special disability citation.

Professor Rajewsky, incidentally, showed me a report he had written concerning a study of the effects of uranium dust on the health of German uranium miners. The study made it clear that under some circumstances inhaling uranium dust could constitute a serious health hazard. I remembered the heavy clouds of uranium particles that I had seen in the machine shops at the metallurgical laboratory of the Manhattan District in Chicago. As far as I know, there were no serious casualties reported among those working in the shops, perhaps because the particle sizes were sufficiently different that they did not penetrate down into the lungs.

In the course of events a major portion of the files of the V2 rocket program was located in a cave in the hills not far from Frankfurt. Robertson and I visited the site. He commented wryly, "A year ago I would have given a fortune for one folder in those files. Today I scarcely have enough interest to pull open a drawer."

General Eisenhower enjoyed playing golf on Sunday mornings, but he also felt duty-bound to deliver some kind of official public address to the German people each week over the radio system.

One day, there came into his hands through a German volunteer a tape deck that Hitler had secretly used for public radio broadcasts to make it appear that he was in one place when he was actually elsewhere, possibly in a distant bunker. Eisenhower saw his chance to spend all of Sunday morning free for golf; he would simply tape his speech earlier in the week. Unfortunately, there were no new, unused tapes so Eisenhower had to record on the same tapes used by Hitler. On one occasion, someone failed to erase a tape properly and as Eisenhower's speech ended, the rasping, shouting voice of Hitler came over the airwaves. Eisenhower was not amused.

Alexander M. Poniatoff used that tape deck as the basis for what became the Ampex Corporation. I had the privilege of serving on the board

Figure 9.19. Four of the principals hoisting the Army/Navy E flag at Los Alamos at the end of World War II. Left to right: Robert Oppenheimer; General Leslie Groves; Robert G. Sproul, the president of the University of California; and Admiral William S. Parsons. (AIP Emilio Segrè Visual Archives. Courtesy of Los Alamos National Laboratory.)

of his company for several years in the 1960s and witnessed the emergence of the system as an integral part of everyday information-processing technology.

Our office, FIAT, continued on well into the U.S. occupation. In fact, one of my scientist colleagues in wartime ordnance work at Princeton, Dr. Merit P. White, served in a post there for about a year starting in 1946.

I returned to the United States in mid-August, arriving home just as we dropped the atomic bombs on Hiroshima and Nagasaki. The end of the war was at hand (Figure 9.19).

Chapter Ten

Post-War Carnegie Institute of Technology

I returned full-time to the Carnegie Institute of Technology in Pittsburgh in September of 1945, soon after leaving the Field Intelligence Agency, Technical in Europe. Inspired by both friendship and curiosity, I made a last visit to the group at the Manhattan District Laboratory at the University of Chicago with whom I had worked; I found that research there had all but ceased, mainly because of concern that the bombs to which their work had contributed had actually been used. Many of the lab staff were deeply distressed by the use of the bomb on Japanese cities and were trying to decide what kind of protest, or policies for the future, if any, should be made. The debates had started immediately after the news of the success of the test at Alamagordo on July 16th. Leo Szilard and James Franck were holding discussions with members of the laboratory in order to get a consensus as to what should be done.

Recalling my discussions with Shockley and others in the Pentagon just before I left for Europe in April, I felt from the start that any action that would terminate the war, without necessitating the full-scale invasion that had previously seemed inevitable, would benefit both the Japanese and the allies.[1] The use of the bombs was a terrible choice, but I could not disagree with President Truman's decision. Moreover, I did not believe that a test on an unoccupied island or in the sea off the coast of Japan could have convinced the Japanese leadership of the need to end hostilities. This view was clearly shared by others at the lab but not by all.

The group did have one amusing story to tell me. A high ranking admiral who had been in command of an aircraft-carrier task force in the Pacific that was being readied for the invasion of Japan, and whose orders had been drastically changed, visited the laboratory to learn what he could about the work of the Manhattan District. Arthur Compton referred him to Wigner's group; there, the admiral asked Gale Young how large the mighty nuclear explosive was, and Young indicated with his hands an object the size of a

---

[1] The intense conflict between those Japanese who wanted to fight on relentlessly after the bombings and those who wished, along with the Emperor, to end the conflict is described vividly in the book by Soichi Oya, *Japan's Longest Day* (Kodansha International, Tokyo and Palo Alto California, 1968).

grapefruit. The admiral flushed in anger and disbelief and said, "I did not come here to have my intelligence insulted." The enormous power of nuclear energy, and its scale, was still literally incredible for the uninitiated.

Wigner, who had purchased a home in Princeton, was preparing to return there and use it as his next base. He was much interested in hearing of my experiences in Europe, but it was clear, not surprisingly, that his roots were now deeply planted in America.

In the autumn and winter of 1944–45, the Carnegie community was electrified by the announcement in November that both Otto Stern and I. I. Rabi had been awarded Nobel Prizes for 1943 and 1944, respectively. The excitement was felt throughout Pittsburgh as well as on campus, and the celebrations went on for weeks with many out-of-town guests and special banquets.

While in one sense the achievement of the Nobel Prize was the glorious climax of Stern's remarkable career, it also caused him, at the age of fifty nine, to withdraw to the fringes of society and become a thoughtful observer of events rather than continue as an active researcher, even though it would now be easy for him to obtain all the resources he could wish for. Certainly the difficulties he had faced at the Institute were partly responsible for this decision, but I believe the problem went much deeper. He actually had achieved his own major life goals when he received his appointment to Hamburg, and there created what, for the time, was a unique research center. To be cast out of his professional home by a ruthless government, which despised the work that meant the most to him, was a near-fatal blow from which he never really recovered. Later, on the eve of his departure to California, came the sudden death of his close friend, G. N. Lewis, the celebrated chemist at Berkeley. This new bereavement could only deepen Stern's sense of isolation.

When Stern was leaving Pittsburgh, he came to see me and said cordially and with a twinkle, "It has been good to know you. I would not change you in spite of your imperfections." He had looked at me somewhat askance when I arrived in 1942, but came to realize that, in my own way, I was no less committed to our science than he. He also said to me, "The world of science and its values are changing rapidly. If I were you, with your interest in mathematics, I would pick an obscure but interesting area in that field at this time and work at it as far away from the centers of future action as possible."

Though light in tone, Stern's proposal was meant seriously, and I believe he was expressing his belief that forces outside of science would inevitably exert pressures to corrupt it. Perhaps he also realized that not all who don the mantle of science can be trusted to protect the rich traditions they inherit, which have sometimes been maintained only with considerable struggle.

Generally speaking, the experiences of the European intellectual refugees of the 1930s depended upon their ages. Those who were much over thirty five years of age, having been compelled to leave well-established positions in Europe, had great problems in adjusting to life in the United States, with its looser and more open social structure. Those who were younger still had the flexibility to adjust and feel comfortable with their new home in spite of what they might regard as its shortcomings relative to the types of special recognitions they would have received as distinguished scientists in a European setting. There were, of course, notable exceptions to both scenarios. Dr. Peter Pringsheim, for example, who came to Argonne National Laboratory after World War II in his sixties to join O. C. Simpson's group, adjusted rapidly and was a cheerful member of a productive team. In contrast, one or two of my colleagues who immigrated well before their thirty fifth year always gave me the impression that they felt they were somehow wasting a part of their lives in the United States, even though their research was successful and they were appropriately honored for it. It is, of course, possible that these particular individuals would have been somewhat unhappy in almost any environment.

The time was soon at hand to begin building up the staff of the Physics Department to accommodate what we knew would be a huge new crop of students, for, in addition to the traditional batch of seventeen- and eighteen-year olds, we would be seeing veterans who were taking advantage of a government-supported education. The latter were soon with us in great numbers. The situation was made particularly acute by the fact that two of our best teachers, who had been in the department when I arrived, accepted chairmanships in four-year colleges, for which they were ideally suited.

The veterans were extremely serious students and were a pleasure to have in the classroom because of their dedication and discipline. Their immediate objectives, however, were very much on the practical side. They wanted to get back into the mainstream of civilian life and to achieve material success and security in some serious technical job. The regular students who came along subsequently provided a contrast, as they were, of course, much younger and much less disciplined. The best of them, however, were excited by science for its own sake, and asked much sharper, more inquisitive questions.

At that time it was unfortunately not uncommon for landlords to take advantage of students returning from military service. A builder promised one of James Koehler's graduate students, Thomas H. Blewitt, a unit for himself and his family in a small apartment house the builder was putting up if Blewitt would help with the construction. Blewitt spent all his spare time at back-breaking labor to help complete the building, and was then turned away by the owner. He should have had a written agreement; unfortunately those of us who might have warned him did not know the details of the arrangement until it was too late.

With the end of the war, foreign students and postdoctoral scientists began to arrive in the United States to catch up on advances in research that they had not had access to during the hostilities. Our campus was soon favored with the presence of two individuals from France: Pierre Aigrain (Figure 10.1) who came to work with Everard M. Williams in the electrical engineering department, and Claude Dugas (Figure 8.3) who spent a year in the physics department. The latter had translated my book of solid state physics into French. We also had an Italian postdoctoral scientist, Fausto Fumi, and a Chinese student, Y. Y. Li. Dugas, Fumi, and Li worked in the field of solid state physics. Both Aigrain and Dugas returned to France to pursue their studies in physics at the École Normale Superieure; Fumi and Li eventually followed me to the University of Illinois. Aigrain had a brilliant career, serving for a period as Minister of Science and Education of France.

The arrival of such colleagues heralded the start of a new era of international scientific cooperation which soon became global in extent.

Since we had a good group in solid state physics, I decided that most of our new staff should come from the ranks of nuclear physicists. We soon succeeded in obtaining an Australian theoretical physicist, Herbert C. Corben, who had studied in the United States with Oppenheimer.

I then talked with Edward Creutz, with whom I had spent much time at Chicago. He had been an experimental nuclear physicist at Princeton, but had joined the Chicago lab when it was formed, to be of whatever service he could. As mentioned in the previous chapter, most of his work in Chicago was closer to metallurgical and chemical engineering than to nuclear

Figure 10.1. Pierre Aigrain as Minister of Science and Education of France. He has been elevated to the position of "National Treasure." (Courtesy of Mrs. Aigrain.)

physics. I hoped to persuade Creutz to join the Institute faculty and resume basic research in high energy physics. His acceptance was assured when representatives from the newly formed Office of Naval Research (ONR) in Washington visited Carnegie and offered to provide the funds and related support necessary to build and run a cyclotron—the equivalent of a gift from heaven. We also obtained suitable local support; Westinghouse offered to let us build on the site of its original broadcasting center for KDKA in Saxonburg near Pittsburgh, and the Buhl Foundation, a generous local organization, provided additional funds. The town of Saxonburg, incidentally, had an interesting history connected with the early career of John Roebling, who would later design the Brooklyn Bridge.

Creutz also succeeded in convincing two young physicists he had met at Los Alamos, Roger B. Sutton and John G. Fox, to come with him. Between Creutz's technical ingenuity and his talent for obtaining gifts in kind, he was soon operating one of the most remarkable cyclotrons in the East.

While visiting the Office of Naval Research in connection with the establishment of the nuclear physics program on campus, I again met E. R. Piore, who had been in uniform in the electronics division of the Navy during World War II, and was now Chief Scientist at the ONR. He told me that the ONR was organizing a section on solid state physics, and introduced me to one of the key members of the staff, Lawson MacKenzie, who became a lifelong friend.

MacKenzie visited our group in the Physics Department in Pittsburgh and helped us enormously by allowing us to expand our research program. He and his colleagues provided us with sufficient funds for new equipment, fellowships for advanced graduate students, and grants for visiting postdoctoral physicists. In the process, we helped MacKenzie create a solid state advisory committee (Figure 10.2) for the Office of Naval Research. Between twenty and thirty committee members visited many laboratories under various circumstances to try to generate some kind of a consensus of viewpoints among investigators. The base to which the committee reported was gradually broadened as other agencies in the Department of Defense and the federal government became interested in solid state physics. As of this writing, the committee still exists, and is attached to the National Research Council of the National Academy of Sciences.

If, during the latter part of World War II, someone had asked me what would ultimately become of the field of solid state physics, I would have guessed that it would become incorporated into what was then the more general field of physical chemistry. Actually, in a sense, the reverse has occurred. Many physical chemists turned physicist and joined physics departments or interdisciplinary groups. The field of physical chemistry was permanently transformed and to a degree diminished. In the meantime, solid state physics has become part of the broader field of condensed matter physics.

During the winter of 1945–46, the scientific community became very concerned about the future development and use of nuclear energy. It was clear that the military-oriented Manhattan District was not the appropriate agency to address issues related to civilian use of nuclear energy, such as the production of electrical power or applications to medicine. In an attempt to remedy the situation, the so-called May-Johnson Bill was introduced in Congress to create a new nuclear agency. The bill proposed a not unreasonable division of responsibility between military and civilian leadership. However, it left open the possibility that the director or administrator of the new agency might be a retired military officer. This prospect was anathema to many members of the scientific community, even though it had the endorsement of individuals such as Vannevar Bush. Led by Szilard and Edward Condon, a substantial opposition was mounted and the bill was defeated. Although I did not have very strong feelings on the matter, I was an interested observer because of my relationships with both Szilard and Condon.

A new bill was drawn up that incorporated the views of those who had opposed the May-Johnson Bill. The new bill was passed by Congress, creating the Atomic Energy Commission. The AEC took over most of the nuclear development program in January of 1947 under the leadership of David E. Lilienthal, who had been head of the Tennessee Valley Authority. A joint Senate-House committee was created to oversee the work of the new agency. Twenty-odd years later, the Atomic Energy Commission, which had functioned reasonably well, was dissolved in response to protests from a new generation of activists who wanted complete separation of civilian and military interests in energy-related issues. The military component was turned over to the Department of Defense and the civilian component to a new civilian agency, the Department of Energy.

The effort to defeat the May-Johnson Bill brought to the fore Szilard's talent for generalship, which Wigner had described to me in accounts of their student days in Berlin. Szilard had succeeded in obtaining financial support for his work from various private sources and had rented an office which was barren except for a desk, a few chairs, and several telephones. It was not uncommon to see him at his desk, carrying on two or three differ-

---

Figure 10.2. The Office of Naval Research Panel on Solid State Physics during a trip to the West Coast in April of 1951. Lawson M. McKenzie is number 23, John Bardeen number 21. (Courtesy of the Office of Naval Research.) In sequence of numbers the individuals are as follows: 1) B. Levin, 2) Clarence Zener, 3) S.B. Batdorf, 4) W.N. Arnquist, 5) CDR Robert Dahllof, USN, 6) E.J. Hassell, 7) John S. Hickman, 8) R. Pepinsky, 9) John Osborn, 10) Brian O'Brien, 11) G.R. Irwin, 12) David F. Bleil, 13) Robert J. Maurer, 14) J.C. Slater, 15) L.P. Smith, 16) R.M. Robertson, 17) I. Estermann, 18) A.H. Ryan, 19) R.L. Sproull, 20) R. Smoluchowski, 21) John Bardeen, 22) Foster Nix, 23) L.M. McKenzie, 24) Frederick Seitz.

Figure 10.2. See legend on page 180.

ent conversations on as many phones. I do not know if conference calls were common then, but Szilard would have preferred his system in any event since it gave him complete control of the discussion.

Once, when I was in Washington on business, he called my hotel quite early one morning and said, "Seitz, it is absolutely necessary that you accompany me to the Hill this afternoon where I have a meeting with several senators." I checked my schedule and found that I could adjust it to accommodate his needs. As we were going to the Hill by taxi I said, "Leo, tell me what's up. Just what do you want me to do and say?" He replied, "Say nothing. I will do the talking. I want you along because you are a native American and tall. I speak with an accent and am short. It is very important for me to have an individual like you along during the discussions to show diversity."

When I went to Washington as President of the Academy some seventeen years later, Szilard had taken up residence in Washington and was living at the DuPont Plaza Hotel. He soon appeared at my office and offered to "help" me in administering the organization. Fortunately he did not press his offer very hard, for which I was grateful. Sad to say, he became terminally ill soon thereafter and I lost the great pleasure of associating with this remarkable man, the true Columbus of the nuclear age.

While he was lobbying for the defeat of the May-Johnson Bill (which activism, I was told later, caused much distress to some of the officers of the Westinghouse Corporation), Condon met Henry Wallace, who was then the Secretary of Commerce. The two became good friends. The post of Director of the Bureau of Standards, which reported to Wallace's Department, became open in 1945, and the Secretary promptly appointed Condon. Condon expressed great faith in Wallace's political future, but those of us in Pittsburgh who were the Condons' good friends were worried about this move since Washington can be unforgiving if one's star begins to decline.

Condon found a large wooden frame house on the grounds of the Bureau of Standards, then on Connecticut Avenue, which was being used for spare office space. He had it renovated and he and his family moved in; it soon became a favorite gathering place for his many friends. Initially, at least, life went on pretty much as it had in Princeton and Pittsburgh, while Condon worked enthusiastically to invigorate the Bureau. To that end, he encouraged frequent visits from scientists, so that he could learn their views. The visitors came from near and far, and some came from Eastern Europe. This indiscriminate hospitality was, unfortunately, not considered acceptable behavior for a high-ranking government employee, particularly as the cold war evolved.

Condon became the target of attacks from a group of Congressmen during the McCarthy era; as a result, he was forced to resign from the Bureau. The ordeal took a heavy toll on him, both physically and emotionally. He was never again quite the same cheerful, carefree man I had known in earlier years.

Following his resignation, Condon first joined the Research Laboratory of the Corning Glass Company, but pressure exerted upon the corporation by his enemies in Washington soon made his position there untenable. He accepted academic positions, first at Washington University in St. Louis and then at the University of Colorado in Boulder.

I owe this wonderful man a great debt but the forces arrayed against him were much too great for me to be able to provide him with more than sympathy. I wrote to his children after his death in 1974, expressing my wish to have done more for Condon. One of them responded sympathetically, saying, "Do not feel too bad. There was nothing that you could have done."

During that same winter of 1945–46, Wigner decided to spend an indeterminate period at Oak Ridge as Director of the laboratory previously run by the University of Chicago, and now under the administration of the Monsanto company. Several of the scientists and staff at Chicago, including Alvin Weinberg, Gale Young, and Katherine Way, decided to join him there. Their objective was to design useful nuclear reactors, such as might be employed to produce electric power. Wigner asked if I would be willing to take charge of a reactor education program to which would be invited about thirty five young scientists and engineers (Figure 10.3) who had not been involved in the work of the Manhattan District on nuclear reactors. The selection was made from applicants who responded to a special public announcement. It was hoped that some of them would be sufficiently inspired by the work as to assume leadership roles in the development of nuclear power for general peacetime purposes. After discussing the matter with the administration at the Carnegie Institute, I agreed to take on the program with a one-year leave of absence.

Sidney Siegel, who had remained at Westinghouse, went on leave from the company to join the group. Another member was John R. Menke, who during the war had worked with the team at Columbia University involved in isotope separation. He stayed on at Oak Ridge and eventually formed a consulting company with Gale Young.

One of the participants, Herbert G. MacPherson, had played a crucial role in the early development of graphite nuclear reactors, such as the experimental unit tested by Fermi and those at Hanford. He was a well-informed research physicist at the National Carbon Company during World War II and became involved in producing graphite for Fermi's experimental group on fission research during its early stages. He noted that the standard graphite produced by the company was formed in furnaces which employed boron-containing fire bricks. Since he knew boron would trap neutrons and serve as a "poison," he influenced his company to build special boron-free furnaces for the graphite used in the research. This step was crucial for the success of plutonium production with graphite reactors. It is assumed that the small German group which failed in its attempt to build a small graphite reactor was unaware of the source of boron in their relatively impure material. For an account of the German and early Soviet work on reactor grade uranium see the book by N. Riehl listed in a footnote on page 168.

I was already familiar with Oak Ridge, having made several visits there during the War. It had three main installations in well-separated geographical areas: the research laboratory, with its one megawatt graphite reactor and associated facilities; an area where the separation of uranium isotopes had been achieved by electromagnetic devices; and an area where the separation was accomplished using a system based on gaseous diffusion. The latter procedure had proved so effective (due mainly to the efforts of Clark Center, a remarkable engineer), that the electromagnetic method had been abandoned, although the equipment was still used to separate other isotopes as a service to experimental groups.

Since Betty and I would require housing and there were other arrangements to make, I went to Oak Ridge a month before my program was actually due to begin. I had planned to stay at the guest lodge in the community center while sorting out our housing. Admiral Hyman Rickover (Figure 10.4), who had decided to take advantage of the special opportunity provided by the presence of the laboratory and lecture program to become intimately familiar with reactor research and development, invited me to take a room at the local Navy bachelor officers' quarters. Rickover[2] had become well-known in Pittsburgh as a naval officer attached to the Westinghouse Electric Company. Immediately after the outbreak of Word War II in 1939, he had persuaded the company to begin manufacturing de-gaussing belts for ships, long before a naval contract had been arranged. Also housed in the

---

Figure 10.3. Our class of earnest students at the Oak Ridge training center in 1947. A complete list of the names of the group is unfortunately not available. Most, however, have been identified with the help of Sidney Siegel, Fred von der Lage, Walter Saeman, and Jordan Markham. The names are as follows in accordance with the numbered diagram. 1) J. Lum, 2) F. Seitz, 3) E. Wigner, 4) M.E. Rose, 5) L. Eisenbud, 6) J.J. Markham, 7) S. Siegel, 8) E. Campbell, 9) L.D. Roberts, 10) unidentified, 11) unidentified, 12) F. von der Lage, 13) S. Lawroski, 14) J.J. Grebe, 15) R.M. Boarts (?), 16) W.C. Saeman, 17) R.C. Mason, 18) unidentified, 19) unidentified, 20) unidentified, 21) S.K. Haynes, 22) J. Buck, 23) unidentified, 24) unidentified, 25) unidentified, 26) S.L. Simon, 27) P.J. Bent, 28) J. Menke, 29) J.R. Dietrich, 30) E.B. Ashcraft, 31) E. Blizard, 32) unidentified, 33) unidentified, 34) unidentified, 35) W.H. Vanata (?), 36) J.W. Simpson, 37) _._. Powell (?), 38) A.V. Masket, 39) H. Etherington, 40) E.J. Wade, 41) W.E. Shoupp, 42) L.P. Hunter.

---

[2] An account of Admiral Rickover's career appears in the book by Theodore Rockwell, *The Rickover Effect* (Naval Institute Press, Annapolis, Maryland, 1992).

Figure 10.3. See legend on page 184.

Figure 10.4. Admiral Hyman Rickover. The father of the Nuclear Navy. (Courtesy of Theodore Rockwell and the U.S. Naval Institute.)

quarters was a group of young officer cadets who had elected to work with Rickover on the development of reactors for ship propulsion. During that month I spent many hours with the Admiral and learned a great deal about him. It was impossible not to admire the tenacity with which he pursued any goal to which he set his mind. Our paths crossed a number of times in later years when he had become very famous. Fame, however, did not alter his personality or his attitude toward his work.

During his early career as a naval officer, Rickover had travelled widely and learned much. He once said, during one of our many conversations, "Of all the people I have seen, the Chinese appear to me to be the most ingenious and resourceful, taking full advantage of their opportunities whether rich or lean. They will prove to be formidable if they gain the freedom to enter into the modern age."

The training program I headed was well-attended by an excellent group of young scientists and engineers. Meanwhile Wigner and his colleagues were laying the groundwork for much of the reactor technology that has been used ever since.

In 1947, the Monsanto corporation, which had held the federal contract for the Laboratory since the end of the war, suffered an enormous disaster at its Texas City plant when a ship anchored in the harbor and loaded with ammonium nitrate caught fire and exploded catastrophically. The blast killed a number of Monsanto executives who happened to be visiting the plant in Texas at the time, leaving the company short on managers. The Laboratory was taken over by the Union Carbide Chemicals and Plastics Corporation.

During World War II, the service staff at Oak Ridge had been built up to accommodate the needs of the time; almost all of the staff were from the local area. Activity decreased after the end of the war, particularly when the electromagnetic separation system developed by E. O. Lawrence was all but closed down. The staff, however, was not reduced accordingly. Since the Monsanto company, which had the contract for managing our Laboratory, was busy building up its commercial business, little attention had been paid to the detailed management of Oak Ridge. It soon became clear to those of us who had come to the Laboratory from the outside that the service staff was, for the most part, serving itself. The scientists and their needs tended to be looked upon as something of a nuisance. When the Monsanto company gave up the contract after the Texas City Disaster, and the lab was taken over by Union Carbide, the new management promptly cut the service staff by about one-third. The efficiency of the operation increased immediately.

During this year, Frank Porter Graham, the president of the University of North Carolina, initiated the forming of a university association that would be linked to and participate in the work of the Oak Ridge Laboratory. Since the Carnegie Institute of Technology was not affiliated either with the eastern group at Brookhaven nor with the midwestern group centered at Argonne Laboratory, I proposed to Graham that Carnegie be invited to join this new association. He was apparently too much of an unreconstructed Southerner to accept the proposal. Two decades later, however, the Institute became a member of the nationally organized Universities Research Association.

Before starting the program at Oak Ridge, Wigner probably wondered if he should spend most or all of his remaining career there. He decided, however, to return to Princeton in the autumn of 1947 to continue with his academic work. Fortunately for all concerned, Alvin Weinberg was named as his successor. Weinberg served in the post until 1974 and succeeded in making the lab internationally renowned as the center of a variety of important activities related to nuclear energy. Weinberg has told me that the laboratory records indicate that I was considered for, and apparently offered, the position before he was. If so, I have completely forgotten the invitation, probably because of the work in which I became immersed upon returning to Pittsburgh.

Although Wigner received the Nobel Prize for his contributions to the theory of the nucleus of the atom, he always felt personally that his greatest contribution to society probably was his work on the theory and technology of nuclear reactors, a part of which he carried out at Oak Ridge.

Betty and I enjoyed our year at Oak Ridge enormously because of the convivial spirit that developed within the laboratory, and our proximity to such beautiful country, including the Great Smoky Mountain Park.

Years earlier, driving through the hills of Kentucky, William Shockley and I had had the terrifying encounter with an on-coming, racing, truck that hurtled toward us in our own lane; we later learned that this was simply the local youths' idea of sport. I found that same playful spirit animated the native drivers around Oak Ridge. In driver training I had been taught to beep my horn in warning before moving out to the left lane to pass a slower car on a two-lane, two-way road. This proved to be a dangerous procedure in that part of Tennessee. Apparently the slow drivers resented being beeped at and speeded up when one came abreast, maintaining exactly the same speed. The idea was that the impertinent beeper would be forced off the road by the next on-coming car. I rapidly learned to use stealth in passing a slow-moving car.

One of the happy events that occurred during our stay at Oak Ridge was a visit from the Dutch physicist, Professor Hans A. Kramers (Figure 10.5), whom we had met before World War II during the summer sessions at the University of Michigan. The Netherlands government had made him a special envoy to the United Nations, and he was taking the opportunity to learn more about our national laboratories, particularly those devoted to nuclear energy. We had him as a house-guest and greatly enjoyed getting reacquainted. The war years had wrought changes in him; he was now very involved in international politics, and devoted his energies to the quest for world peace.

The sale of liquor in the county where the Oak Ridge Laboratory was located was nominally prohibited. Actually, a form of raw corn whiskey that was, according to its label, "guaranteed to be less than one year old," was freely available on the local black market. It was widely believed at the lab that the local prohibition law had been enacted primarily to protect this

Figure 10.5. Professor Hans A. Kramers (right) with General Chapman during the former's period of diplomatic service with the United Nations after World War II. (AIP Emilio Segrè Visual Archives. Photograph by Leo Rosenthal.)

cottage industry. Since a modest amount of social drinking was common within the scientific community, anyone travelling out-of-state felt obliged to bring back a few bottles of standard liquor.

Driving back to Oak Ridge from a trip to Washington, Betty and I passed through her ancestral hometown of Bedford, Virginia. We wanted very much to visit her maiden aunt Roberta, catch up with Betty's old friends, and see the ancient family home—with its inevitable shrine to Robert E. Lee in the foyer. As might be imagined, Betty and I created a minor scandal when, without thinking, we went to the local liquor store and loaded up with a case of bottles. Much clucking ensued among Aunt Roberta's friends about her wild relations.

The policy of importing small amounts of good liquor into Oak Ridge had actually started during World War II when members of the Chicago scientific community visited Oak Ridge to carry on experiments, or obtain samples from the nuclear reactor. One member of the chemistry division who went to Oak Ridge every two weeks or so usually carried a bottle of whiskey for his numerous co-workers in Tennessee—a gift that required the contribution of ration point tickets from colleagues in the Chicago laboratory. At some point the guards at the entrance to Oak Ridge began taking it upon themselves to search visitor's luggage and to confiscate any alcohol they found—and proceed to enjoy it themselves. Our chemist colleague became annoyed at this ongoing persecution, and one day seasoned a bottle with Croton oil, a potent but otherwise harmless emetic. Half an hour after passing through the gate he found himself under arrest, charged with "attempting to poison the guards."

While at Oak Ridge, I was asked to participate in a meeting at Hanford, Washington, where the large nuclear reactors were operating, to discuss issues related to radiation damage. I decided that as long as I was that far west I would take the opportunity to spend ten days or so with my parents, since I had not been with them for an extended visit since 1941. They were then in fairly good health and were enjoying the relative calm of the post-war period. We took a number of trips into the countryside, as in the old days when I was young, and enjoyed many talks together.

As an intended surprise treat, my father took me to a place in the Santa Cruz mountains known locally as the Mystery Spot, on which was a small cottage. Upon entering the cottage, one immediately found oneself in a strange world in which water seemed to run uphill, other laws of physics were defied, and one's sense of direction was completely confused. These phenomena, of course, were common carnival illusions produced by devices such as tilted floors and blind windows. I had long since been familiar with these tricks, but apparently my parents thought they were genuine natural mysteries, and when we left the cottage my father, sure that he had amazed me, said, "Well, what did you think of it?" When I disappointed him by trying to explain the illusions, he simply pointed to a sign that said that there was no known explanation for the effect and that even the professors at the University of California at Berkeley were baffled. He said, "How could you possibly understand it if they don't?" As far as he was concerned at that moment, the investment in my education had clearly been wasted.

One evening my parents hosted a gathering for a number of their oldest friends. Some of the guests who were older than my mother had memories of the Bay Area going back to the 1870s, and they enjoyed reminiscing about what California had been like when the population was well under one million. It was a marvelous discussion that took one back to the period following the Gold Rush when the state was just beginning to develop its agricultural base.

Just as I was preparing to leave Oak Ridge, I was asked to join a special group, the Einstein Committee, whose goal was to try to generate a sensible plan for the future international use of nuclear energy. Most meetings were held in Princeton. Although Albert Einstein served as chairman, the two most active members were Leo Szilard, with whom I had spent many hours at the University of Chicago discussing the fate of the world, and Harrison Brown (Figure 10.6), a radiochemist, then deeply involved in geological dating.

Figure 10.6. *Harrison Brown during his period as Foreign Secretary of the National Academy of Sciences. (Caltech Photograph, from Archives of the National Academy of Sciences.)*

The first order of business for the committee was attempting to meet privately with scientists from the Soviet Union, if that were possible, to try to find practical ways to head off an arms race. Harrison Brown succeeded in scheduling meetings with Foreign Secretary Andrei Gromyko, but was unable to arrange a convocation of scientists as Moscow vetoed the plan. The news of the Soviet fission bomb in 1949 made it clear that it was already too late to prevent a buildup of arms. The notion of holding cooperative meetings survived, however, and led to the creation of the Pugwash conferences nearly ten years later.

During the winter of 1946–47 I received two requests. The first was to give the 1947 summer lectures on solid state and nuclear physics at the University of California at Berkeley, which invitation I was most pleased to accept since it would give us another opportunity to see our families. Then Donald Mueller, an old friend from graduate school days at Princeton who had been at Los Alamos during the latter part of the war, asked if I would stop off there on the way to Berkeley to discuss our possible cooperation on a problem.

The six weeks at Berkeley turned out to be even more stimulating and pleasurable than we had anticipated. Ernest Lawrence's Radiation Laboratory was returning to basic research and had assembled a brilliant group of investigators. Robert Oppenheimer, however, was in the process of leaving to become head of the Institute for Advanced Study. We saw many old friends as well as family members and had a thoroughly enjoyable time. Unfortunately it was the last occasion when I was able to spend appreciable time with my beloved mother. She died soon after of a sudden heart attack.

The prior stop-off at Los Alamos opened a very interesting doorway. Mueller was working on a miniaturized system for determining the equation of state of materials when placed under compression. I became a part-time consultant to the lab for this work, forming a fruitful and stimulating relationship that lasted for almost a decade. In addition to making a number of new friends, Betty and I had an opportunity to renew our relationship with Jack Clark (Figures 4.5 and 4.6); he had gone to Los Alamos after leaving Aberdeen Proving Ground, where he had served as one of the uniformed members of the scientific staff. Jack held one of the important positions at Los Alamos, being in charge of the technical arrangements for certain tests of nuclear bombs that were to be carried out on the Pacific Islands. Off-duty he enjoyed exploring the local terrain, and showed us many interesting sites in northern New Mexico. Through him we acquired very precious pieces of pottery made by the celebrated Zuni potter, Maria of San Ildefonso. With the Muellers we explored the verdant wonders of the local Valle Grande, the crater of a gigantic extinct volcano, and the many beautiful canyons associated with it.

In the first decade after the war, Los Alamos became a regular gathering place during breaks in the academic year for a number of notable individuals. Norris Bradbury was the director at this time, having succeeded Robert Oppenheimer in 1945. Among the many visitors were John and Klari von Neumann, the George Gamows, Enrico Fermi, who usually came with his son, and the Tellers. The visitors tended to congregate at the homes of the permanent staff in the evenings. The Stanislaw Ulams were usually part of the group, adding wit and wisdom. Lois and Norris Bradbury ran a more or less continuous open house where all of us met frequently. Another favorite gathering place was a duplex whose two units were separately occupied by Jack Clark and Nicholas Metropolis, a mathematician with a variety of interests including the newly developed electronic computers.

I have rich memories of those times, of exciting work and delightful social gatherings. Our convivial evenings had considerable educational value in their own right. For example, Von Neumann's broad interests and eternal curiosity were always in evidence on these occasions. He was prepared to discuss any topic, whether related to the past, present, or future, dealing with science, technology, economics, or sociology. His insights were unfailingly keen; his detailed knowledge of history was awesome, and his discussions of economics were usually quantitative and enlightening. He would have been an outstanding figure in almost any field of intellectual endeavor. Because of his interest in electronic computers, he was excited by the announcement of the invention of the transistor. He undoubtedly saw it as a more reliable replacement for the vacuum tube, which was limited in frequency response and very wasteful of energy. Unfortunately, he did not live to see the age of the integrated circuit.

Von Neumann's creative interest in electronic computers and the logic involved in their operation inspired a number of applied mathematicians to seek to work with him in his university setting. While all admired his magnificent mental gifts, those of his colleagues who were committed Marxists were puzzled by his strong dislike of Soviet communism; to them this was an illogical attitude to be held by an individual so comprehensively brilliant. One such colleague, in a biographical work, linked von Neumann's feelings to some endemic xenophobia, inculcated in Hungarians from childhood, which particularly emphasized a dislike of Russians. The biographer could not have been more mistaken in his case. The simple truth is that the von Neumann family knew the face of communism well. They had experienced first-hand the tyranny of Bela Kun who brought communism to Hungary in 1919, just after World War I. The entire von Neumann family fled the country as refugees.

To the deep regret of everyone who knew him, von Neumann died of cancer in 1957 at the age of fifty three while serving as a member of the Atomic Energy Commission. The knowledge of his impending death, and worse, the loss of his creative powers had a devastating psychological effect upon him in his final days. I hope he found some solace, as we did, in reflecting on the gifts he had already given us.

George Gamow (Figures 10.7 and 10.8) was also a stimulating guest at the Los Alamos gatherings. He was full of creative ideas related to physics and cosmology, and he also enjoyed discussing some of his adventures while still a Soviet citizen.

Returning from Niels Bohr's institute in Copenhagen soon after wave mechanics had been developed in the 1920s, Gamow was invited to give a lecture on the subject at the University of Moscow and did so in all innocence. Both he and the director of the institute were arrested after his talk, as it was deemed heretical by the authorities. The Head of the Institute was sent to Siberia, but Gamow's punishment was merely a year of exile in Georgia since he offered the defense that he only did what he was asked. There he was welcomed by the Georgian Academy of Sciences as a valued new member of their community. A special dinner was held in his honor, accompanied by many toasts to him. When Gamow's turn came to respond, he wobbled to his feet and expressed the wish that Georgia should become independent of the Soviet Union. No sooner had he spoken than he was overwhelmed with fear that he had misspoken again, and that there might be reprisals. Instead, his companions leapt to their feet, drained their glasses, and threw them into the fireplace.

Jack Clark was not only an excellent host but, being involved in the nuclear

Figure 10.7. George Gamow at Bohr's Institute in the early 1930s. He is in the front row on the right followed successively by Wolfgang Pauli, Werner Heisenberg, Niels Bohr, and Oscar Klein. Felix Bloch is directly behind Gamow (partly hidden). The toys are probably Gamow's. (Niels Bohr Institute and the AIP Emilio Segrè Visual Archives.)

Figure 10.8. A more formal picture of Gamow taken in the 1950s. (AIP Niels Bohr Library, Physics Today Collection.)

tests, was always a welcome guest at any informal gatherings since he provided colorful first-hand accounts of the work in progress.

For several years after World War II, it was believed that useful deposits of uranium ore were very rare and that the element would always be in limited supply. On one occasion Clark described a camping trip near Monument Valley. A member of the party found a small outcropping of carnotite, one of the ores containing uranium, and joked, "If we could find enough of that we would be rich!" In fact, that campsite would later be part of a major mining operation. They should have staked out a claim.

Betty and I spent the entire summer of 1950 at Los Alamos in a rented home. My recently widowed father joined us there for a month and took pleasure in visiting the beautiful canyons and the abandoned Indian cliff dwellings, which he was inclined to compare to New York City apartment houses. He was then seventy four years of age, but clambered up and down the cliffs with the agility of a far younger person.

Edward Teller returned to Los Alamos full-time in 1949 to pursue the development of a fusion weapon, on which he had started research during World War II. He was ultimately joined by a theoretical group working under the leadership of John A. Wheeler. Operations at Los Alamos had become multi-purpose by that time, as the lab was running regular tests along with the exploratory work on fusion weapons. Teller decided the atmosphere was not compatible with his own interests. While there during the summer of 1950, I worked diligently as an ombudsman along with Frederic de Hoffmann, a friend of both Bradbury and Teller, to see if we could find a rational, and not too radical, solution to the difficulties that had developed. The problem at Los Alamos turned out to be insoluble; however, the answer eventually came with the Atomic Energy Commission's creation of the Lawrence Laboratory at Livermore, California. The new lab was designed to have a great deal more flexibility and could satisfy Teller's needs.

We returned to Pittsburgh in the full expectation that we would be there for the remainder of my professional career. The Physics Department was doing very well and we enjoyed our many friendships. Moreover, my close proximity to Washington made it possible for me to participate in the work of several agencies there, such as the Office of Naval Research and the Atomic Energy Commission.

It soon became evident that Creutz's nuclear program at Pittsburgh was sufficiently successful that, in fairness to both him and the program, all appointments in the department in the immediate future should conform as closely as possible to his needs which would inevitably grow as his accelerator began to function. After this had been established, however, I was faced with an entirely unexpected problem.

President Doherty was heavily involved in civic affairs and it was at this time that the city of Pittsburgh decided to take action to rid itself of the terrible smoke and fumes that had become its infamous trademark. What today would be called a major environmental cleanup was initiated, and Doherty was chosen to serve as chairman of the city commission charged with directing the program. Faced with this new responsibility, he decided to appoint a provost to act as the inside administrator on the campus, and he named to the post a professional educator who had been a colleague of his at Yale. For reasons which I completely failed to understand, the provost decided to change the physics curriculum drastically without either our input or approval; I suddenly found much of my time and energy being taken up in tense discussions with him. He was quite headstrong and tended to use his authority arbitrarily to put across his own point of view.

It was just at this point that Louis Ridenour, who had become Dean of the Graduate School at the University of Illinois in 1947, asked me to meet with him and F. Wheeler Loomis, the Head of the large Physics Department, to discuss the opportunities there. In brief, they offered me a research professorship, at an excellent salary, that would also give me the opportunity to make a number of departmental appointments in the field of solid state physics. I was, moreover, assured of a substantial research budget. I had known Loomis as one of the prominent senior members of the American Physical Society and as Associate Director of the Radiation Laboratory at M.I.T., where he had played a major role as advisor to Lee DuBridge and as confidant to the entire laboratory. He had undoubtedly been instrumental in having Ridenour appointed dean.

I found their proposal difficult to refuse, although Betty was quite unhappy at the thought of leaving Pittsburgh. As mentioned, she had spent the war years teaching physics and mathematics at the Pennsylvania College for Women, and she had a wide circle of beloved friends. Moreover, she had decided long ago that she would rather live almost anywhere but in the Midwest. Fortunately, as things actually worked out, our years in Illinois were among the happiest of her life.

When I returned to Pittsburgh and explained the situation to Creutz, he expressed his sincere regret at my decision, but was sympathetic to my reasons. He was promptly made Head of the department. Soon after these matters were settled, Robert Doherty decided to retire, and J. C. Warner was appointed to succeed him; Warner immediately limited the provost's authority to matters more appropriate to his experience. I, however, was looking forward to new challenges by that time.

Creutz stayed at the Institute for ten years and then, in 1956, joined Frederic de Hoffmann in creating a new laboratory, General Atomics, in La Jolla, California under the auspices of the General Dynamics Corporation. There again his accomplishments were many and extraordinary, including the development of small, safe nuclear reactors that could be used for educational purposes.

In 1970 General Atomics was acquired by the Gulf Oil Company which replaced the lab's entire senior staff. Creutz then received a Presidential appointment to the National Science Foundation as an Assistant Director in charge of a large area of physical science. With the change in administration in 1977, he was unfortunately replaced but then became the Director of the Bishop Museum in Hawaii. He had acquired a vast knowledge of the South Pacific, as a result of a lifelong amateur interest in the region, that provided an excellent foundation for his work there.

Frederic de Hoffmann became President of the Salk Institute in La Jolla and served it exceedingly well. Unfortunately, he underwent major surgery and received blood transfusions before it was known that the available blood supply could be tainted with the AIDS virus. He acquired the disease and suffered an untimely death in the 1980s.

I was invited to the Michigan summer school again for the early part of 1949, but by this time competing programs at other universities had made the summer session there less unique. Betty joined me on that occasion and we had a wonderful time with our Michigan friends.

In the winter of 1948–49, Robert Maurer and I received invitations from Professor Robert Pohl (Figure 9.18) at Goettingen University to attend a conference there the following August dealing with color centers in the alkali halides. We both attended, going by way of Holland and Hamburg and returning by way of Switzerland, France, and England, taking a few days for a rare visit to Ireland. While in Hamburg we visited Stern's old institute, which had survived the massive bombings experienced by the city but which still looked sad and destitute. Currency reform had just taken place in West

Germany, providing coinage that finally supplanted packs of American cigarettes as everyday specie.

Goettingen University, which had been greatly helped by the British oc-
cupying authorities, was very active, although it showed the physical drab-
ness of the period. Among our fellow guests there were Nevill Mott, Rudolf
Hilsch, who held a post at the University of Erlangen, and Walter Schottky,
a brilliant, innovative physicist who was employed by the Siemens Company laboratory in Erlangen. I also had the pleasure of meeting the great
generalist-metallurgist Professor C. Tammann whom I had always pictured
as a forbidding *Geheimrat*. Instead I found a kindly and sympathetic individual. Otto Hahn appeared at one of our small receptions.

It was at Goettingen that I first met in person Heinz Pick (Figures 10.9
and 10.10) with whom I had corresponded prior to World War II. He was
Pohl's assistant and was responsible for the excellent organization of the
meetings, which he arranged during a difficult transitional period. He, incidentally, had never joined the National Socialist Party and, as a result, earned
his living at menial, routine factory work during the war.

When Pohl retired, many on the faculty wished to have Pick take his
place, but the appointment eventually fell to Hilsch, who had worked with
Pohl early in the 1930s. Pick instead was given a position at what was then
the Technical University of Stuttgart (now Stuttgart University), where he
inherited a very badly damaged portion of a bombed building. Fortunately,

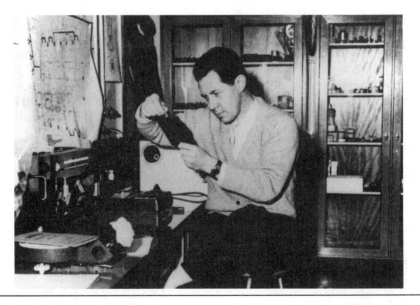

Figure 10.9. *Heinz Pick in the laboratory during his Goettingen period. (Courtesy of
Mrs. Pick.)*

Figure 10.10. *Professor and Mrs. Heinz Pick during his period as director of the Physics Institute at the University at Stuttgart. He proved to be a brilliantly successful and inspired leader. Many students and others spent happy evenings in their cheerful home. (Courtesy of Mrs. Pick.)*

Stuttgart recovered economically much more quickly than many other German cities. A decade later Pick had a marvelous new laboratory and associated lecture hall, and began hosting many international meetings. It was my sad privilege to give a memorial lecture in his honor in November of 1985.

In 1943, when our group at the Carnegie Institute was working on the development of the so-called dark trace tubes using cathode ray screens composed of evaporated crystalline layers of potassium chloride, I revived a long-held interest in the nature of the color centers developed in such crystals during electron bombardment—a subject which had received much attention in Pohl's laboratory.

The structure of the monatomically dispersed color centers, or "F-centers" as they had been designated, was well understood at that time. They, in fact, were responsible for the primary, desirable darkening. It was not clear, however, how the atomically dispersed centers coagulated to form the relatively dark, persistent deposits which prevented practical use of the screens. Since the available, relatively crude, information on ionic mobilities near room temperature suggested that the centers could not migrate alone at such temperatures, I had developed an elaborate mechanism to account for the coagulation. Being more empirically minded, Heinz Pick proposed an alternate scheme that was much simpler than mine. We became friendly rivals in our discussions of this issue. Detailed measurements of the coagulates eventually demonstrated that his scheme was the correct one—a resolution that gave him great personal satisfaction. Much later, in the 1960s, Robert Maurer and a student, Robert G. Fuller, demonstrated in a delicate series of experiments that the old experimental data on migration rates that I had been using were wrong. The atomically dispersed centers could in fact coagulate by direct migration near room temperature.

Chapter Eleven

The University of Illinois, 1949-1965

In the late summer of 1949 we moved into a fully furnished home in Urbana that belonged to John van Horne and his family. Van Horne, a professor of Spanish with a fine personal library, was on sabbatical leave for a year. The post-war boom was on; the university faculty was growing; and housing, whether new or old, was in short supply in the area. We were fortunate, thanks to some research on the part of our friends the Loomises, that we had a fine, if temporary, home in which to begin our new life.

Later, we purchased a relatively small, older house on Iowa Street near the campus and close to our friends, the Ernest Lymans. With the aid of an architect we renovated it to suit what we thought were our short-term needs, as it was our plan to soon build a new home. Sixteen years later, when we left Illinois, we had excellent architectural drawings and a fine site but the new house remained unbuilt. Those years were stimulating ones, however; our home on Iowa Street saw us through the Korean War; the creation of two new research laboratories on campus; a major University crisis involving President Stoddard; and even a time away from it, when I was on leave at NATO headquarters in Paris.

The community consists of the twin cities of Urbana and Champaign, which have grown into one another while retaining separate municipal governments. When we arrived there were still many veterans on campus and the student population was over thirty thousand—somewhat larger than the full-time population of the community.

In spite of her concerns, Betty promptly fell in love with the Midwest—in fact, we both did, conceiving an affection for our new homeland that almost transcended both the excellent opportunities afforded me by the university, and even the many fine friends we made. The great American prairie has many of the qualities of the sea; there is the same kind of rich interplay between sky and rolling land as is seen between sky and undulating water. Brilliant sunsets and magnificent storms hover over the vast spread of the land. And, like the sea, the land presents markedly different aspects at different times of day and at different seasons. In the early spring, the new plantings emerge as if by magic from the rich black soil. During the autumn, or in snow-covered winter, the pheasants move like gleaners through the grainfields, often whole flocks at a time. In the summer the whole world

200

seems to be bursting with vigorous life of all kinds. Fortunately for us, air conditioning became commonplace very early in our stay, mitigating the occasional periods of high humidity. Midwesterners took to this new convenience quickly, as they have always enjoyed comfort, both in their homes and in the workplace. We never had a happier or more fulfilling period in our lives.

The President of the university was George D. Stoddard (Figure 11.1), who had spent a number of years at the University of Iowa and had helped in the wartime creation of the State University of New York. He was a man of substantial intellectual stature who had specialized in educational psychology. Appointed President in 1946, he was doing his best to transform an institution which had become somewhat lethargic during the Depression into a vital intellectual center. We became friends with the whole Stoddard family and developed a great admiration for both Dr. and Mrs. Stoddard. She was an intelligent, caring woman who was much loved by all who came to know her. Stoddard himself, alas, had two weaknesses which brought him to grief in the long run. He could act impulsively on occasion, without appropriate preparation, and he was often openly contemptuous of those who would move more slowly and cautiously than he.

The Stoddards lived in the President's official house on campus, which was designed to accommodate large receptions and which had a number of guest rooms for the use of visiting notables. Someone complained to the Board of Trustees that the Stoddards were living a life of oriental splendor at the university's expense and that they should be investigated. With some reluctance, the Board appointed a special committee to visit the Stoddards

Figure 11.1. George D. Stoddard as President of the University of Illinois. (Courtesy of the University of Illinois.)

and look into the matter. The committee arrived while Margaret Stoddard was in the basement doing the laundry for her large family. She greeted the group in her work clothes and invited them to look around the house while she finished her work in the basement; she could be reached downstairs if there were questions. The committee shuffled about in the foyer for a few moments, and then quietly left.

The Physics Department was in the Engineering College, which was headed by William L. Everitt (Figure 11.2), an electrical engineer who was greatly respected within the college. Unfortunately, Louis Ridenour, Dean of the Graduate School, who at times exhibited more than a trace of arrogance, tended to disparage Everitt because he had come from a different technological culture and was a more deliberate thinker and planner than Ridenour himself. Since I had been brought into the University through Ridenour's efforts, Everitt's understandable sensitivity to Ridenour's attitudes rubbed off in part on me in an unfortunate way. As a result, Everitt was never fully at ease with me, even though I did my best to compensate since I admired him fully as a wise and steadfastly dedicated leader.

As was typical of large state universities, the Illinois Physics Department was much larger than those at most private institutions, having at least fifty senior professors on staff at any given time. The Department Head, F. Wheeler Loomis (Figure 11.3), had been trained at Harvard in atomic and molecular physics. He was appointed to his position at Illinois in 1929 after serving on the physics staff of New York University. The Depression for some time hindered the development of the department, but Loomis, calling on his considerable resourcefulness, had maintained it in the top ranks of the field. One of his graduate students, Polykarp Kusch, later won a Nobel

Figure 11.2. William Everitt, the Dean of the College of Engineering. (Courtesy of the University of Illinois.)

Figure 11.3. F. Wheeler Loomis, the head of the Physics Department. (AIP Gallery of Member Society Presidents.)

Prize for work carried out in cooperation with I. I. Rabi at Columbia University. And just prior to World War II, Donald W. Kerst, a young staff member, had invented the betatron, an important device for accelerating electrons.

In the immediate aftermath of World War II, the State of Illinois provided the University with sufficient funds to enable Donald Kerst to construct a large betatron. The machine attracted a number of visitors as well as several new permanent staff members who were interested in working on nuclear physics. One of these last was Gilberto Bernardini, a brilliant Italian physicist who was also drawn to the University by its proximity to Enrico Fermi in Chicago.

Loomis went on leave from the school during World War II to serve as Associate Director of the Radiation Laboratory. On returning, he filled some departmental vacancies with young individuals who had distinguished themselves at that lab, taking care to be highly selective in spite of the pressure to build up the faculty as quickly as possible. In the group was Albert Wattenberg who had been a colleague working with Enrico Fermi at the Chicago Laboratory of the Manhattan District. All of the new staff members were interested either in nuclear physics or in some form of advanced atomic physics. Most were in their early thirties; however, one, Andrew Longacre, had been an advanced physics student during my Princeton days and had spent some time at Oak Ridge immediately after World War II. There were a few members of the research staff, in addition to Kerst, who had joined in the 1930s. One, Gerald P. Kruger, was a nuclear physicist who had served as Acting Head of the Department during the War. Another, Gerald M. Almy (Figure 11.4), was a spectroscopist.

Figure 11.4. Professor Gerald M. Almy, a major figure in the life of the Department of Physics. (Courtesy of the University of Illinois.)

The Physics Department possessed a remarkably congenial environment, considering its size. The members worked together well, in close cooperation and in close quarters, and I do not recall any significant feuds developing among us. Wheeler Loomis's commanding presence and sense of fairness was undoubtedly responsible for this healthy atmosphere. Moreover, the community, being what is sometimes termed centrally isolated, tended to attract individuals with the appropriate temperament, as also seems to be the case at Cornell University in Ithaca, New York.

Upon arriving in 1929, Edith and Wheeler Loomis had purchased and renovated a large, handsome wooden-frame house that dated from early in the century. Their home became a central gathering place for faculty members throughout the University, not least those in the Physics Department, and Betty and I spent many enjoyable hours there.

The Chemistry Department, which was in the College of Liberal Arts and Sciences, was headed by Roger Adams (Figure 11.5), a close friend of James Conant. That Department was particularly outstanding in the field of organic chemistry and had developed a close relationship with the DuPont Company in a way that enhanced rather than impaired the quality of its research. While Loomis and Adams were, in a sense, friendly rivals on campus, I developed a warm friendship with both. Adams was an exceptionally good academic politician, as well as a good scientific politician on the national scene, and had great influence on campus when he chose to use it. He succeeded, for example, in keeping chemical engineering within the Arts College against conventional practice.

In 1954 Adams was succeeded by Herbert E. Carter, who appropriately

Figure 11.5. Roger Adams, the head of the Chemistry Department and a national leader. (National Academy of Sciences.)

turned the focus of the Department towards the rapidly evolving field of biochemistry; he and I also became good friends.

A wonderfully durable figure in the Chemistry Department was Carl S. Marvel, popularly known as "Speed." He was on a first-name basis with every organic chemist in the country and had been a Ph.D. advisor, either directly or indirectly, to most of them. He was greatly loved by a wide circle of chemists.

The state of Illinois pointed with pride to the University's Agricultural College which dated back to the founding of the University in 1868 and carried on a valuable extension program. The college worked a large tract of land on the southern edge of the campus, complete with animal pens and other wonders, but it also maintained, as part of its proud tradition, a two-acre cornfield in the very center of the campus, much to the annoyance of the campus planners. I came to know the Dean, Henry P. Rusk, reasonably well and found him remarkable in many respects. He told me a story one day about himself and his counterpart at the University of California at Davis; they had been classmates years ago and had remained friends. When Rusk took the position at the University of Illinois, his friend recommended that he focus the work of the college on basic research rather than on the extension program. Rusk had not followed this advice, but now, after all these years, had reached the conclusion that his friend had been right. It takes real greatness for an individual to so honestly acknowledge fallibility, especially at the peak of a very successful career.

The Library of the University was then, and remains now, one of the great scholarly centers in the country. Soon after arriving, I asked if the

Library could possibly get hold of a special volume published by the Academy of Sciences in Goettingen. It turned out that there already were three volumes in the University stacks. Some of my happiest hours were spent browsing through those stacks for general reading material.

The University's Music School was excellent and held many attractions for both of us, especially for Betty who rapidly made a number of friends there, among them Ludwig Zirner (Figure 11.6) and his wife, Laura, Austrian refugees who ran the opera workshop. The school also had a resident string quartet—the Walden Quartet. As usual, the mathematicians and physicists established close connections with the music school.

Betty, who had enjoyed playing the piano since childhood, took this opportunity to resume her practice under the expert guidance of Professor Zirner, and soon was the principal pianist in an excellent amateur group. One of my fondest memories is of one evening at the Stoddards' home when the group gave a beautiful rendition of the Mozart Piano Quartet in G Minor. I have often wished since that today's common home recording equipment had been available that night.

At the time we arrived, the Music School was under the leadership of Professor John M. Kuipers, a Dutch composer-musician with great talents. Unfortunately, he became dictatorial in his attempts to form a university orchestra, and completely alienated his entire department, which proceeded to oust him. He was replaced by Professor Duane A. Branigan a much more diplomatic figure. We had come to know Kuipers well and valued his friendship a great deal; unfortunately, his difficulties with the department caused him to leave a year or two later.

Figure 11.6. Ludwig Zirner, Director of the Opera Workshop and a valued friend.

Another interesting member of the Music School was Soulima Stravinsky, the son of the great composer. He and his wife enjoyed an association with the physics community so that we were together frequently, at both musical and non-musical events.

During our early years in Illinois, the Metropolitan Opera still came to Indiana for a week every May, giving performances at the university in Bloomington and at Purdue in Lafayette. We drove to those towns through the rich countryside in midwestern spring, just as the crops were beginning to burgeon. We were always welcomed by friends like the H. Y. Fans and their young family in Lafayette; the Fans usually gave us a sumptuous Chinese dinner before the opera. When Rudolph Bing became head of the opera company, the singers' salaries began to mount, and to our grief, he found that the midwest tours were no longer cost-effective, so they were discontinued.

In arranging the move to Illinois, I succeeded in having transferred there several federal research grants which had been in effect in my name in Pittsburgh. Robert Maurer and James Koehler, who came with me, were able to do the same, thanks largely to the efforts of Edward Creutz, who remained in Pittsburgh. Harriet Koehler, incidentally, was an accomplished pianist and added to the local amateur music world.

Dillon E. Mapother (Figure 11.7), one of Maurer's students at the Carnegie Institute, had completed his doctoral work there in 1949 and had become involved in helium temperature research. He agreed to join us when, at our suggestion, the university offered to provide him with a Collins helium cryostat. Mapother had been wondering if he should investigate opportunities in industrial laboratories, since, before starting graduate work, he had en-

Figure 11.7. Dillon E. Mapother.
(Courtesy of the University of Illinois.)

joyed a three-year stint as an engineer at the Westinghouse Research Laboratories during the time Edward Condon was Associate Director. I arranged for Mapother to be interviewed by William Shockley for a possible position at the Bell Telephone Laboratories. Unfortunately Shockley, who previously had been a quite reasonable, and even inspiring, interviewer had just developed a complex questionnaire that he believed would accurately reveal the capabilities of a candidate. The interview was held while both men were attending a meeting of the American Physical Society in Cleveland; after being subjected to what he regarded as an inquisition for several hours, Mapother decided to join us at Illinois, where today he continues to pursue a very productive career.

Two other new Ph.D.'s of interest to our group joined the Department at this time: Charles P. Slichter (Figure 11.8), who had been a student of Edward M. Purcell's at Harvard, and David Lazarus (Figure 11.9), who had worked at the University of Chicago under Andrew Lawson, my former associate at the University of Pennsylvania. Both Slichter and Lazarus have spent their entire careers at the University of Illinois, greatly enriching both the Department and the University. Sidney Dancoff, who had taken a leave to work on Wigner's theoretical staff at the University of Chicago during the war, and with whom I very amicably shared an office there, had re-

Figure 11.8. (Left) Charles P. Slichter in the 1970s. (National Academy of Sciences.)
Figure 11.9. (Right) David Lazarus, probably taken in the 1980s when he was editor of the Physical Review. *(Courtesy of the University of Illinois and the AIP Emilio Segrè Visual Archives.)*

joined the staff at Illinois. Unfortunately he died of a form of lymphoma in 1950.

Another major addition to the young faculty was Hans Frauenfelder, a Swiss physicist educated at the technical university in Zurich (ETH). He not only expanded research in the field of electron and nuclear resonance but was able to induce many of his former Swiss colleagues to spend a period in the Department. Later in his career, he focussed his talents on the study of molecules of biochemical interest with great effectiveness. He also served a productive term as Chairman of the Governing Board of the American Institute of Physics during which a decision was made to move the Institute office to Maryland.

Y. Y. Li and Fausto Fumi, who had worked under me in Pittsburgh, both decided to move to Illinois with me. Li obtained his doctoral degree there, and after a few more years in the United States, returned to mainland China to join the Physics Institute of the Academy of Science in Peking. Another of my graduate students of this period, Russell C. Casella, eventually became a broad-ranging "House Theoretician" at the National Bureau of Standards, now the National Institute of Science and Technology (NIST). Fumi's experience inspired a long succession of Italian postdoctoral fellows to come to the university. Among them were Franco Bassani and Gianfranco Chiarotti, both of whom became prominent in Italian academic physics in later years. Fumi met his wife on the Illinois campus; she was an Italian exchange student.

We also had the privilege of hosting, on extended appointments, a remarkable group of young English physicists who had worked with Mott at Bristol, or were otherwise interested in solid state physics. Almost without exception, the foreign fellows returned home to successful careers in academic or governmental institutions.

While expanding our research programs, we developed close relationships with colleagues at the University of California at Berkeley and the University of Rochester. Charles Kittel, an ingenious physicist, had gone to Berkeley from the Bell Labs in 1950 to put together a group to study the various forms of resonance in magnetic fields in solids. Several of his Ph.D. students spent some time at Illinois. A similar close relationship developed with an excellent group in Rochester.

In 1950 a tempest erupted on the Berkeley campus in California over the requirement that faculty members sign a loyalty oath. The university community was deeply divided, and a number of faculty decided to accept appointments elsewhere. One of these, Geoffrey F. Chew, an excellent young physicist, came to Illinois, and greatly stimulated the Department; a year later he was followed by his colleague Edwin L. Goldwasser (Figure 11.10). Unfortunately, Chew was lured back to Berkeley in 1957. Goldwasser, however, remained on. Eventually, Goldwasser took an extended leave to serve as Deputy Director of the Fermi National Accelerator Lab near Chicago. He then resumed his position as a major figure on the Illinois campus.

Figure 11.10. E. L. Goldwasser, hurrying home from the laboratory on his bicycle.

Laura and Ludwig Zirner of the music school introduced Betty and me to yet another remarkable set of new friends. They were Central Europeans, mostly Austrians, who had come to the United States either as refugees or as part of the so-called paper-clip group of scientists from Central Europe who were encouraged to come to the United States after the war by our government.

One member of the group, Henry Quastler, who had worked as an x-ray therapist at a local clinic, joined the Physics Department for a time and eventually moved to Brookhaven Laboratory. He had earned his medical degree in Vienna and subsequently served as personal physician to King Zog of Albania; Quastler was forced to flee Europe at the outbreak of World War II. Another Austrian, Willibald K. Jentschke, came after the war and worked in our nuclear physics program. He went to Hamburg in the 1950s and established a world-renowned accelerator laboratory there. One of our most enduring friendships was with the Heinz von Foersters, who also arrived after the war. They had three teenage sons who rapidly made themselves popular with the young people. Heinz was soon joined by his sister, Erika de Pasquali, and her husband, Giovanni, who joined the physics staff.

Betty and I were concerned as to how our son, Jack, who had spent his childhood in a much different environment, would adjust to the small, midwestern community that was his new home. To our delight, he had no problem; on his new bicycle, he was soon riding around town making friends. He was quickly accepted by his classmates at the local public grade school and felt at home everywhere.

The University High School which, Jack entered in 1956, was trying an innovative method of secondary school mathematics education. Introduced by a Dr. Max Beberman, the new course focussed on point-set theory rather than algebraic manipulations. While Jack easily picked up the "new math," as it was called, I observed with some dismay that he was learning none of the standard algebraic manipulations. To round out his education, he spent the latter half of high school at Andover Academy in Massachusetts, which then had a more traditional curriculum to which he responded easily. He ended up in the class of 1964 at Harvard. I must confess that I was never sure at any given time just what his major was; students were permitted a lot of latitude in the normal curriculum and Jack took full advantage of it.

In the meantime, Betty's brother, Lauriston, who had remarried, accepted a post as Chief Scientist at the Link Belt Corporation in Indianapolis. He and his friendly and warm-hearted wife, Lucie Sewell (Figure 11.11), arrived with an infant daughter and promptly had two more children. We soon found ourselves involved in a regular round of family affairs (Figure 11.12).

Whatever complicating personality traits Ridenour may have possessed, he had instincts that bordered on the psychic when it came to predicting the future growth of technology. From the time he arrived in 1947, he pushed computer research, guided in part by Abraham Taub, a classmate of mine at Princeton who had joined the mathematics staff in 1946.

Before Ridenour even arrived on campus, a brilliant civil engineer, Nathaniel Newmark, had chaired a study committee on computers, and

Figure 11.11. Lauriston and Lucie Marshall in a relaxed moment.

Figure 11.12. *A family gathering in Indianapolis at Thanksgiving time, 1952. Right to left: Lauriston Marshall, Betty's older brother; Nancy Sewell (later Woollen), Lucie Marshall's first cousin; Lucie with her daughter Clarice, future ballet dancer; Betty; Robert Marshall (with a necktie); Jack age 10 kneeling before the table; me.*

had convinced the university to acquire one of the newly developing electronic computers. Newmark was supported in his campaign by two other members of his committee, namely Taub and Ralph Meagher, a young experimental nuclear physicist. An order was placed early in 1948 with an instrument company for what was to be a copy of the EDVAC machine developed at the University of Pennsylvania. Within six months progress on the machine was lagging considerably. At that point, John von Neumann, probably encouraged by Ridenour and Taub, wrote a letter to Newmark suggesting that the University consider building its own computer, following a design that von Neumann himself had developed. One such machine was already under construction at the Institute for Advanced Study at Princeton. Permission was obtained and the previous contract was canceled. By the following winter, the University of Illinois computer group had contracted with the Army Ordnance Corps to build two machines based on the von Neumann design instead of one. The Digital Computer Laboratory was created to handle the project, and Ralph Meagher was named Chief Engineer. Newmark remained Chairman of a reconstituted computer committee until September of 1957 when he was succeeded by Ralph Meagher. At the same time the Digital Computer Laboratory was made a separate department.

These vacuum tube computers were being completed when I arrived in 1949 and were causing much excitement, both locally and nationally, since they were the first machines of their kind to be put into regular operation. One of the machines (ORDVAC) was sent to the Ordnance Laboratory at Aberdeen, Maryland. The other (ILLIAC) stayed at the university. Von Neumann was rather annoyed at how quickly and efficiently construction had been completed at Illinois, because he had had a team working on the same task at Princeton since the end of the war, with less effective results. He took comfort, however, in observing, "You people may be first, but I am the only person who knows how to use these machines properly."

We enjoyed many visits from this remarkable genius, who clearly relished the opportunity to come and play with his brainchild. In the question period after one of his lectures, someone asked him if he thought we would eventually match the capabilities of the human brain with these machines. He grimaced and responded, "I am afraid that the brain shows the really professional touch."

Although von Neumann's interest in computers is often considered to be a relatively late development, from 1944 onward when he first learned about the work of Eckert and Mauchly at the University of Pennsylvania in connection with the development of the ENIAC, as related to him by Herman Goldstine, the start actually was nearly twenty years earlier. Even in his Goettingen days, he wondered about the logic of the operation of the human brain. He broadened this interest in the 1930s by inviting Alan Turing, the English mathematical logician, to work with him on the logic of what he then termed thinking machines. This activity and wartime research on such matters as shock waves convinced him that slow, purely mechanical computers would never be adequate to handle the non-linear equations one meets in many fields of science and engineering; the best machines should be electronic. The news of the work at the University of Pennsylvania stimulated him to renewed action. His own concepts of the logical design of such machines were soon far ahead of those of anyone else.

There is little doubt in my mind that Ridenour's interest in building up a solid state group at the University was inspired by his keen appreciation of the work that had been done on silicon diodes during World War II, and of course was fueled by the recent invention of the transistor. He foresaw a brilliant new era of electronics and wanted the University to be at its forefront. His legacy is still very strong on the campus many decades later, although perhaps under-recognized. He died in 1959, and so did not live to see the full development and application of the integrated circuit, which was invented in 1958 by both Jack St. C. Kilby of Texas Instruments Company and Robert N. Noyce of Fairchild Semiconductor Company.

Early in the history of the University, a small group of earnest thinkers had formed the Philosophy Club. They met monthly to discuss selected intellectual matters, usually taking a famous book or treatise as their starting point. The Provost, Coleman Griffith, was President of the Club when Betty and I came to the university. He asked me to become a member and I was pleased to do so since the topics under discussion were of personal interest to me. Moreover, the gatherings gave me an opportunity to meet a diverse group of people from different disciplines. Some of the best scholars in the University were members, including the Head of the Astronomy Department, George C. McVittie, an Englishman whose family had lived in the Near East, and Gordon N. Ray, who was then Head of the English Department. Ray eventually became President of the Guggenheim Memorial Foundation in New York City, where we crossed paths again years later.

One prominent figure on campus was Professor Henning Larsen. He was also in the English Department and replaced Coleman Griffith as Acting Provost during the interregnum period after President Stoddard had been dismissed under conditions to be described later. Larsen and Wheeler Loomis joined forces at that time to make it clear to the Board of Trustees that a new, and qualified president must be found soon.

Anna and Eugene Rabinowitch lived in a large old house in Champaign with their sons, Victor and Alexander. Actually, our friendship with them had begun at the University of Chicago during World War II. Eugene, a biophysicist, was in the Chemistry Division of the Metallurgical Laboratory of the Manhattan District, working closely with James Franck, whom he had known since their European days when he was a student of Franck's in Germany. Eugene became very concerned with international affairs, having witnessed with alarm the dawn of the nuclear age. In 1945 he and Hyman Goldsmith established the *Bulletin of the Atomic Scientists* to provide a forum in which to discuss issues arising from the successful creation of nuclear bombs. Eugene was also very active in organizations like the Federation of American Scientists and Pugwash. Anna's commitment to this cause was equal to his own, and together they worked tirelessly to raise global awareness of the hazards of the nuclear age and to help improve international relations during the period of the cold war.

Eugene, who had grown up in St. Petersburg, and Anna, who had been raised in Kiev, enjoyed entertaining their large circle of friends during the holiday season in accordance with old Russian traditions. At year's end, festive parties filled the Rabinowitch home to overflowing. Their son, Victor, followed in their footsteps and became in his turn deeply immersed in international scientific affairs, eventually serving as Director of the Office of The Foreign Secretary of the National Academy of Sciences and later as a senior officer in the MacArthur Foundation.

Soon after arriving in Illinois I succeeded in arranging visits to the campus, at different times, for Nevill Mott, Robert Pohl, and Heinz Pick. Each stayed for several weeks. We also had the privilege of hosting extended visits by Rudolf Hilsch, Pohl's successor at Goettingen, as well as Kessar Alexopoulos of the University of Athens and Vaino Hovi of the University of Helsinki. The Swedish physicist Professor G.Borelius and his wife visited us several times during the winter months, glad to see what they regarded as the "southern" sun. Shorter visits were made by many other Europeans. One of our frequent English visitors, Alan H. Cottrell, later served as Science Advisor to his government for a period, on leave from Cambridge University. Mott had an indefatigable curiosity and energetically poked into various parts of the campus, turning up in the process areas of research that were of interest to him.

A week or two before Pohl arrived, I received a call from James Franck, now at the University of Chicago. Franck told me that he had heard Pohl was to be a visitor on our campus and he wished me to inform Pohl that there would be no welcome for him at Franck's laboratory. Apparently they had been close friends as young scientists but had had a serious falling-out while at Goettingen, presumably over matters of campus politics. Moreover, they had been residing on opposite sides during a long and bitter war.

Shortly after Pohl arrived, he told us that he had friends in Chicago whom he wished to visit, and he announced his intention of paying a call on Franck while he was there. I tried as best I could to persuade him that that might not be a good idea, since Franck had very strong opinions about what had transpired in Germany after 1933. Pohl merely laughed and said that he was certain it would work out. A week or so after Pohl had returned from Chicago Franck called to thank me for "insisting" that Pohl visit him! Apparently the two had fallen into one another's arms as soon as Pohl appeared. Decades of animosity were dissolved in the memories of their old comradeship.

The Korean War broke out in the spring of 1950 when the North Koreans violated the truce that had held since 1945, and invaded the South with the support of the Soviet Union and the Chinese Communists. To those of us for whom World War II was still fresh in our minds, it appeared certain that the two major Communist countries were on the brink of a major effort to expand their powers globally. Betty and I spent the summer at Los Alamos, and on returning home, found a group in the Physics Department who felt stirred to take some type of action. In the meantime, Ridenour had accepted a temporary post as Chief Scientist of the Air Force, and Wheeler Loomis had agreed to help establish an advanced, multi-service research laboratory—the Lincoln Laboratory—in Cambridge, Massachusetts. Gerald Almy became Acting Head of the Physics Department, and Joseph R. Smiley, Ridenour's deputy, was made Acting Dean.

To clarify our position at Illinois, the staff had several meetings with Ridenour. We agreed to establish a laboratory to study the ways in which digital computers could be used for military purposes, with the additional goal of introducing redundance into electronic systems in order to achieve increased reliability. Since Ridenour's office was in the Air Force, it was also agreed that the program should be monitored by the Army Ordnance Corps in order to avoid a conflict of interest. The new University lab was designated the Controls Systems Laboratory (CSL for short) and was set up in an old power house on campus after suitable renovations had been made. I was selected to be the first director, and we set to work quickly and with a will.

During the Korean war, in addition to advancing computer technology, our laboratory achieved two major accomplishments. First, it demonstrated that even in that ancient vacuum-tube era, digital computers, when coupled to appropriate memory devices, could manage complex problems. The research focussed on the management of a cluster of defensive aircraft within radar range of the fleet. Abraham H. Taub and Arnold T. Nordsieck played a major role in this work. Second, following the lead of Andrew Longacre, who had been struck by a chance observation of radar operation during World War II, the Laboratory developed a back-pack radar system for detecting moving objects in a static background using a Doppler shifted frequency — a moving target indicator, or MTI. Both of these systems evolved into standard equipment, granting that aircraft controls became vastly more sophisticated with the evolution of both radar and computers.

Our exciting work was somewhat marred by one unfortunate occurrence, the result of Stoddard's impetuosity. An individual in the Ordnance Department in the Pentagon wanted to develop a gun that would fire nuclear explosives, and he attempted to divert the relatively modest funds supporting our laboratory to his own interests. Stoddard, without consulting me, complained to his friend the Secretary of Defense — but Stoddard for some reason put the blame on individuals in the Chicago office of the Ordnance Department rather than on those in the Pentagon. In one sense our problem was solved as we kept our funding, but meanwhile we needlessly lost the friendship of what had been a highly supportive group in the Chicago office. One of the consequences of the fracas was that a leading general in the Pentagon branch of the Ordnance Department designated me a "money-mad SOB."

During the Korean War, Lee DuBridge decided to establish a new military-oriented research program at Caltech, Project Vista, as he wanted to give the faculty there an opportunity to serve, if they wished. Our group, which included many who had worked under DuBridge in Cambridge during World War II, participated in several meetings there and offered to cooperate with Project Vista, but no very significant degree of overlap in our programs ever developed.

In 1953 President Stoddard was forced to resign by the Board of Trustees. The board member who instigated this move was a politically ambitious businessman who supported the worst excesses of the McCarthy era. He claimed without foundation that the University was strongly under the influence of the Communists, and that Stoddard had behaved irresponsibly in ignoring this trend. The accusation concerning Communism was nonsense, particularly on that reasonably conservative campus. Most of the faculty regarded themselves as Governor Adlai Stevenson Democrats. The trustee involved had previously been a great friend of the University in general, and of the scientific community in particular, having supported the funding of Kerst's new accelerator. Stoddard's strengths as a leader were many, but he undoubtedly had alienated some of the Board's members with his condescending manner. Our excellent Provost, Coleman Griffith, was also dismissed in the process.

The school went through a two-year hiatus in its leadership, during which the former Controller, Lloyd Morey, functioned as Acting President. He had served the university for many years and was more than a bit reactionary. To restore what he regarded as normalcy, he terminated some of the more imaginative programs in communications and economics that Stoddard had initiated. The non-tenured instructors for these programs were terminated as well, while a number of Stoddard's appointees left voluntarily. All of these well-qualified faculty went on to good positions elsewhere, so it was the University that came out the loser. In fact, Stoddard himself promptly received an appointment as Dean of one of the colleges at New York University in New York City, followed by an advance to the position of Chancellor, where we crossed paths with him in later years. Morey was undoubtedly successful in his own field, but he was not the right leader for the University in a period of expansion and change.

The truculent board member lost his bid for public office, and shortly afterwards his interest in policing the University. After a two-year search, the Board offered the position to David D. Henry, who was, ironically, a colleague of Stoddard's with the position of Vice Chancellor at New York University. Henry proved an excellent President. He was sufficiently diplomatic to maintain good relations with the board and with various university factions, and he worked tirelessly to enhance the quality of the institution. He was widely respected in educational circles, not just throughout the state, but nationally, as well.

One of Henry's greatest challenges was the creation of a second university campus in Chicago. There already was a branch campus, at the so-called Navy Pier on the shores of Lake Michigan close to the central city, that offered a program for the first two college years, but this was inadequate for the post-war needs of the state. It was therefore decided to create a new campus in the city of Chicago under the same administration as the campus at Urbana/Champaign. We of the old campus feared that the attention of the legislature would be focussed primarily on the new campus, and that our school, which in a real sense had been the pride of the state, would suffer accordingly. Happily, this did not occur, just as the creation of the University of California at Los Angeles did not detract, in the long run, from the status of the campus at Berkeley.

During this turbulent period, I received four or five offers of employment from other institutions that might have been very attractive but for one thing. At none of those other schools would I have been able to develop a group of any significant size in solid state physics. So I remained at Illinois, and fortunately never came to regret it.

When the Stoddard trouble started on the Illinois campus, Ridenour, still on leave with the Air Force, decided to resign from his position. He was replaced temporarily by Robert E. Johnson, a physiologist, while Joseph Smiley became Dean of the College of Liberal Arts. After serving as Chief Scientist of the Air Force, Ridenour moved to the West Coast, held a series of positions in the local aerospace industry, and, on the whole, continued his creative but tumultuous career until his untimely death.

When Loomis returned from New England in 1952, I turned over the directorship of the Control Systems Laboratory to him. I was amused to learn that some of the young people who had occupied prominent positions at the Radiation Laboratory during World War II, and whom he had brought into Lincoln Lab, felt that they could run the Lincoln Laboratory better than he. I had noticed similar attitudes among some of the individuals at CSL.

During the winter of 1950–51, I had the pleasure of crossing paths with John Bardeen at a meeting. At the end of the war he had joined the Bell Telephone Laboratories and had settled in New Jersey with his family. When Shockley first tried to develop what is now called the field effect transistor, and only achieved indifferent results, Bardeen and Walter H. Brattain made a new attempt, using a different method that employed point contacts, and succeeded in obtaining a positive triode effect. Additional developments soon followed, and a new era of electronics began. Unfortunately, relations within the triumvirate fell apart, with Shockley doing his best to make life difficult for his colleagues.

In the meantime, an experimental study of the superconductive transition temperature in metals having different isotopic composition strongly indicated that normal superconductivity in metals was related to the coupling between the conduction electrons and lattice vibrations. Bardeen decided to focus his attention in that field, which was his first love. He also concluded that he could probably work most effectively in an academic environment, where he would have students and postdoctoral fellows to work with. Naturally, the management of Bell Labs was anxious to keep him, and offered him anything within reason that he wished.

At our chance meeting, Bardeen asked me for any leads I might have on available academic positions. Hiring had slowed considerably at the University of Illinois and elsewhere; the veterans were gone, and the regular student body was smaller than usual because of the low birth rate during the Great Depression. Dean Everitt, in a heroic effort, using every means at his disposal, managed to bring together enough resources to be able to make Bardeen an attractive offer—which Bardeen accepted. He requested, and got, joint appointments in the Physics and the Electrical Engineering Departments. He remained at the University for the rest of his career, emerging as a gigantic figure on campus.

David Pines, an imaginatively creative and enterprising theoretical physicist who had recently completed graduate work at Princeton University, and was deeply interested in the collective behavior of electrons in metals, joined Bardeen in 1952 as a research assistant professor, and eventually became a senior member of the faculty. He and John formed a close, lifelong professional relationship that was rewarding to both. In his research and related professional activities, Pines was highly effective in focussing the interests of physicists who had previously concentrated their activities on crystalline materials more broadly on what came to be called "condensed matter"—a major generalization of the field. He proved to be a brilliant leader in both national and international activities.

The announcement that Bardeen, Brattain, and Shockley had won the Nobel Prize for the invention of the transistor came in 1956 when Bardeen was well-established at the University, and caused great jubilation among friends and colleagues. He and Jane were given a great send-off to Sweden (Figure 11.13).

Figure 11.13. Jane and John Bardeen in happy a mood — on their way to Stockholm. (Courtesy of Mrs. Bardeen and the AIP Niels Bohr Library.)

In 1951 the Solvay Institute in Belgium decided to hold one of its celebrated conferences on the physics of metals, and invited leading scientists in Europe and the United States to a meeting in Brussels. Lawrence Bragg was the designated convenor of the symposium. This was an unusual meeting for the Solvay Institute, which had previously focussed on such issues as the frontiers of quantum theory and high energy physics. Apparently, the administration had decided to diversify its agenda and include some topics that were of more immediate practical interest. My strongest memory of the meeting, apart from that of the pleasure of seeing old friends like H. A. Kramers, is of the food that was continuously offered to us. It was amazing both in its variety and in its richness, and I came to realize that I had more food allergies than I had ever suspected. The conference ended with a magnificent banquet from which it took me several days to recover. Belgian hospitality left an indelible impression upon me in more ways than one.

In the decade following the end of the war, much of the government funding that universities received for basic research came from either the Atomic Energy Commission or one of the military agencies; the National Science Foundation was not created until 1951. Since the Department of Defense still maintained a regular transportation operation between the East Coast and Europe at that time, it was not uncommon for scientists to travel abroad in military transports.

In 1951, following the caloric Solvay Conference in Brussels, I went to Paris to travel home on military planes. The plane left from Orly Field, and those of us taking it checked in at a special booth in the corridor leading to the main terminal. When I arrived, I joined about thirty other travelers who were signing in on a flight ledger, which involved writing down the details of one's authorization for the flight. When my turn came, I was bent over the ledger for perhaps a minute-and-a-half, filling out the particulars. When I finally lifted up my head, everyone had gone, including the service personnel behind the counter. The corridor was completely empty of everyone who had been there just a minute earlier. I stared around me in amazement until one of my friends came puffing back to tell me, "Jane Russell passed through the corridor and we all decided to follow her." Jane Russell, of course, was a major movie star of the day, well-known for her pulchritude and perhaps less for her acting. Whatever she had, she had enough of it to empty airport corridors almost instantaneously.

In the early 1950s, physicists in Japan decided to end the long period of isolation from the international community which they had endured as a result of the war, and they arranged to hold a month-long meeting of the International Union of Pure and Applied Physics in Japan in 1953. Most of the funding for the meeting was raised through public subscription. Contributions came from many sources, indeed even from school children to the government. Bardeen and I travelled there together through the courtesy of the U.S. military air transportation system.

The Japanese people had been made aware of the meeting and of its significance as a healing gesture, so all of the participants were warmly welcomed. The Nobel Prize winning physicist, Hideki Yukawa (Figure 11.14), served as Chairman of the conference.

I certainly had no complaints about my own reception but, as might have been expected, Bardeen was given very special attention, as Japan was now greatly interested in the transistor — an interest that foreshadowed the amazing developments to come. I attended the meeting for only two weeks, but in that short time I met many Japanese physicists who were anxious to spend time in the United States. This led to the arrival of a long series of Japanese postdoctoral fellows whom we were able to accommodate at the University of Illinois. Loomis had very mixed feelings about this influx of visitors, for reasons of his own. Eventually, much to my regret and embarrassment, I had to withdraw one or two invitations to Japanese scientists, because he felt we had pretty much reached our quota for the time being. Among others, we lost the benefit of a closer association with a distinguished young physicist, Dr. Koichi Kobayashi, as a result. Perhaps because I was raised in San Francisco with its mixed population I never seemed to have developed any particularly adverse sensitivity toward our Asian cousins on a collective basis. Again, most of our Japanese visitors returned home to prominent positions, forming lifetime links with their U.S. friends. One of Koehler's

Figure 11.14. *Dr. Hideki Yukawa and Mrs. Yukawa in the company of Niels Bohr and J. Robert Oppenheimer. (The Niels Bohr Institute and the AIP Emilio Segrè Visual Archives.)*

visitor-colleagues, Masao Doyama, who rose to important academic posts in Japan, was a major factor in retaining such associations.

One consequence of my friendship with Roger Adams was a meeting he arranged for the two of us with Alfred P. Sloan in New York City in 1954.

In addition to being a brilliant entrepreneur who succeeded, in cooperation with the DuPont Company, in making General Motors a giant among automobile manufacturers, Alfred Sloan had a profound understanding of the delicate interrelation between basic science and technology. There are far too few individuals of his kind active in public affairs in the United States today.

Sloan had known Adams for many years and referred to him as his "intellectual son." The philanthropist was very concerned because the federal government was funding so much of the country's basic research. He feared that this could be harmful to both science and technology in the long run if the government began to exercise too great an influence on the kinds of research that would be conducted. He hoped that private foundations would continue their own support of scientific inquiry, as far as their means would allow, both to provide a balance of funding sources and to assure the maintenance of high standards. For his part, he had decided to add a sum to the Sloan Foundation endowment for this purpose and was wondering how best to distribute and administer those funds.

A small group of us met with him on several occasions to review this matter. He was quite deaf by this time and had a complicated, now antiquated, microphone-amplifier system to help him hear. From our discussions emerged the plans for the Sloan Research Grants to young postdoctoral investigators, intended to help them get their projects underway. A chemist, Richard T. Arnold, was selected to be the first Director (1955–1960) of the grants program and I was given the assignment, for a three-year term, of making recommendations as to who should receive the first grants in the field of physics. Unexpectedly, my most difficult job initially was finding anyone to apply for a grant. The young investigators I approached had become so accustomed to federal grants by that time that it seemed disruptive to them to use an additional mechanism to try to get funding. The program soon caught on, however, and applications began pouring in. Later on I had the dubious pleasure of being accused of favoritism in developing the early grants, because many went to young friends of mine in solid state physics — they were the only ones I could persuade to apply.

It would be a great boon to basic science if major foundations would follow Sloan's example and set aside part of their income to support research in the basic sciences. Most federal legislators today seem to lack an understanding of the history of science and the importance of basic research in the development of new technology. This is evident in their insistence that the National Science Foundation concentrate on supporting applied work as opposed to basic science—a requirement that is bound to damage both science and technology, in the long run.

In 1954 I was asked to serve as Chairman of the Governing Board of the American Institute of Physics. Located in New York City, the AIP had been created in 1930 to coordinate the activities of a number of societies promoting the physical sciences. It had been noted in the 1920s that many fields of applied science of industrial interest, such as optics, acoustics, and rheology, were spinning off from the core field of physics with the creation of new societies; this centrifugal process was expected to continue in the future. Farsighted individuals, such as Paul D. Foote, the head of the Gulf Research Laboratories in Pittsburgh, and K. T. Compton (Figure 11.15), then head of the physics department at Princeton University, agreed that an organization should be created to serve the common interests of these related sciences.

I had been elected a member-at-large of the AIP's Governing Board early in the 1950s and was asked to serve as Chairman when George R. Harrison, who had represented the Optical Society of America, left the post in 1954. I served for five years. The Director at that time was Henry A. Barton, who had held the position since the early 1930s. His Manager-Treasurer was Wallace Waterfall, a remarkably well-informed and capable individual who had been an acoustical expert.

I was privileged to take part in three events of historical significance during my term as Chairman of the AIP. First, in 1955, we held a celebration to

Figure 11.15. Karl T. Compton. (AIP Emilio Segrè Visual Archives.)

commemorate the twenty fifth anniversary of the institute's founding.

Second, early in my term, we found that our original office building, a beautiful townhouse located at 57 East 55th Street, was unfortunately no longer adequate for our needs, and we set about finding a new home. After a search, led by Waterfall, we acquired a building at 335 East 45th Street, which continued to serve the institute well for nearly forty years. Raising the money to finance our transition proved to be an interesting exercise. Aside from the normal frustrations of fundraising, we had to cope with a group of physicists in Cambridge, Massachusetts, who for reasons known only to themselves, began lobbying their fellow physicists *not* to support the financing of the institute in its plans.

Arrangements went ahead as scheduled, however, and, as a third significant event, we had the great good fortune to have Prince Philip (Figure 11.16), who happened to be visiting the United States, join with us in dedicating the new building in the autumn of 1958.

When we celebrated the AIP's twenty fifth anniversary, it was my honor to introduce Robert Oppenheimer (Figure 11.17) as the key speaker at one of our symposia. This was his first formal public appearance following his security hearings the previous year. We had also asked him to serve as chairman of the planning committee for the science program section of the cel-

Figure 11.16. *Prince Philip, Duke of Edinburgh, at the dedication of the Compton Memorial Room of the American Institute of Physics in 1957. I was chairman of the session. (AIP Emilio Segrè Visual Archives.)*

Figure 11.17. J. Robert Oppenheimer in the 1960s. (AIP Emilio Segrè Visual Archives. Courtesy of CERN.)

ebration. He graciously accepted this duty, and fulfilled it to perfection, showing a keen understanding of developments in the various fields of physics.

During an informal dinner following a meeting of his planning group, composed for the most part of former Oppenheimer associates, the discussion turned to current events in Communist China. As was to be expected, many scientists who were sympathetic to the advance of socialism had great hopes that the transformations that were taking place in China might lead the world to a new level of technical, economic, and intellectual progress. Someone at the dinner observed that the ongoing cooperation between Communist China and the Soviet Union might prove to be a great boon to the world. I had the nerve to suggest that it was not unlikely that the two nations would ultimately have a falling out, since they had a long-standing tradition of mutual animosity. To my surprise, Oppenheimer disposed of my conjecture with a rapier-like thrust, saying, "That can never happen now that they are both communist countries." Considering both his brilliance and his knowledge of history, I cannot believe that he was expressing his true opinion. Rather I think that he, as host of the gathering, probably was trying to please the others in the group.

In any event, my prediction came true three or four years later although it is not clear why the rupture occurred just when it did. My own guess is that the Chinese demanded more access to Soviet weapons technology than the Soviet leaders were willing to give, and probably proposed a military alliance that the Soviets considered too risky to themselves.[1]

Although Robert Oppenheimer died in 1967, more than a quarter of a century ago, I note that to date no one has been willing to write a biographical sketch for the Memoirs of the National Academy of Sciences. Instead, his friends and former students paid homage to his memory at a session of the American Physical Society in 1967 (*Oppenheimer*, Scribners, New York, 1969). The absence of a memoir is not surprising. Of all the brilliant and great scientists with whom I was fortunate enough to be associated, he was by far the most complex, protean, and undecipherable. The Oppenheimer I knew had several different personalities; he was like a very great actor who could throw himself with complete immersion into a number of roles and play each out with flawless consistency. Whether this was a skill developed by training or a gift of genetic origin others may judge better than I.

Some who were close to him claim that he was even more brilliantly creative than he appeared, or than his truly great contributions to science would indicate. They say that he did not care for the labor of putting his best ideas down in writing, and was content to have his inspirations recorded by his students and postdoctoral colleagues, in their own words and freely mixing in their own ideas.

Oppenheimer's awareness of his intellectual superiority, while natural enough, unfortunately caused him at times to dismiss the views of others when more discussion would have been of value to all concerned. In this way he made enemies of people whom he really could not afford to alienate, whether his opinions were right or wrong. He made himself vulnerable to attack by being unwilling to compromise, or even to consider compromising, in important situations. He seemed, particularly after World War II, to live above his fellows, secure in a personal envelope of splendid isolation.

That he was brought low by charges that he was a security risk was more than unfortunate—a great man was maligned. Whatever flaws he had, he clearly was never a risk to national security as a result of his associations; and the review of his clearance should never have been used against him as it was. This point was made exceedingly clear by John von Neumann in his testimony at the hearings in 1954—probably the most lucid and important item that emerged from any of the hearings. Of course, as soon as Oppenheimer's reliability was brought into question, the result was almost foregone as a result of his earlier associations, prior to the war.

---

[1] Paradoxically, in 1994 the entrepreneurial Chinese on the Russian border are busy selling badly needed consumer goods to the free-enterprise Russians for Russian rubles. The Chinese invest the Russian money in Russian real estate which they then lease out or otherwise utilize to stimulate business on the Russian side of the border. It is possible that Chinese entrepreneurs may serve as one of the critical catalysts for the evolution of free enterprise democracy in Russia.

Oppenheimer's flaws were of a much more personal nature. He insisted on staying outside and above the normal process of rational discussion. In opposing the exploration of fusion weapons, he assumed that the Soviet Union was capable only of following and duplicating results we obtained first. Subsequent events showed otherwise. Even while he was voicing his opposition, capable Soviet scientists were using their ingenuity to prove him wrong. Their first, primitive, hydrogen bomb was tested in 1953, the year following our very powerful bomb.

Oppenheimer's spectacular reversal of fortune made him a tragic figure. The next dozen or so years could not have been happy ones for him, in spite of the support of his many devoted friends who rallied around.

I do not know whether the Greek gods were ever put on trial by their human subjects. I feel however, that the physics community, which had enjoyed god-like status for a decade of so after World War II, underwent its own trial in 1954 when Oppenheimer was held up to such public disapprobation. The scientific community has not been quite the same since, nor will it quickly reclaim that prestige and public confidence which it had at its peak.

Edward Teller, through both his own actions and the accident of circumstance, became the focus of much of the resentment that was engendered in segments of the scientific community by the outcome of the Oppenheimer trial. He was, in any case, an object of animosity among many intellectuals, including scientists, as he shared neither their admiration for the doctrines of the Left nor their deep distrust of the so-called "Establishment" in the United States.

When Wigner was elected President of the American Physical Society in 1956, he and Harry Smyth decided that the activities of the American Institute of Physics should be critically reviewed by the member societies. While such a review can be a reasonable and salutory exercise, their approach was needlessly accusatory, particularly since Barton and I were old friends and associates of theirs. Our conflicts reached crisis proportions when the Society of Exploration Geophysicists applied for membership in the Institute. In the main, this group, many of whom were physicists, used acoustical methods in their oil exploration projects and, in the opinion of the Governing Board, deserved serious consideration for membership. Most members of that society came from industry as did many others in the member societies of the institute, in keeping with its broad initial purpose. The furor raised by Wigner and Smyth over this issue turned out to be a tempest in a teapot, but it demonstrated to me how irrational even the best of scientists can be under certain circumstances. I was also made acutely aware of the hazards of attempting to serve the public or a profession when those gaining the benefit do not appreciate the service they are receiving. Most unfortunately, the incident convinced Barton that it was time for him to retire as Director of the Institute. I was urged to take the post but after only a little reflection decided against it. Elmer Hutchisson, who had moved from the University of Pittsburgh to the Case Institute of Technology in 1944, accepted it and served as Director without major incident until 1964.

I met Wheeler Loomis shortly after this ordeal and said to him, "It is clear that our physicist colleagues can be as crazy and irrational as anyone else." He laughed and said, "You are right, but can you think of a better gang to be associated with?" I did a quick mental survey of the various professional types that I had dealt with, and decided that he had a point.

My period as Chairman of the Governing Board was marked by one other misadventure. Two distinguished American scientists had created an important journal which was being printed in another country. A number of difficulties associated with the press and the transfer of material across borders developed. It appeared that a valuable, well-conceived publication might collapse. At my suggestion, the staff of the Institute undertook the arduous task of rescuing the journal, with the permission of the founders and with the intention that it would become one of the Institute's special journals. A year or so later, however, when matters were finally in the clear, Captain Robert Maxwell, who was just beginning to appreciate the special opportunities that the field of scientific publishing might offer him, succeeded in persuading the founders to turn the journal over to him. Maxwell took action abruptly, over a weekend, without discussing the matter in any way with the administration of the institute, much to my personal embarrassment. Our friends at Academic Press, with whom David Turnbull and I established a series of review volumes in the field of solid state physics, had already experienced business relationships with Maxwell and had warned me somewhat earlier to be cautious about any dealings with him. This incident served to underscore their admonitions. He had become an expert at hostile and other takeovers long before the process was commonly experienced in the United States.

Betty and I took great pleasure, during my years with the Governing Board of the American Institute of Physics, in our frequent meetings with Elizabeth and Karl K. Darrow (Figure 11.18) who lived in an apartment on New York's Riverside Drive. Darrow was the nephew of the renowned trial lawyer, Clarence S. Darrow, who had defended John T. Scopes, as well as the teaching of evolution in schools, in the famous "Monkey Trial" in Tennessee in 1925.

A research investigator for Bell Telephone, Karl was also a popular writer and lecturer, and served as Secretary of the American Physical Society for a quarter-century beginning in 1941. He did his best to maintain the dignity and formality of the Society as it was first envisioned by the earnest handful of physicists who had founded it in 1899. Unfortunately the generation of physicists who came up in the post-war era had a different image of the society from Darrow's. Once he sensed the change in attitudes, Karl resigned his post, perhaps painfully, in 1967; he and Elizabeth retreated to a more private existence, but remained our good friends. The Physical Soci-

Figure 11.18. Karl K. Darrow, the anchorperson of the American Physical Society for many years. (AIP Emilio Segrè Visual Archives. Photograph by Fabian Bachrach.)

ety has not been the same since, in spite of the good work of his successor, William Havens; great change, for the better or worse, was inevitable, however, given the enormous growth it has had to sustain in the meanwhile. Efficiency of a kind has replaced Darrow's standards of gentility.

During this period, I received a telephone call from Dr. Mervin J. Kelly (Figure 11.19), the Director of the Bell Telephone Laboratories. He asked if

Figure 11.19. Mervin J. Kelly, the Director of the Bell Telephone Laboratories. (AIP Emilio Segrè Visual Archives. Photograph by Werner Wolff.)

we could meet to discuss the commemoration of Oklahoma's fiftieth year of statehood, which would be in 1957. He was widely known, both nationally and internationally as a public-spirited individual and a creative organizer; he was often asked to provide leadership on occasions such as this. Kelly and I had known one another since early in the decade, having met while serving together on several government committees. The state was planning to mark its anniversary with a huge celebration, and Kelly had been asked to organize a scientific symposium for the occasion. He wanted help both from me personally, and from the American Institute of Physics.

I met with Kelly to lay out plans and to recommend speakers and other participants, drawing from various disciplines. He then suggested that we go to St. Louis to meet with the chairman of the group overseeing the entire operation in Oklahoma. On the way he said to me, "You are going to meet the most dynamic man I have ever known. Don't be surprised if he changes the course of your life." Coming from Mervin Kelly, who was the most dynamic individual I had met up to that time, this was a remarkable statement indeed.

The person he referred to was James E. Webb (Figure 11.20), who did, in fact, have a major impact on my life, and whose remarkable qualities had not been overstated by Kelly. Webb was in St. Louis on that occasion seeking some help and advice from James S. McDonnell of the McDonnell Aircraft Corporation. Webb had been head of what was then called the Bureau of the Budget under President Truman and, being a staunch Democrat, had left Washington during the Eisenhower administration to work for the Kerr-McGee Oil Company interests in Oklahoma. He regarded this position as an interim job while he waited for the return of what he then regarded as a

Figure 11.20. James E. Webb, the administrator of NASA as he appeared in the 1960s. (Courtesy of NASA.)

congenial administration; it also gave him the opportunity to replenish his bank account. He had served in the Air Force of the Marines in World War II.

Soon after we met, he said to me with a laugh, "People tell me that listening to me talk is like trying to get a drink of water out of a fire hydrant." One of Webb's favorite anecdotes was about the night following the election of 1948. President Truman had gone to his family home in Independence, Missouri to await the outcome, while Webb waited at his own home in Washington. When it became clear, late that night, that Truman had won despite all the predictions to the contrary, Webb assumed correctly that the President would head for Washington immediately in his official plane and go directly to the Oval Office. Webb arrived at the White House well before dawn and waited there. When the President strode in at daybreak to what he thought was an empty office, Webb was there to greet him with, "Well Mr. President, I guess we won that one!"

The celebration in Oklahoma City was a rousing success, with all the state's notables and celebrities taking the opportunity to hob-nob in the limelight. It was a particular pleasure to see so many citizens of Cherokee descent playing a major role in the proceedings. Science was very popular with students at that time, so our meeting, which we held in a large sports arena, was, to our delight, full to the rafters.

One rather amusing incident occurred in connection with the events that was not so amusing for those directly involved. A group of scientific leaders, including Lee DuBridge, Julius A. Stratton, and William V. Houston, as I recall, were flying to Oklahoma City together on a commercial plane that made a refueling stop of about half an hour in St. Louis. Our heroes, along with a number of other passengers, decided to stroll around the airport to pass the time. When they returned to the gate the plane was gone, well before its scheduled departure time. It turned out that Dave Garroway and his crew for the "Today Show" had been on board and had used their special influence to talk the pilots into taking off as soon as the refueling was completed, leaving those on the ground to make their own way as best they could. As it happened, our friends were able to get on the very next plane and so were not too grievously delayed—but both the airline and the Today Show lost a number of fans that day.

In 1954, the United States Air Force decided to establish an air defense laboratory in Europe under the aegis of the North Atlantic Treaty Organization. A team of U.S. scientists was sent to Western Europe for several weeks, to meet with officials in the various NATO countries, to review needs and options and to make one or two specific recommendations. Carl F. Overhage, at that time with the Kodak Company and formerly of the Radiation Laboratory, was selected to lead the team; Loomis and I were asked to join and accepted. Another member of the team was Ragnar Rollefson, who had also served at the Radiation Laboratory during World War II, and who was an advisor to the government. We spent a rewarding four or five weeks journeying from Norway to Italy, meeting with a wide range of officials and scientists, and observing thankfully as we went the resurgence of normal human activity in continental Europe.

It soon became clear that the government of the Netherlands was most anxious to have the defense lab based in Holland, and it made a very attractive offer to help provide scientific leadership and facilities. Our recommendation was based on that offer and was accepted by the Air Force.

Betty accompanied me on this trip, meeting me in Paris where we were greeted by the Pierre Aigrains and other old French friends who had been students at Pittsburgh and Illinois. It was Betty's first visit to Europe, and as the great wave of tourism that was to flood the continent in later decades had not yet begun, we had an enviable opportunity to visit many of the treasures of Europe in relative peace.

Our visit to Oslo gave us the opportunity to meet Dr. Robert Major, a Norwegian scientist who headed a government office responsible for scientific research, particularly research that could have applications for technological development. Major's mission was to help his country move as rapidly as possible into the modern technical age. To that end, he was making a thorough study of international patterns of development, and he had made a series of agreements with the Dutch government for mutual cooperation and exchange of information. The friendship we developed with Major on this occasion would flourish over the following two decades.

In 1958, the Secretary of the United States Navy decided that the Navy would help NATO create a naval research unit complete with seagoing research ships. I was asked to join the committee that would decide where the unit would be based. It was understood that the preferred site was on the Mediterranean coast in the south of France; unfortunately, the French government which was involved in the Algerian war, removed its navy from NATO while we were touring several NATO countries seeking advice. After some deliberation, we selected a site near the Italian navy base at La Spezia, which has served its purpose well. It was on this trip, while in Bonn, that I first met Dr. Rudolf Schrader who was to be my deputy later on at the NATO.

In 1954, Nevill Mott accepted the Directorship of the Cavendish Laboratory in Cambridge. He organized a symposium at the University of Bristol, partly as a farewell gesture, and partly to commemorate his quarter-century there. The meeting was a bittersweet experience; wonderful in that it again brought together so many good friends and colleagues, and melancholy in that it marked the end of the many happy visits I had made to the city.

President Eisenhower and Chairman Khrushchev held a summit meeting in Vienna in 1954, in an attempt to alleviate some of the tensions of the cold war, which had intensified during the Korean War. They agreed that an "Atoms for Peace" meeting would be held in Geneva in 1955. This developed into a large international gathering that brought together physical scientists, engineers, and biologists who had been involved in any way with peaceful uses of nuclear energy. At the meeting I was presented with a Russian translation of my book, *The Modern Theory of Solids*, by a leading metallurgist with many compliments, but no mention of royalties. I never raised the issue.

One of the amusing aspects of the royalties occasionally paid by the Soviet Union for translations of foreign books lay in the level of the uncertainties involved in the process. My physical chemist friend and colleague, Eugene Rabinowitch, who had been a major partner in the creation of the *Bulletin of the Atomic Scientists* and was a devotee of the Pugwash movement, made frequent trips to the Soviet Union on behalf of Pugwash. He and his wife, Anna, felt quite at home in Russia as they both had been born there. The Russians had translated a two-volume series he had written on photosynthesis. He suggested that he deserved a royalty. They demurred at first, but finally, on one of his visits, he was handed a suitcase packed full of rubles. The joker of the situation, he was informed, was that he could not take the rubles out of the country. He found himself in considerable difficulty, because he did not know what to do with the contents of the suitcase. He hid it under the bed and in the closet, but always felt quite insecure. Finally, Anna said, "You give me those rubles, I know just what to do." She went to a pawn shop and bought a great deal of semi-precious jewelry, which they were allowed to take out of the country.

The Soviet exhibits at Geneva were of great interest to all of us from the West. I noted two significant items. For one thing, the notations the Soviet scientists used in papers dealing with reactor theory were identical to those used in our previously classified documents, a "coincidence" that showed us the efficiency of their espionage system. For another, a listing of Russian scientific publications in the field of nuclear and high energy physics in the years from 1945 to 1953 showed a very strong dip, making it clear that the Soviet scientists had been drafted away from normal research to work at the nuclear bomb centers. The dip in publications abated after the successful test of their simple hydrogen bomb in 1953.

Atoms-for-peace meetings continued to be held over the years, but became less general; they focussed increasingly on special topics.

In the early 1950s, the nuclear physicists at mid-western universities had become deeply concerned about developments at Argonne National Laboratory. The University of Chicago had continued to administer the lab, and Walter Zinn became its Director after 1945. Whereas the laboratory created by Ernest Lawrence at the University of California prior to World War II returned rapidly after the war to work on high energy physics and the scientists on the East Coast succeeded in converting Camp Upton on Long Island to a center for basic nuclear and high energy physics, Zinn decided to make Argonne a center for the development of nuclear reactors for applied, including military, purposes. In this he was encouraged by Admiral Hyman G. Rickover, who saw an excellent opportunity to begin developing a nuclear-powered submarine. Much of Argonne became a high-security lab, where a relatively high-level clearance was needed if one wished to visit.

By the mid-1950s, the mid-western universities had formed their own association—the Mid-West Universities Research Association (MURA)—with the intention of establishing an additional national laboratory, if need be. Gerald Kruger served as President of MURA. In the meantime, Donald Kerst, who had invented the Betatron, began working with his brilliant younger colleagues to develop another invention, namely the Fixed-Field Alternating Gradient Accelerator, a machine that would produce a very high flux of charged particles. The organization applied to the Atomic Energy Commission for funding for the machine for which they proposed to establish a new open national laboratory near Madison, Wisconsin, to promote high energy physics.

A national committee of physicists was appointed by the AEC to review the overall situation and to decide upon the best step to take next in machine construction. The decision, reached in a reasonably impartial way, was that it was more important to emphasize higher energies rather than higher fluxes, and that the next large machine should do the former. That decision was undoubtedly the right one for the time. It would be difficult to deny, however, that regional competition entered significantly into the discussions. At one point I visited the offices of Thomas H. Johnson, the Director of the Research Division of the AEC, on an unrelated matter. I noted the following message scrawled in chalk on his blackboard in huge letters: "KILL MURA." Johnson had been head of the Physics Division at Brookhaven Laboratory on Long Island and had joined the Atomic Energy Staff in Washington from there. The message on the board was not exactly that of an impartial judge. In spite of such evidence of sectional rivalry, it must be admitted that the high energy physicists have behaved about as rationally as is humanly possible.

Walter Zinn tried to compensate for the criticism which led to the formation of MURA by obtaining funds to construct an accelerator at Argonne. The result was a machine, called the Zero Gradient Synchrotron, which was built under the direction of John J. Livingood, one of Ernest Lawrence's former colleagues. It was by no means a trivial piece of equipment, but it was probably funded at the wrong time for the wrong reasons. It never really became part of the mainstream of accelerator research.

Admiral Rickover eventually shifted his interest from Argonne to the Westinghouse Nuclear Power Center in Pittsburgh where he helped create the Bettis Laboratory. He had been Inspector of Naval Materiel at Westinghouse during World War II, so his move was motivated in large part by practical considerations, including Pittsburgh's proximity to Washington. Zinn later resigned from Argonne to form a private consulting company, and the laboratory became much more accessible to the scientific community at large.

In 1957 the Atomic Energy Commission decided to accept applications for two high energy accelerators of intermediate size to be constructed on university campuses. I was then Head of the Physics Department at the University of Illinois, as described below, and convened a series of meetings of the staff in the department to see if it made sense for us to enter into the competition. The group wisely decided against making such a bid since it felt that the machines would be out of date for frontier research in high energy physics almost as soon as they came on-line. The machines were actually constructed on the Harvard and Princeton campuses, having affiliated links with the Massachusetts Institute of Technology and the University of Pennsylvania, respectively, but were terminated after a relatively short period of operation, just as the Illinois staff had concluded would probably be the case.

When William Shockley left Bell Laboratories in 1955, he and his new wife began driving westward across the country with the intention of establishing a new electronics company somewhere in California. The route they chose took them through Urbana, where they were our houseguests for about two weeks while Bill formulated the details of his plans. He had a long list of names of West Coast entrepreneurs, and he tied up our phone almost continuously as he negotiated with his prospective business partners. He and our son, Jack, formed a warm friendship, as Bill, with his parlor tricks and tales of adventure, was a fascinating figure to a twelve-year-old boy.

Bill eventually decided to set up shop in the Bay Area of San Francisco, where he proceeded to become involved in many complex business adventures which culminated in his getting an appointment on the faculty of Stanford University. His great contribution was that he sowed the seeds for a new industry in an area of the San Francisco peninsula that had formerly been mainly residential and agricultural. He was indeed the Moses of Silicon Valley, although he never personally achieved as much success in business as he had hoped for.

Leo Szilard's idea of holding regular meetings between Soviet and American scientists, in the hope that they might advance the cause of arms control, came to life again in the mid-1950s through the work of Harrison Brown and the philanthropy of Cyrus S. Eaton, a Cleveland capitalist hoping to gain prominence on the international scene. The first meeting took place in the community of Pugwash, Nova Scotia, in Canada, and included groups from Canada and England as well as from the Soviet Union and the United States.

A much more prominent meeting of the Pugwash organization took place in Kitzbuhl, Austria in 1957; the issue of arms control had commanded increased world attention as a result of the Atoms for Peace meetings which had been initiated in 1955. The Austrian meeting turned out to be a truly gala affair, with much productive open discussion as well; about one hundred scientists attended. I was invited to dine at Mr. Eaton's table on one occasion and was amused to learn that his opinion of the general reliability of the Soviet leaders was on a par with my own; that is, it was not exceedingly high. Nevertheless, he enjoyed being the focus of so much attention and there is little doubt that the meetings helped lay the foundations for a number of lasting friendships among scientists on both sides of the Iron Curtain.

The Pugwash movement lapsed for a period when the Soviet Union had some highly contaminating open-air tests of megaton weapons in the early 1960s. The leadership of the Pugwash group persisted in its endeavor, however, and the organization had a long life with a carefully controlled membership. I attended several meetings when I was President of the National Academy of Sciences, and continue to feel that the organization deserves credit for encouraging international connections that might not otherwise have been formed. Unfortunately, many of those who eventually assumed leadership of the organization seemed to blind themselves to the dangers inherent in the Soviet system. Some were inclined to regard the United States as the source of world tensions.

An interesting incident occurred in connection with a Pugwash meeting in Udaipur, India in the mid-1960s. The meeting took place in a small palace on an island in the middle of a lake; attending were some of the leading political and scientific figures in India, including Indira Gandhi. To broaden our knowledge of Indian culture, we were all taken on a half-day trip to a justly famous Jain temple about thirty miles away from Udaipur. I was assigned to a car with Professor C. F. Powell, the Nobel Prize winning physicist who had discovered the pion and whom my NATO office had supported in connection with the tracking of balloon flights, as described in Chapter 12. He had been a very strong friend of the Soviet Union in 1959 and 1960, but on this occasion I noted that he spoke of our Soviet Pugwash colleagues in quite a derogatory way. At first I thought that he might have undergone some miraculous form of enlightenment akin to that of Saint Paul on the way to Damascus. It turned out, however, that he had been wooed by the Chinese communists while on a journey through China the year before. He now felt that the Chinese held the true key to the future of humanity and that Soviet communism was an aberrant and vulgar sideshow.

When the Office of Scientific Research and Development (OSRD) and the

National Defense Research Committee (NDRC) were closed down at the end of World War II, it was recommended that their activities be taken on by the armed services. As a consequence, Vannevar Bush and James Conant proposed that an organization be established within the Department of Defense to serve as advisor to the service branches in selecting programs for research and development. The result was the Research and Development Board, an advisory board with a full-time active staff. The organization never operated effectively, even though Bush and K. T. Compton served successively as full-time Directors. Apparently the services felt that the full-time, centralized organization merely interfered with their own plans and programs.

The Research and Development Board was eventually dismantled and replaced by the Defense Science Board, a purely part-time advisory board composed of individuals from industry and academia. H. P. Robertson served as the first Chairman. Because of his stature, and because the Board carried out comprehensive reviews of activities of the services throughout the Pentagon, the Board started its life with most of the upper echelons of the Pentagon participating in its meetings. I had the privilege of serving on one or two advisory panels early on, and was impressed with the Board's capabilities.

The three services had created similar advisory boards, with more limited scopes to serve their own interests. Of these, the Naval Research Advisory Committee was in a sense the most prestigious; it had statutory standing, having been created by an Act of Congress. The Science Advisory Board of the Air Force, however, was considered the most glamorous, particularly during the buildup of jet aircraft and long-range rockets. Toward the end of the decade, I had the pleasant task of serving as Chairman of the Naval Research Advisory Committee.

In 1955, I received an invitation from Professor Willy Dekeyser (Figure 11.21) of the solid state science laboratory of the University of Ghent in Belgium to spend a summer at the university, lecturing and reviewing the research programs underway there. The invitation was made with the cooperation of The Institute for Industrial and Agricultural Research, a general research funding agency in Brussels directed by Dr. Louis Henry, who had once studied at Yale. I accepted the appointment for the summer of 1956, and Betty and I spent about six weeks in residence in Ghent. We had the privilege not only of becoming conversant with many aspects of scientific and industrial research in Belgium, but of gaining familiarity with the rich history of the Netherlands, particularly that of Flanders and the Flemish.

In retrospect, I can see that Belgium was then at a critical point in its contemporary development. It was clear that the country could not expect to hold its African colonies much longer and was attempting to make the transition as smooth as possible for both sides. Sadly, this was not to be; the

Figure 11.21. Professor Willy Dekeyser, the Director of the Solid State Laboratory at the University of Ghent. (Courtesy of the University of Ghent.)

colonies broke away abruptly, and with an unfortunate amount of violence, in 1960.

It was also evident in 1956 that the relationship between the French-speaking Walloons and the Dutch-speaking Flemish was deteriorating. The bitter feeling that developed between the two groups of Belgians was a painful consequence of the all-too-typical human tendency toward irrationality, inasmuch as the land and its people have a glorious common heritage going back to the period prior to the Spanish oppression of the Netherlands in the sixteenth and seventeenth centuries. The separation of Holland and Belgium in the wake of the struggle with Spain gave rise to something of a caste system, where one's cultural status became attached to one's language, and a rift developed within a previously unified people.

It became fashionable for many prominent individuals from Flemish-speaking families to "cross the line" and turn their backs on their roots. Meanwhile, the Flemish-speaking population of Belgium, which was a relatively small minority in 1830 at the time of the separation—following a "reunion" with Holland that emerged out of the Peace of Vienna in 1815 that ended the Napoleonic Wars—became a majority because of a higher birth rate.

I once asked Secretary General Paul Henri Spaak of NATO, a Belgian and one of the "Wisemen" of Europe at the time, if he thought that the situation could be resolved without some form of drastic action. He shook his head sadly and said, "Because we are a small country, we must think small." In the meantime, some form of federalism is in the offing if the status of Brussels can be resolved.

Louis Henry, who did not normally speak Flemish, had served in the Belgian underground resistance in World War II and had been captured by the Gestapo during one of his clandestine forays. He was badly injured when he was tortured while a prisoner of the Nazis. Whenever Henry visited Ghent, Professor Dekeyser, who admired him enormously both as an individual and as a national hero, gave special orders to the Flemish staff, "Today we will speak French."

In spite of the problems the natives were struggling with, Betty and I were treated with gracious hospitality wherever we went. Professor Dekeyser and his wife proved to be extraordinary hosts, providing food for the mind as well as the body. He modestly claimed that it was sheer accident that he had become head of the laboratory. Others, he said, more deserving, had suffered ill fortune through bad health or the war. Actually, he was the perfect leader for that time and place, and he was widely appreciated and respected.

One of the saddest trips we took while in Belgium was along the World War I battle line that extended from Tournai to Nieuwport, where so many died for the sake of a few meters of ground.

In the latter part of the summer, we took the train to Vienna. Austria had just been declared a neutral country upon the withdrawal of both the Soviet and Western armies of occupation. Vienna was in the process of returning to normal life, although the opera house was still badly damaged. The great museums, which had been in the zone occupied by the American and British forces, were open and in good repair; it was a rare privilege to be able to visit them then while they were still free from crowds.

During our stay in Ghent, I received a telegram from a Physics Departmental Committee at the University of Illinois asking if I would be willing to serve as Head of the Department starting in 1957, when Loomis planned to retire. Betty and I talked it over and decided to accept. It was not an appointment I would have sought at that moment; I did, however, deeply appreciate the confidence shown by the members of the Department, and I felt I should do whatever I could to promote our common enterprise. The drawback for me was that I could expect to lose much of the professional freedom I had enjoyed since Loomis had returned from Lincoln Laboratory in 1952.

My term in the departmental office was marked by the launching of the satellite Sputnik by the Soviet Union in 1957. Needless to say, this caused great consternation in Washington, and support for science was promptly increased. President Eisenhower replaced his Secretary of Defense, Charles E. Wilson, when the latter short-sightedly attempted to hinder the advent of the satellite age.

Soon after becoming Department Head, I was elected to the Council of the National Academy of Sciences, its governing body, for a three-year term and became more intimately familiar with the workings of the institution and the individuals in it—both members and staff. As will be described in Chapter 13, the Academy has a complex structure, being a private organization with a federal charter which commits it to advise the government "on request." Detlev Bronk, whom I had known since Philadelphia days, held the post of President on a half-time basis, devoting the other half of his time to the Rockefeller Institute for Medical Research. He was proving to be an inspired leader, with an exceptionally fine intuitive sense, and a respect for quality in all fields of science. He was also an elegant showman. He had taken office in 1950 after what turned out to be a competitive election, so he was in every sense popular with the membership. The term of the office was then four years and he was in his third term. I had served on the nominating committee that had put him up for his third term. While there was inevitably some slight objection to one President serving for twelve years, he was reelected by an overwhelming majority.

The problems the Council faced at that time were on the whole routine and did not cause serious controversy. There was some stir when Sputnik was launched, but the Academy had risen to the occasion and had offered the government sound advice, particularly once James R. Killian, the head of the Massachusetts Institute of Technology, replaced I. I. Rabi as President Eisenhower's personal Science Advisor, Rabi's own recommendation.

Rabi, who had known President Eisenhower since the days when the latter was President of Columbia University, once remarked that at their first meeting after the launching of the Soviet satellite, Eisenhower said, "You seem a bit annoyed with me." Rabi responded, "You seem to have ignored the scientific community up to this point. I trust you will make up for it." Rabi also emphasized the importance of having someone more physically active than he to replace him as Science Advisor.

Wernher von Braun, who had been waiting for several years at the Army Materiel Command Center (later the George C. Marshall Space Flight Center) at Huntsville, Alabama to launch a satellite, soon allayed national fears to some extent by launching his somewhat makeshift but entirely adequate American satellite. He had been prevented from doing this earlier by U.S. advocates of arms control who felt, rightly or wrongly, that such a step would be provocative in the Cold War environment.

After Sputnik, the Pentagon determined that it should create twelve new laboratories devoted to materials science, in order to strengthen its position in the coming space age. Actually the concept of creating such government-sponsored materials laboratories had originated with John von Neumann during his period as Commissioner on the AEC, starting in 1954. He had decided that advances in such research were as important for both science and technology as any other; much more concentrated attention should be devoted to the field. In the process he elevated the field of research in materials science to a state of dignity and respect which it has continued to enjoy since. One of his first actions had been to ask me to prepare a proposal to the AEC for establishing such an interdisciplinary lab at the University of Illinois. I complied at once with the support of the University. He also urged other agencies of the government to follow this lead. His death from cancer early in 1957 derailed the initial proposal to the AEC, but the plan took root at the Department of Defense. In any event, the bidding was opened and the individual in charge of selecting which academic institutions would be provided the new labs was Charles Yost, an old friend and supporter of our laboratory. At first he felt that we were already so strong that a new lab on our campus would not add significantly to the national strength. Fortunately for us, he changed his mind, recognizing that many of our best people might very well be offered positions at new labs elsewhere, and would leave us. He relented.

Unfortunately, Senator Everett M. Dirksen of Illinois had incurred the wrath of one of his Senate colleagues from Missouri. Dirksen, who was exceedingly powerful at that time, had rewritten a Senate bill so that a federal prison destined for Missouri ended up in Illinois. The Missourian's wrath was not quenched until our laboratory was removed from the bill which would have authorized it. To compensate, a friend of ours at the AEC, Donald K. Stevens, found a new solution involving the cooperation of his agency, the DoD and the University administration. A laboratory was finally constructed, using funds from a state construction authority which was reimbursed over a ten-year period by the federal agencies. All of the participants including Charles Yost had to overcome a number of obstacles before this result was achieved. By the time the laboratory was operating, my colleague, Robert Maurer, was the lab's Director—a post he held until his retirement in 1981.

My attempt to establish some kind of routine life after becoming the Head of the Department was abruptly derailed in 1959 when Bronk approached me to ask if I would be willing to serve as Science Advisor to Secretary General Paul Henri Spaak of the North Atlantic Treaty Organization. Soon after Sputnik was launched, the NATO Council began to wonder if it was devoting sufficient attention to the basic sciences. To answer that question, the Council created an advisory committee, of which I. I. Rabi was a strongly influential member, and which told NATO that the answer was no. Rabi recommended the establishment of an office devoted to basic science, and the installation of a Science Advisor who would have a rank equivalent to that of a cabinet officer. The NATO Council agreed and Norman F. Ramsey accepted the post for the first year (1958–59). After thinking the matter over and discussing it with the University administration, I agreed to succeed Ramsey, with the understanding that Gerald Almy would take over the Department in my absence.

Taking on the NATO position meant that I had to be made a Foreign

Service Officer in the Department of State. H. P. Robertson was of great help in ensuring that the formalities such as security clearance and health checks were handled expeditiously. I was grateful to him on two scores. First, I was delighted to see that he was genuinely interested in having me undertake the assignment. Second, it was well-known that the State Department was not very efficient in handling matters dealing with science and might, under normal conditions, have taken a long time to settle the affair. At that time Wallace R. Brode, a well-known chemist, was the Science Advisor in the State Department—a post that has been all but eliminated since. He, however, had developed a great distaste for anything connected with NATO and would not have been of much help. My friends in the Pentagon made the same special arrangements for me that they had for Ramsey, which would make it possible for me to return from Paris to Washington every two months or so. (See Chapter 12 for details of the NATO years.)

When I returned from NATO in the autumn of 1960, I was approached by both the Convair Division of the General Dynamics Corporation and what was then United Aircraft with offers to join their staffs as a senior science advisor. I decided to remain at Illinois, but did accept a consultantship with United Aircraft.

That advisory work provided me with great insights into several newly emerging areas of science and technology. The research group, for example, was working on ion propulsion for rockets destined for long interplanetary missions, as well as on both high-powered and highly refined lasers. The first type of laser provided a possible means of transferring a large amount of energy at a high rate; the second proved useful for the development of laser gyros. Jet-engine development was moving ahead rapidly for a variety of applications such as civilian and military aircraft, as well as the lunar landing modules. I felt as though I were in front of a wonderful window, watching the evolution of some of the most exciting technical developments of the decade.

My association with United Aircraft enabled me to form friendships with some remarkable engineers. Among them were Wesley A. Kuhrt (Figure 11.22) and John G. Lee (Figure 11.23). Wesley Kuhrt had joined the company in 1941, early in his career, with an advanced degree in aeronautical engineering from M.I.T. He was put in charge of a number of research projects and then became Director of Research in 1963 when all of those activities were consolidated into a formal laboratory. In 1968 he was made the head of the Sikorsky Aircraft Division, which was involved in the development and production of helicopters. One of Kuhrt's prized possessions was a letter, which he had framed, offering him his first position at the company. The letter congratulated him on his successful interview with the employment office and offered him the company's top starting salary of fifty cents per hour.

Figure 11.22. (Left) Wesley A. Kuhrt. He rarely wore this intensely serious expression. (Courtesy of United Technologies, Inc.) Figure 11.23. (Right) John G. Lee, aerodynamic engineer – a pioneer. (Courtesy of United Technologies.)

John Lee was also a graduate in aeronautics from M.I.T., but had completed his work there twenty years earlier. He was truly one of the aviation pioneers, having seen all the phases of research and development from fabric-winged biplanes to the jet age. His ingenuity in the field was apparent in many ways. Among his more ubiquitous inventions are the asymmetrical, teardrop-shaped metal stubs one sees on the upper wing surfaces of many commercial airplanes. These devices help maintain laminar flow over the wing to sustain lift.

One of Lee's great-uncles was the Confederate engineering general, Stephen Lee. Stephen was involved in the bombardment of Fort Sumter and initially in charge of defenses in the siege of Vicksburg; he was also with General Robert E. Lee (no close relative) at the surrender at Appomattox. He was given a few gold coins as his pension on departing from R. E. Lee. John Lee had his great-uncle's dress uniform mounted in a glass case in their large home in Farmington, Connecticut.

A relative on Lee's mother's side had been involved in one of the first attempts at heavier-than-air flight. The machine ran on rails and its propellers were powered by a relatively light-weight steam engine. John showed me a photograph in which it seemed the machine had actually gained sufficient lift to rise off the rails. Unfortunately it was unstable and soon crashed. Any meeting with John was a special occasion.

At just about this same time, I was asked to serve as scientific consultant and board member of the Ampex Corporation. The company had been formed by Alexander M. Poniatoff to explore the potential of the magnetic tape recorders that had been invented in Germany, and that had been turned over with special licensing privileges to American corporations. This was the period during which that form of magnetic recording and its descendants came into their own for countless uses, ranging from audio and video systems to computing devices.

I also served on the board of the AMF Corporation when Morehead Patterson was Chief Executive Officer. The company was then in the process of helping to revolutionize manufacturing through increased automation. Following Patterson's death in 1964, however, the research lab was virtually shut down and the company concentrated on marketing its leisure-time products; I left the board at that time.

Provost Gordon N. Ray, a former member of the Department of English Literature and one of our much-admired friends at the University, resigned in 1960 to become President of the Guggenheim Memorial Foundation in New York. Campus rumor had it that the faculty committee had nominated me as a replacement for the position but that Dean Everitt had demurred — perhaps a hangover from the Ridenour period. As a result, Lyle H. Lanier, a psychologist, was selected. I was actually pleased at the choice, since I greatly admired Lanier and knew the confining burdens of the position. Lanier was indeed an excellent Provost rendering great service both to the University and to President Henry during a major transition period. Lanier held the position until his retirement in 1972.

In 1963, Fred Wall, the Dean of the Graduate School, left Illinois to accept a position as Chairman of the Chemistry Department at the University of California at Santa Barbara. President Henry asked me to succeed Wall as Dean and also to accept the title of Vice President for Research. I moved into the Dean's office and Almy became Head of the Physics Department. In retrospect, it might have been wiser for the continuity of departmental affairs if Almy had been chosen in 1957, but one never knows what lies ahead. When Almy retired in 1970, he was succeeded as Department Head by Professor Ralph O. Simmons (Figure 11.24), a fine physicist and former Rhodes Scholar. Simmons maintained a highly productive research program while carrying the administrative burdens of the department.

I served as President of the American Physical Society from 1961 to 1962. Traditionally, the society held its major annual meeting in Washington at the end of April. Betty, always a good hostess, had helped the local organizing committee to arrange a visit to the Supreme Court Building for both the woman members and the wives of the members, complete with a luncheon at which the guest of honor was Ladybird Johnson. Betty had the privilege of sitting next to her.

Some members of the academic community tended to scoff at the Johnsons

Figure 11.24. *Ralph O. Simmons.*
(Courtesy of the University of Illinois.)

because of their "down-home" Texas background. After the lunch Betty told me, "That was a remarkable experience for me. Mrs. Johnson is an exceptionally fine and intelligent woman. The people who joke about the Johnsons are completely wrong." We saw a great deal of Mrs. Johnson during the five years that Lyndon Johnson served as President, starting in 1963, and admired her enormously, not least for her keen understanding of people and situations both personal and political. We both believed that over the holidays in 1967, it was she who persuaded her husband not to run for another term in 1968, a decision he announced in March. She probably felt that the complex burden was undermining his spirit and health. His demeanor in public was quite different before and after that Christmas.

In 1961, I was elected Chairman of the United States Delegation to the International Union of Pure and Applied Physics and was also chosen to serve on the Executive Committee of the Union. That Union was one of the first of the scientific unions founded in the decades following World War I to promote regular communication between members of the professions.

The Executive Committee meets annually and the entire Union convenes every three years. The meetings are held at different sites each time to give members an opportunity to observe first-hand the activities and facilities of colleagues in the various member countries. The one-month meeting held in Japan in 1953, described above, was one of the first meetings of the entire Union in the period following World War II. I served on the Executive Committee for six years, attending meetings in such places as Australia, Canada, India, Switzerland, and Sweden.

In the immediate aftermath of the war, there apparently had been some pressure exerted by our State Department to prevent the election of a Soviet member as President of the Union, should one be nominated. One effect of this "request," if indeed it had been made, was that the other member nations agreed informally among themselves that they would not elect a President from the United States unless one from the Soviet Union was elected first. This political intrigue was completely contrary to the aims and ideals of the Union; our activities were specifically meant to be non-political, and our purpose simply to serve the advancement of good scientific research and standards. To neutralize the conflict, I nominated Professor D. I. Blokhintsev, a long-time Soviet member of the Executive Committee, for the Vice-Presidency with the understanding that he would become President in three years. He was elected.

Then, in 1966, I nominated Robert Bacher Vice-President, with the same understanding. My selection of Bacher was well-calculated. Bacher was a very distinguished scientist who had served his country as a member of the Atomic Energy Commission. He was now a major figure at Caltech, which had initially been brought to national distinction through the efforts of Robert A. Millikan, the first scientist from the United States to serve as President of the Union. Millikan had been a staunch supporter of the International Unions and it pleased my sense of continuity to bring in a distinguished successor from the same institution.

I had not conferred with Blokhintsev before nominating him, as I had hoped to pleasantly surprise him. In fact, he seemed rather dismayed for a minute or two and said a bit anxiously, "I am not at all certain that my country will allow me to serve in the office. I must check with the authorities." Apparently "the authorities," who held their citizens on short leashes, were agreeable.

Early in 1962, William Houston and Julius Stratton approached me on behalf of the nominating committee of the National Academy of Sciences to ask if I would be willing to serve as President of the Academy starting in July of that year. The Presidency had traditionally been a part-time position, really almost an honorary office over part of its span. However, having served on the Council under Bronk, as well as on other committees, I knew that by now the office entailed real and increasing responsibilities. I was not sure that I could adequately fill that post while continuing my activities at the University of Illinois. Houston and Stratton said that the nominating committee, which apparently felt that I would combine adequate judgment and enterprise, understood the situation; in fact, they hoped I might help the academy clarify the future role and involvement of the president. With that understanding, I agreed to accept the nomination. The offer came as a complete surprise to me because I had assumed that Bronk would be followed by George Kistiakowsky, who had followed Killian as Science Advisor to President Eisenhower. It was never clear to me whether Kistiakowsky had been offered the position and refused it, or whether the nominating committee had decided that it preferred a somewhat less volatile individual. In any event, the election at the end of April went through uneventfully. Three years later, when I agreed to become the first full-time president of the academy, Herbert Carter occupied my position at the university on an interim basis until it was filled by Daniel Alpert (Figure 11.25), an old and valued friend, who had served as the third director of CSL.

Soon after my appointment to the Academy had been announced in mid-1962, Illinois Governor Otto Kerner asked if I would help him create a science committee for his office. I did so with pleasure, serving as chair for two years. Some of the major issues we addressed included land and water management, the status of medicine in the state, and the stimulation of business. The governor arranged for me to speak about our advisory work at one of the National Conferences of Governors, which experience gave me a fascinating glimpse of national politics at that level. I gained a whole new respect for Kerner, seeing how myriad were his responsibilities, and how complex the environment in which he tried to fulfill them.

Governor Kerner came to national prominence during the social turmoil of the 1960s when he was chairman of a group appointed by President Johnson to attempt to analyze the source of the troubles and to offer possible remedies. While the efforts and intentions of what became the Kerner Commission were significant, it served mainly to mirror the problems rather than solve them. The level of disruption was too great to be resolved by a direct, rational approach.

Looking back, I realize how idyllic were our years in Illinois, in spite of the difficulties of President Stoddard. Faculty and students alike comported themselves with dignity and were deeply committed to their work, while the state expressed its pride in our institution by funding it generously. The

Figure 11.25. Professor Daniel Alpert. The third director of CSL. A most valuable and profoundly thoughtful addition to the university. He later served as Dean of the Graduate School.

Chapter Twelve
NATO (1959–1960)

By 1949 it was clear that the Soviet Union had no intention of entering into an open democratic partnership with the Atlantic community of nations and would follow its own agenda of aggressively seeking to gain political and military control wherever it could. Therefore, the Atlantic nations decided to form a defensive alliance, which came to be called the North Atlantic Treaty Organization (NATO). Lord Ismay of Great Britain was Secretary-General; he was aided by many of the most brilliant political minds in the Western community. By 1956 the European community had recovered economically and the organization had established a stable structure; its civilian headquarters were in Paris, although still in temporary buildings at the Palais de Chaillot awaiting the completion of a new home. The military headquarters were nearby at Rocquencourt. Each of the fifteen member-nations appointed a full-time NATO ambassador.

In order to strengthen non-military relations between NATO and its member-nations, the Secretary General appointed an internal committee composed of the ambassadors from Norway, Italy, and Canada—"The Three Wise Men"—to determine the best means of achieving that goal. After reviewing the matter, they made two important recommendations: first that NATO keep in close touch with the political affairs of its member-nations, and second, that NATO establish its own science office to help increase the effectiveness of scientific research within the NATO family.

The persuasiveness of I.I. Rabi and the launching of the first Soviet satellite were both powerful inspirations in the creation of the science office. Up to that time, technical activities in NATO had focussed on the work of the Armaments Committee, which was the design and procurement of military equipment.

Both of the Wise Men's recommendations were adopted by the member-nations and Professor Norman F. Ramsey (Figure 12.1) of Harvard arrived in Paris in the spring of 1958 as the first full-time Science Advisor; his first task was to oversee the development of the Science Office and its program. Ramsey had been a student of Rabi's in the 1930s. Rabi, in turn, was selected to be the U.S. Representative on the NATO Science Committee, the part-time advisory board for the office on which each member-country would be represented.

Ramsey quickly won the admiration and respect of the leaders of the

Figure 12.1. Norman F. Ramsey, Jr., the *first Science Advisor to NATO. (AIP Emilio Segrè Visual Archives, the* Physics Today *Collection.)*

NATO family, and in a remarkably short time, had succeeded in convincing the NATO council to pass the detailed legislation required to make his office operable. He emphasized three major goals: the creation of a NATO fellowship program to promote exchange visits among scientists of different NATO countries; the support of study institutes and summer schools involved in the review of topics of current interest for the advancement of basic scientific research; and the cultivation of collaborative research programs involving NATO scientists and facilities.

Ramsey assembled an excellent full-time staff to help him pursue his goals. His deputy was Anthony Sargeaunt, a seasoned operations research scientist. Also recruited to the team were Dr. Rudolf Schrader, a physicist who had been a member of the Civilian Science Corps in the German Defense Ministry in Bonn, and Dr. C. Klixbull Joergensen, a Danish physical chemist.

Ramsey had agreed to stay for only one year and I was asked in the late autumn of 1958 if I would be willing to succeed him in the spring of 1959. After discussing the matter with the University administration, I was able to accept the invitation. In order to become familiar with the situation in NATO, I made several brief trips to Paris before settling in there full-time. Betty was somewhat apprehensive initially about being involved in a diplomatic post in an international organization but faced the challenge with cheerful good will.

I was to have the privilege of spending most of the money that Ramsey had negotiated through the NATO Council, as it became available only as he was leaving. Anthony Sargeaunt, who wanted to write a book on operations research, stayed with me only a few months, whereupon I promoted Rudolf Schrader to Deputy and took on two new individuals: a retired French Air Force general, Jean Truelle, and a Portuguese administrator, Mr. Lopes.

The United States Ambassador to NATO, W. Randolph Burgess, was not a professional foreign service officer as is often the case with American ambassadors; he was a wealthy businessman as well as a public servant and a scholar. He had extensive experience in both public and private banking and monetary planning, and had served as an officer in the Federal Reserve Bank and the Treasury Department. He was, incidentally, a good friend of President Eisenhower. Betty and I had a good relationship with him and his wife, the former Helen H. Woods, who had been a general in the Women's Auxiliary Corps of the Army. Burgess' right-hand colleague, Frederick E. Nolting, was a dedicated and highly effective foreign-service officer, who later served as our Ambassador in Vietnam during the war there.

I found an easy answer to our housing question in Paris. A number of faculty members at the University of Illinois had successively taken sabbatical years in Paris and had sublet, one after the other, an apartment on Avenue de Lamballe in the Passy District, 16th Arrondissement. The primary renters of the unit lived in Southern France and held on to the lease as the fixed rent was very low, but were happy to rent out the apartment through an agent to what were then regarded as rich Americans. I was pleased to take over the unit when a family from our French department at Illinois finished their sabbatical and returned to the United States.

Betty and I were not embarking on a leisurely sabbatical, but would be part of a diplomatic corps. Since our position would require us to entertain from time to time, and since the previous tenants had mainly been interested in the bedroom and bathroom facilities, I took it upon myself to spruce up the entire apartment before Betty arrived in June. Fortunately, I had access to American cleaning equipment with which I was familiar at the Post Exchange. These supplies made my task more manageable, and with an investment of sufficient time and elbow grease, I was able to get the place into reasonable shape, although much remained that required Betty's finer touch. The apartment, which dated from the 1920s, was of a classical French upper-middle-class design, with a number of well-mirrored walls. When tidied up, it was a wonderful space both for daily living, and for guests and occasional receptions. From a corner balcony, we had an excellent view of the Seine and of the nearby bridge that featured the diminutive original version of Bartholdi's Statue of Liberty. On the ground floor of the building was one of the best restaurants in the district, run by a genial, effervescent individual who went under the name of the Ours Martin—the Bear Martin. His wife, who bore a striking resemblance to Madame La Farge, presided at the cash register.

Betty who had contemplated her complex new life—running a household in a foreign country while dealing with the diplomatic community—

with more than a little apprehension, as mentioned earlier, rose magnificently to the challenge, and quickly became a popular figure among both our diplomatic colleagues and among the shop-keepers of Passy. Fortunately she had had a fine start in learning French at the Shanghai-American School. With the help of the local merchants, she soon felt at home everywhere. I rather envied her this opportunity to immerse herself in the French language, because English remained our official business language at the office. So although I was able to exercise my ears for French, building on the foundation I already had, my speaking skills were only modestly improved.

Paul Henri Spaak (Figure 12.2), a Belgian diplomat who was highly respected throughout Europe, had taken over as Secretary-General from Lord Ismay in 1957. He had spent the war years in England as part of his government-in-exile. He was an exceedingly eloquent spokesman and had little difficulty in becoming the leader of the NATO Council of Ambassadors, who were inspired to support his proposals to the extent their governments permitted. In addition to the weekly meetings of the NATO Council, which were open to the staff, Spaak occasionally held more exclusive meetings (séances privées) to discuss special political issues that arose. One such meeting took place when, in 1960, Chairman Nikita Khrushchev abruptly walked out of the summit meeting in Paris with President Eisenhower, as a result of the U-2 overflights of the Soviet Union.

Figure 12.2. *Paul Henri Spaak carrying the weight of the world. This is an official photograph. He was very eloquent and had a great sense of humor. (Courtesy of NATO.)*

The Deputy Secretary-General was F. D. Gregh, a brilliant member of the French Civil Service who had spent a number of years in Washington. He was indispensable to Spaak. Although his credentials were of the finest, he was somewhat out of favor with the de Gaulle administration, so that his post in NATO was something of a haven for him. While I could not help but admire his brilliance, I was a member of a substantial group of his colleagues who felt that he was continually scheming to acquire more personal power.

The NATO Ambassadors were, on the whole, an elite, hand-picked group who were veterans of their respective countries' diplomatic service. The senior Ambassador was Mr. Michel Melas, who had represented Greece in Washington. Lord Frank Roberts, who was already a distinguished member of the Foreign Office of the United Kingdom, went on from NATO to serve as Ambassador, first to the Soviet Union, and then to the Federal Republic of Germany. Ambassador Salim Sarper of Turkey was called from his post in 1960 to become President of his country. Turkey was struggling through a period of great turmoil, and as Sarper was widely respected by the entire citizenry, he was able to exert a strong stabilizing influence on the situation. Ambassador André de Staerke of Belgium, often a counselor to Spaak, eventually became his country's leading diplomat. The Dutch Ambassador, Dirk U. Stikker, who left a highly successful business career to serve his country, followed Spaak as Secretary-General when the latter left office at the end of 1960.

During this period General Lauris Norstad held the important post of Commander of the NATO forces, headquartered at Rocquencourt, and was frequently present when the Ambassadors met. His office provided the link between the various military forces of the NATO countries stationed in Europe.

Rudolf Schrader (Figure 12.3) had been in an advance party when Germany invaded Norway in 1940. He was captured by the Norwegians, turned over to the British forces and shortly afterwards found himself in a Canadian prison camp, where he remained for nearly six years. The camp inmates developed their own internal organization centered around daily chores and educational programs. Schrader became proficient in both English and in the basic sciences. A talented amateur musician, he often entertained his campmates on the piano and the accordion; in the process he acquired a large repertory of American jazz.

When he returned to Germany immediately after the War, he entered the physics program at the University of Goettingen with the intention of completing his education as rapidly as possible and moving on with a normal life. Each Saturday afternoon a large group of students, almost all veterans, took a local train into the countryside where they haggled with the farmers over their next week's food supply. The trains were jammed well beyond capacity and the students were accustomed to climbing in and out of the windows as well as the doors. One day an elderly woman who could stand

Figure 12.3. Rudolf Schrader who became my deputy. (Courtesy of Mrs. Schrader.)

the pandemonium no longer stood up in Rudi's car and shouted at the top of her lungs, "You awful men need discipline! You should serve time in the army!" A stony silence enveloped the car as the men digested the full implication of what she had said.

The American Science Attaché in the U.S. Embassy was a well-known chemist, Edgar Piret. He had been appointed by Wallace Brode, mentioned in the previous chapter, and was somewhat apprehensive about association with me at first. We were brought together on so many occasions, however, that his concerns soon vanished.

Relations between NATO and President de Gaulle, who had taken office in the late spring of 1958 at a time of great crisis in France, were very complex. He had had a bad time in England during World War II as head of the Free French government, and he saw NATO as the creation of those who had demeaned him. Moreover, the Algerian War was in progress, further straining the French government. One of President de Gaulle's first acts in office had been to pull the French Navy out of NATO. Then, in the mid-1960s, he ordered the entire NATO organization to leave France, whereupon it was invited to move to its present headquarters in Belgium and received a warm welcome there.

My successor, Dr. William A. Nierenberg (Figure 12.4), had his own problems with President De Gaulle. Nierenberg had succeeded in gaining a consensus among most of the other NATO members regarding the creation of an international institute of science and technology that would be attached to NATO and that would be an important unifying force with respect to international educational policy and research. An initial survey of the vari-

Figure 12.4. Dr. William Nierenberg, my successor. (AIP Emilio Segrè Visual Archives.)

ous NATO Ambassadors made it seem that unanimous approval could be expected. The committee was chaired by Dr. James R. Killian, who was then Chairman of the Corporation of M.I.T. He led a distinguished working group, and I was privileged to sit on an advisory council. A crucial meeting was held in London that would determine the fate of the institute. Shortly after the meeting started, Dr. Pierre Aigrain, who was both the French representative and my old friend of many year's standing, arrived and took one of the remaining seats across the table from me. In a jocular vein reflecting the optimistic mood of the group, he said to me, "At last, Fred, we are on opposite sides of an issue." We all laughed and the meeting proceeded. Half an hour later a member of the secretariat entered the conference room with a message for Aigrain. As he began to read it, his face grew tense and he blanched. Then, with some hesitation he said, "President De Gaulle has decided to veto the plan for the institute." Our meeting ended since NATO actions required unanimous consent.

NATO was still housed in the temporary quarters in the gardens of the Palais de Chaillot when I arrived. 1959 was a great vintage year and the temperature in my office did its best to reach 40 degrees centigrade (104°F) almost every day during that summer. We moved to our new headquarters at the Porte Dauphine on the edge of the Bois de Boulogne late in the autumn, taking over an attractive suite of offices overlooking the Bois. The building has since been turned over to the French educational authorities and the site has been renamed Place du General de Lattre de Tassigny.

When I arrived in May of 1959 to begin my full-time stint, Schrader and Joergensen were enormously helpful in dealing with the not unpleasant problem of allocating the money Ramsey had acquired for us. Both were experienced, indefatigable workers, and Joergensen especially was full of practical, imaginative ideas. Ramsey had already started to distribute fellowships, and we were soon advertising the availability of money for both scientific conferences and research grants. All three forms of funding were eagerly and gratefully received in communities that were starved for support.

One of the questions we faced regarding scientific conferences was whether scientists from Eastern Europe could be allowed to participate in the meetings. In keeping with long-standing traditions in the scientific community, we decided to place no restrictions on who could participate—a decision which did much to popularize the support within the international scientific community.

There were a few amusing incidents regarding the research grants. Professor C. F. Powell (Figure 12.5), the English cosmic ray physicist who had discovered the pion with the use of photographic emulsions, requested both money and the use of the NATO navy to track cosmic ray balloons released in the Eastern Mediterranean. Powell was a well-known member of the Communist Party in England. We decided that his program was a sound one and agreed to fund it. There proved to be no difficulty in getting NATO ships to leave port and undertake the required sea duty in his service.

There were one or two awkward moments. For example, the organizers of the Pugwash meetings asked me to stay away from a conference to which I had been invited prior to my appointment at NATO, presumably because they felt that in my new position I could not support their disarmament goals. In another instance, we invited a group of European astronomers to a meeting at NATO headquarters to see if there were ways in which we could

Figure 12.5. Professor C. F. Powell. (AIP Meggers Gallery of Nobel Laureates.)

support astronomical research. Some of those invited became noisily indignant when they found they had to sign in and wear the customary visitor's badge.

Dr. Nierenberg has enjoyed pointing out that one of the remarkable features of the Science Office over the years has been the small size of its staff, especially when one considers the many tasks it undertakes. These people have always aimed to serve the needs of NATO's clientele rather than the interests of the NATO staff. This policy stands in refreshing contrast to the standard practice at many philanthropic foundations, including a large number of foundations in the United States. Generally speaking, I have noted that those that spend a large portion of their money internally eventually become detached from the real world and lose a great deal of their influence for good. Eventually, they are reduced to serving only limited causes of special interest to the staff.

The members of the NATO Science Committee were also a select lot. In addition to Rabi, there were Baron Zuckerman of the United Kingdom (then Sir Solly); Dr. André-Louis Danjon, the leading French astronomer; and Dr. P. B. Rehberg, one of the leaders of the Danish resistance in World War II. The Portuguese representative, who was head of the Atomic Energy Commission in his country, had the surprising name of J. F. Ulrich. He told me that he was from the Azores, where his family had lived for ten or twelve generations. Many of the original settlers on those islands were from northern Europe; some had probably fled the Netherlands because of the Spanish persecution; others may have come from the Rhineland because of the Thirty Years' War in Central Europe. I have read that the Azores were once called "The Flemish Islands."

All of our work, then as now, required approval from the Science Committee which oversaw our activities. The Committee gave us little trouble as long as we prepared our presentations carefully before the meetings and kept an eye on the geographical distribution of funds. Occasionally the Committee had trouble getting along with itself, but that was another story.

Rabi and Zuckerman tended to dominate the Advisory Committee. They sat next to one another in accordance with an alphabetical seating tradition based on nation names, and usually joined forces when one of them saw fit to disagree with something another member had put forward. Such disagreements were invariably focussed on matters of general policy and not on the details of distribution of funds. Most of the members of the Committee did not seem to resent this high-handed behavior, although they may have been suffering in silence somewhat as I did. On one occasion, however, Professor Danjou, our French representative, could stand it no longer and vehemently denounced both Rabi and Zuckerman, claiming that the whole program was under the sway of what he termed "the Anglo-Saxons." He also announced that he was resigning and that he would inform his government of his opinions he had just expressed.

Remarkably, Danjou came to me after this outburst and said that he realized that I was chairing the Committee under difficult circumstances and was genuinely trying to do my best to maintain a balance between the various national groups. He thanked me for my efforts and wished me good fortune both for the remainder of my term and when I returned to the United States. I was deeply moved, since he was obviously sincere and need not have gone out of his way to reassure me of his good opinion of me.

In addition to visiting scientific and cultural institutions in the various NATO countries, we followed many research activities within NATO itself. Most prominent among these at that time were the Air Research Laboratory in The Hague, established in 1954, and the Naval Research Laboratory at La Spezia, which was newly established in Italy in 1959. Trips to the former gave us an opportunity to return to the Mauritshuis Museum in The Hague with its wonderful Vermeer, "A View of Delft," whereas visits to the latter allowed us another opportunity to explore Florence.

Our staff also kept up with the work of the Armaments Committee. Their meetings were usually much more tempestuous than our own, since the committee recommendations greatly influenced the choice of design and sources for the armaments used by NATO forces. It was chaired by an American, Robert B. Fiske, who found the burden a heavy one. Betty and I formed a close relationship with him and his wife Lenore.

The fairly regular visits of Theodore von Karman (Figure 12.6) to Paris were always exciting for us. He was the chairman of the Advisory Group for Air Research and Development (AGARD), affiliated with NATO, and had been instrumental in reviving in the continental NATO countries a serious interest in aeronautics, an art and science which had lapsed since the war. His efforts were highly successful.

Figure 12.6. Dr. Theodore von Karman, the aerodynamic wizard – and much else. He was the first individual to receive the U.S. National Medal of Science. (AIP Emilio Segrè Visual Archives.)

Von Karman was of Hungarian origin. He had studied physics in Germany prior to World War I, and had then became one of Europe's leading aerodynamic experts. He left Germany to accept a position at Caltech in 1930 and made great contributions to the advancement of American aviation during World War II, not least by advocating the use of rocket jets to assist in the take-off of heavily loaded planes. He had a sly sense of humor which he used to great advantage when chairing the AGARD meetings. He was fairly deaf as a result of years of open-cockpit flying in World War I, and used a microphone and earphone during meetings to improve his hearing. Someone noted that if the speaker was unusually dull or had little new to say, von Karman would covertly turn off his microphone and sit through that part of the session communing with his own thoughts. In attempting to describe the unique capabilities of his fellow countrymen, von Karman asserted that if a Hungarian follows a normal person through a revolving door, you can expect the Hungarian to emerge first.

The AGARD office in Paris was run by von Karman's faithful staff executive, Dr. Frank Wattendorf, who had been one of his students at Caltech. Wattendorf lived during the summer months on a houseboat in the French canals; every year he moved his home a few miles further along so that he always had a new landscape to look at.

I first encountered von Karman in 1933 at a meeting of the American Physical Society, which I was attending with Condon. An American-made dirigible, the Akron, operating out of Lakehurst, New Jersey, had crashed into the Atlantic Ocean a few days earlier on one of its trial runs. A science reporter for *The New York Times*, William L. Laurence, approached Condon to ask if he had any opinions to offer as to what might have caused the crash. Condon pointed out that von Karman, an expert on aerodynamics, was attending the meeting, and that, he would be worth talking to. The four of us met for a discussion session (I was a silent but interested party). The concepts of air mass analysis, developed by the Norwegian J. Bjerknes, were very new at the time, but von Karman was completely familiar with the theory. He suggested that the Akron had encountered a strong weather "front" with high shearing forces and had probably been broken in half. This scenario fitted in well with the structure of the weather map at the time of the accident. Von Karman's comments were featured in *The New York Times* the next day in an article by Laurence.

Present-day information suggests another possibility: that the airship encountered a powerful down-draft in the turbulent region of air just before the front, and that the down-draft forced it into the sea.

One of the problems put before our office dealt with the technical design of Field Headquarters for the operations of NATO military forces, which would have to withstand substantial overpressure, of the order of two hundred pounds per square foot when under bombardment. Since I knew that this problem had been addressed previously in the United States, I was able to find an engineering expert, Commander Christiansen of the U.S. Navy, who had been successfully effective in this field of engineering. We put together an international advisory committee for the study and invited the group to attend what I hoped would be the first of only a few meetings. Christiansen would describe his work and I hoped that his designs for construction would be accepted as the NATO standard.

When at the start of the meeting I asked for questions or statements before we got underway, the hand of a middle-aged French officer shot up. He said, "You are very fortunate. We have here three graduates of the École Polytechnique. We will solve all of your problems. Our experience is vast. We were responsible for the design and construction of the Maginot Line." Christiansen, who was sitting next to me, said incredulously, "What's this stuff about the Maginot Line?" This was a magnificently designed structure meant to thwart a possible German invasion, but its purpose had been defeated because it was not extended to the North Sea and hence was easily by-passed. In the United States, the Maginot Line was a symbol of failure.

It soon became evident that my short series of meetings was extending itself indefinitely as the highly imaginative members of the committee in turn demonstrated their prowess at coming up with new ideas. Christiansen's efforts were in vain. I turned the meetings over to Schrader, and as I recall, they were still going when he was called back to Germany several years later. I never learned the ultimate result of the study.

With the help of Dr. Shepard Stone of the Ford Foundation, whom I met through Ramsey, we succeeded in obtaining a grant to carry on a study of the status of scientific research in the NATO family and to offer recommendations for its further advancement. We felt particularly fortunate when Louis Armand, a scientist-engineer who had been instrumental in rebuilding the French railroad system after World War II, agreed to lead the work of the study committee (Figure 12.7). Unfortunately, he soon became involved in helping the Mainland Chinese develop their own railroads and the leadership role fell back on our office and some borrowed staff.

The final report from our study was a highly cooperative effort, involving not just members of our science committee, but many others from throughout Europe as well. The study may seem bland and even primitive by present-day standards; but, one must keep in mind that in its time it represented the breaking of entirely new ground for many of the participants, and provided a starting point for many more sophisticated European studies that would follow.

The few months spent with the Sargeaunts, Ramsey's deputy and his wife, were most rewarding. Anthony Sargeaunt was not only a highly skilled operations expert who had been deeply involved in important work in World War II, but had as well many other attributes that made him a fascinating companion. He and his wife and family lived and entertained guests on a large motorized boat moored on the Seine. When the Seine was in flood, which it is several times a year, Sargeaunt had to take a leave from the office

Figure 12.7. *The core of the group studying ways of increasing the effectiveness of western science. Those in the front row, left to right are: Professor G. Puppi; Professor Paul Bourgeois; Sir John Cockcroft; M. Jean Willems; Sir Solly Zuckerman; and I. I. Rabi. Second row: first to the left, General J. Truelle of our office; fourth from the left, Professor H. B. G. Casimir, I am to the right of him. Rear from the left: Mr. E. McCrensky; third and fourth from left, Captain B. Bennett and Dr. I. Estermann; seventh from left, Rudolf Schrader; far right, Dr. C. K. Joergensen also of our office; Dr. William Nierenberg is on the rear right. (NATO photograph by Dominique Beretty. Courtesy of Mrs. Margaret Schrader.)*

and attend to his boat full-time, while the office staff waited tensely for assurances that it had not been washed out to sea. He came of a family that had raised race horses in Ireland for many generations; visits to Longchamp racetrack with him and his wife were educational and modestly rewarding in a monetary way.

We were all sorry to see the Sargeaunts leave, although Betty and I did have a chance to visit them later when they were moored on the Thames near Hampton Court.

Just before departing, Sargeaunt said to me, "NATO is a complicated place. If you need any solid advice, go to Lord Coleridge. I have already spoken with him." The man he referred to was the grandson of the poet Samuel Taylor Coleridge and was head of one of the divisions of NATO—a wonderfully warm yet aristocratic Englishman who was indeed a good friend to our office.

During most of the year, the NATO staff was on the whole a jolly, friendly group of people who worked together in a spirit of great camaraderie. Relations deteriorated abruptly, however, when the annual negotiations began concerning raises for the support staff. Then the department heads showed no mercy in denigrating the quality of each other's staff while ferociously defending their own. Their discussions had all the gentle courtesy of a barroom brawl since the total fund available was fixed beforehand.

General Jean Truelle was greatly inspired by the existence of our science office and the role that it could play in promoting a sense of cultural unity among the NATO members. He wrote a number of lengthy and glowing essays on the subject which I was pleased to see in some of the French publications to which he had access.

Once our office was running smoothly and Betty and I felt settled, we invited our office staff, and some others with whom they worked closely, to an informal evening party at our apartment. The kitchen was stocked with supplies of all kinds, and Rudi Schrader brought his accordion. The Truelles were apparently expecting a rather different kind of evening, and arrived in formal dress complete with gloves, silks, and cane. The Truelles, Betty, and I sat stiffly in our formal, mirrored living room, more or less discussing the weather, while the rest of the office staff noisily took over the apartment, tramping back and forth to the kitchen regularly for drink refills and nourishment. The entire apartment house reverberated to Rudi's accordion and to the group singing that accompanied it. In the living room, Betty and I shifted in our seats and smiled politely.

Beginning early in December, the various embassies began a two- to three-week marathon of black-tie evening receptions. They must have been coordinated by some master planner because they were all held without embarrassing competition or overlap between the various embassies. These receptions, of course, were far more than social occasions. In between holiday toasts, individuals from different embassies could hold brief but important conversations, and settle more issues than could be resolved in months of negotiations through official channels.

Following Ramsey's practice, Betty and I always hosted a large reception at the time of the meetings of our Science Committee held every four months; we usually had between one and two hundred guests from the NATO circuit. This not only helped to cement good relations, but again gave staff members that special opportunity to chat across national lines.

One of the most rewarding aspects of my job with NATO was the chance to visit the various member countries, to meet with individuals involved in their government's science administration or in the institutions where we provided research or study grants. Betty and I were especially eager to visit Greece, Turkey, and Portugal, which were new to us, although we enjoyed our trips to Italy and Scandinavia no less. We were in friendly hands wherever we went thanks to arrangements made by members of our science advisory committee. It was a wonderful time to travel since the tourist population in Europe was not yet much greater than it had been in the mid-1950s. I recall a visit to Athens when Betty and I were all but alone both in the archaeological museum and at the Parthenon. I also enjoyed a quiet picnic with Professor Kesar Alexopoulos and a group of international students on a deserted Corfu Island beach. Today on that beach there is scarcely standing room on a typical summer day.

Our visits to Turkey also seemed magical; we were especially thrilled by the wonders of Istanbul. As an amateur archaeologist, I relish the memory of two days in Ankara with its historical artifacts and structures going back to the Romans, and even farther to the Hittites.

In Portugal, the gentleman from the cultural bureau who was our guide had apparently been trained in the social sciences and maintained an attitude of disdain, real or feigned, for hard science and scientists. While driving along one of the handsome boulevards in Lisbon, I commented to him that Dr. Ulrich, our Portuguese committee member, must have been highly esteemed in his homeland. He frowned a bit and said, "How so?" I replied, "This beautiful boulevard is named after him." He answered dismissively, "One boulevard does not mean a man is famous."

Perhaps our most fortunate experience as tourists came on a trip to the Dordogne Valley over the Christmas holidays when NATO headquarters were closed. We drove to the celebrated Lascaux Cave, arriving there on the morning of the 31st of December. The attendant, who had been one of the boys who discovered the cave in 1940, remained outside after cautioning us in a humorously bantering way not to contribute any graffiti to the walls. So Betty and I spent an hour and a half exploring all on our own. It was awe-inspiring to realize that one was viewing the spectacular artistry of individuals who had resided in the cave during the last ice age, nearly twenty thousand years ago. The very next week, the cave was permanently closed to the general public because of the rise in humidity and its accompanying mold associated with the presence of crowds.

One incident that occurred during my travel for NATO showed me that when we hold an official post of any kind, we may cast a shadow much larger than ourselves. Rudolf Schrader, who succeeded Sargeaunt as my deputy, had been employed as a scientist in the German Defense Ministry, and he was eager to have me visit there at my earliest convenience in order to meet his former colleagues.

The visit finally took place when I was attending meetings at some laboratories in the Federal Republic. I was met in Bonn and escorted to military headquarters, which was closely tied into the NATO military structure. On entering the main gate, I noticed two identical flag poles, one flying the flag of the Federal Republic and the other the NATO flag. I asked my escort if this was routine practice. He replied, "No, only when we have an important visitor from NATO." Curious, I asked, "And who is visiting now?" He answered: "You!" At that moment I understood the temptation of assuming for oneself the significance of one's office. Or, as the justly celebrated scientist-author Lewis Thomas once said in relation to being honored: "It is not very dangerous if you do not inhale!"

Thirty years later I visited the German Embassy in Washington and noted that the flag of the Federal Republic and that of the European Council were set up as counterparts in the lobby of the embassy. This time when I asked I was told that both flags were permanent fixtures. Europe seems to be evolving, slowly but perhaps surely.

The Science Office was allocated about five thousand dollars for office entertainment. I parcelled it out among the staff in what I thought was an equitable fashion. Some of the other NATO officers used that money for personal purposes, but Betty and I decided to take the opportunity to get to know our NATO colleagues better, and to entertain some official guests. We quickly learned that ambassadors, who must do so much entertaining themselves, loved to have the roles reversed and to be the guests. In the more private and relaxed atmosphere of our parties, ambassadors from several different NATO countries would enjoy exchanging tales about international incidents that in their time were volatile but now simply wonderful dinner stories. On these occasions Betty and I were rapt listeners. We made one or two gaffs early on in the seating arrangements for such august groups, but suffered no ill will and quickly learned to discuss such matters with protocol officers.

I had the great privilege of meeting two individuals whom I had admired from afar, namely Raymond Aron (Figure 12.8) and Jean Monnet (Figure 12.9). I had become acquainted with Aron through his journal articles, and was particularly affected by his book on Communism, *The Opium of the Intellectual*. He served with the French Air Force and had joined De Gaulle in England as editor of a patriotic journal. He returned after the war and became a writer. He wrote brilliantly in both French and English on many issues, but was most impressive as an opponent of Communism. He was a potent advocate for his cause, despite opposition from the left-wing corps of the French intellectuals whom he had initially supported. I wrote to him from my office at NATO and had the great good fortune to have him to lunch on several occasions, so that we developed something more than a nodding friendship.

Fortunately the tide of opinion shifted among French intellectuals following the brutal putdown of the Prague spring in 1968 and the invasion of Afghanistan in the 1970s. As a result, Aron ultimately found a form of national acceptance among the intellectuals for his views. In his own way he was as influential as the Austrian Friedrich von Hayek, who wrote the equally famous book, *The Road to Serfdom*. Hayek had the great fortune to live to see the collapse of Communism in the Soviet Union, and thus gained international renown very late in life.

In the hopes of stimulating a productive discussion of trends within the

Figure 12.8. (Left) Raymond Aron, the patriotic Frenchman who followed General De Gaulle into exile. He returned from England to join the French left, but ended up as a liberal, outspoken opponent of the communist system. (Photograph by Roger-Viollet, Paris. Courtesy of Mrs. Pierre Aigrain.) Figure 12.9. (Right) Jean Monnet. He and Robert Schumann laid the groundwork for the Common Market. (Photograph by Roger-Viollet, Paris. Courtesy of Mrs. Pierre Aigrain.)

Atlantic community, and elicit suggestions for promoting further collaboration, I invited Aron to a luncheon meeting with I.I. Rabi and Solly Zuckerman. The encounter almost immediately got out of hand and turned into a disaster. Rabi and Zuckerman started right in by attacking the French plan to develop an independent nuclear arsenal (Force de Frappe). Aron, as a French patriot, found himself having to defend the program regardless of his own reservations about it. It would have taken a diplomat with the talents of Talleyrand to get the discussion under control. Not surprisingly, Aron completely changed his public attitude toward the nuclear weapons program after that lunch, becoming one of its strong supporters. I concluded that scientists on the whole are very poor politicians outside their own spheres of self-interest.

I wrote similarly to Jean Monnet, one of the first exponents of European Community. Soon after World War II ended, he began working vigorously to develop relations between France and other European countries. He invited me to a small luncheon party at his apartment and expressed his concerns about the way President De Gaulle was catering to French nationalism and straining the carefully cultivated ties with other countries. Monnet told us that he had asked the President, "But what will happen to our France when you are no longer with us to lead?" De Gaulle replied, "It will fall back into the mud hole from which I rescued it." Unfortunately Monnet did not live to see the dawn of the new age when internationalism reemerged as a guiding spirit in Europe.

While serving in Paris, Betty and I were frequently invited to receptions at private homes. On one occasion, at a gathering hosted by Madame Adrianne Weyl, we met the distinguished Finnish sculptor, Eila Hiltunen, whose work was being exhibited in Paris. She was pleased to see us there as English was her second language, and we had a delightful time discussing her life and her work. When Betty and I visited her exhibit, she sold us the working model for a bronze statue she was showing. Entitled "The Priest," the figure is a robed, hooded man standing as in a formal ceremony. It has been one of our most treasured possessions.

Hiltunen won a Finnish competition to create a memorial to the great composer, Jan Sibelius. Betty and I were invited to Finland in 1967, soon after the unveiling. It was an honor to meet her again and to be among the first to view her creation, which has won world acclaim.

Our office hosted many visitors, both from NATO member-countries and from those outside the "family." It was not uncommon for an American official to stop by our office, since, among other things, such a visit provided a good excuse to spend an extra day in Paris.

On one occasion I had a very informal visit from a Swiss official who wondered if it really was wise for Switzerland to remain outside NATO as its traditional neutrality dictated. As he put it, "It might be good for us to know the places from which those missiles we may occasionally see flying over our country originated."

An Israeli official who paid a call invited me to stop over in his country the next time I traveled to Greece or Turkey to observe Israel's scientific community. When I attempted to do this, my plans came to the attention of Secretary-General Spaak, who asked me to cancel the trip. He was concerned about the difficulties my visit might cause among the Moslem countries, the Algerian War with France being in progress at that time. I then appreciated the intense seriousness with which Spaak took his position as guardian of the organization.

I was also visited, again informally, by a Swedish official. Sweden was then developing nuclear reactors using uranium derived from their own very low-grade granitic ore. They knew that I had worked on the problem of radiation damage to reactor materials and asked if I would give an open talk in Stockholm on the subject, which I was happy to do.

Another visitor, with Swedish family connections, was Professor C. Richard Soderberg (Figure 12.10) from M.I.T. We were to see a great deal of one another in the next decade in connection with advisory work for what was then the United Aircraft Company in Hartford, Connecticut. He had

Figure 12.10. Professor Carl R. Soderberg, a distinguished engineer and a valued friend. He played a catalytic role in the development of jet engines in the U.S. (Courtesy of the Massachusetts Institute of Technology Museum.)

not only played a major role in numerous developments during World War II, but had encouraged the company to enter successfully into the development of jet engines, overcoming the resistance of well-entrenched engineers committed to piston engines. He had learned along the way that, with a combination of patience and wit, he could overcome most obstacles.

Back home after my term with NATO, I had a conversation with my father in which I commented that the European nations still showed pronounced differences and that one might well despair of their ever coming together. He said, "You must give them time. They have no other way to go."

In retrospect, I feel that the concept behind the creation of the NATO Science Office—to support programs that strengthen communications between scientists in different countries—was a brilliant one, for which Rabi deserves enormous credit. With relatively modest resources, the office has managed to remain in operation for nearly four decades and has had substantial and positive impact. Moreover, it was established at a time, barely a decade after the end of World War II, when it could provide a unique service to the vast segment of the scientific community not included in the programs that were developed so effectively for the nuclear and high energy physicists at CERN in Geneva, Switzerland. I hope it does not develop the view that its primary goal should be to support applied research.

In 1989 Ramsey, Nierenberg, and I had the pleasure of attending a thirtieth birthday celebration of the creation of the office (Figure 12.11).

Now that great changes are taking place in the Soviet Union, some individuals have proposed that either NATO be dissolved or that the United

Figure 12.11. Ramsey, Nierenberg, and I at a conference in Brussels in 1989 commemorating the thirtieth anniversary of the creation of the NATO Science Office. I was convalescing from virus pneumonia. (Courtesy of William Nierenberg.)

States withdraw from the organization. I do not believe that either action would be wise until the economic and political unification of Western (and possible Eastern) Europe has proceeded much further. At the least, NATO provides a base for establishing common policies among a group of nations that have common economic and political interests, and that will certainly continue to have some common defense interests as well.

In promoting a united policy of arms development and procurement, NATO's influence can forestall a potential arms race. This is a critical function, and one that, in my opinion, the United Nations is not yet sufficiently evolved to handle.

Chapter Thirteen

The National Academy of Sciences,[1] 1962-69

In June of 1962, after school activities had quieted down on the Illinois campus, Betty and I went to Washington to look for housing and to familiarize ourselves with the staff and activities of the Academy. Our old friends, the Leonard Lee Bacons, were a great help in getting us oriented. After some searching, we found a simple but elegant apartment in the "Cox Row" on N Street in Georgetown, just a few doors away from where the Kennedys had lived when JFK was a Senator; the area was a tourist attraction because of that. Cox Row dated from the first decade of the nineteenth century and, it was said, had been a way station on the Underground Railroad for runaway slaves fleeing North.

The apartment house was owned by Mr. and Mrs. George DeVeer. He was of English origin and had been in the trading business in Brazil. She was American, and had decided that retirement in the United States would suit them best. We became very good friends during the three years we lived there.

In that apartment, we not only started our term with the Academy but also lived through a number of memorable events: the Cuban Missile Crisis; the campus unrest that began with the Berkeley riots; the centennial celebration of the 1863 founding of the Academy; and lamentably, the assassination of President Kennedy. A month to the day before he was killed, Kennedy had addressed the domestic and foreign membership of the Academy in Constitution Hall as part of the centennial.

We also had congenial informal visits at home from Soviet Ambassador and Mrs. Dobrynin and Dr. and Mrs. M.M. Lavrentyev from the Research Institute in Novosibirsk, during a time when relations between the United States and the Soviet Union were relatively cordial.

Once it was determined that the Academy Presidency should be a full-time position, we looked for a somewhat more substantial official academy residence for the president where we could accommodate overnight guests

---

[1] There appear to be two comprehensive histories of the National Academy of Sciences. One, by Frederick True (National Academy of Sciences, Washington, D. C., 1913), was written for the semi-centennial celebration of the academy in 1913, the other by Rexmond C. Cochrane (*Ibid*, 1978) at the time of its centennial in 1963. The latter was initiated at my request.

and host receptions. By a most fortunate coincidence, John A. McCone, the Director of the CIA retired at that time and offered to sell his home on Whitehaven Street behind the British Embassy to the Academy. The house was ideal for our purposes and we lived there for our remaining four years in Washington. As might be expected, our guest room was seldom unoccupied. The house had a small garden that backed onto the British Embassy, and we soon became good over-the-fence friends of Ambassador and Lady Dean, whose hulking bulldog found our petite poodle Ginger an intriguing neighbor.

Our later years in Washington were marked by the great riots that exploded following the assassination of Martin Luther King. Even the Embassy District became unsafe at night; a walk of a few blocks along Massachusetts Avenue from the Cosmos Club to Whitehaven Street could be hazardous, as I learned from personal experience.

Although the house suited our needs well, our successors, the Philip Handlers, preferred apartment living. I therefore took responsibility for disposing of the real estate when we left Washington, and was able to realize a small profit for the Academy.

The National Academy of Sciences is a complex organization, as it has three very strong but very different objectives. They are to recognize and support good science through formal studies; to recognize excellent work by individual scientists; and to give advice to the government "on request," as stated in its charter of 1863. The Academy is both private and public — its relationships with scientists, particularly regarding membership in the Academy, are privileged and private, but the results of its work with the government are paid for by the government and may therefore be made public by mutual agreement.

Most scientific academies in Western Europe are now primarily honorific, although they may have originally been created to serve their governments. The National Academy of Sciences might have gone the same way; however, it created a "working" adjunct, the National Research Council (NRC), during World War I. This led to the establishment of a large full-time staff. In addition to calling on its members, the NRC avails itself of the services of many non-members, finding experts wherever they may be. This advisory work was expanded in the wake of World War II under the guiding hand of Detlev Bronk.

By the time I took office, the National Research Council had nearly five hundred advisory committees. Some were created to handle special problems on an *ad hoc* basis, and were short-lived. Others had long-term goals. The Highway Research Board and the Food and Nutrition Board were of the second type. The former board was particularly important since it provided guidance for the Federal Highway Program which was underway at that time. Most of the committees met in Washington, but there were summer studies conducted at Woods Hole on Cape Cod. Bronk succeeded in obtaining the use of a mansion owned by John Hay Whitney for a nominal rent. Most of the longer-term studies held there were supported by the Department of Defense.

Having a long-standing interest in historical matters, I spent much of my spare time reading about the background of the Academy and its membership, devoting particular attention to the two dozen or so volumes of members' memoirs then available. I was aided in this endeavor by a sprained back that confined me to our apartment for a month in 1964.

The Academy was originally created in substantial measure to serve the military purposes of the Union during the Civil War, as well as to recognize the importance of science for its own sake. It almost died with the termination of that war, but was rescued by Joseph Henry, a highly creative scientist who was then head of the Smithsonian Institution. He had not played a role in the creation of the Academy, but saw that it could be a valuable instrument for promoting good research. He nurtured it for the next thirteen years, giving it a home at the Smithsonian and thereby assuring its preservation.

The Academy gained national prominence in World War I through its applied work, and thereby generated sufficient public support to acquire both an endowment and a home of its own on Constitution Avenue. It languished somewhat between the wars, but then came to life again in World War II when then-President Frank B. Jewett, recently retired from the Bell Laboratories, brought the National Research Council into the center of action in areas ranging from medicine to undersea warfare. On retiring as Academy President in 1947, Jewett, in a brilliant address to the membership, reiterated Joseph Henry's desire that the Academy should keep its focus on the future of science, and should continue to promote basic research.

Bronk (Figure 13.1) had taken over as President of the Academy in 1950 after a brief period when A. Newton Richards, a research physiologist at the University of Pennsylvania, had held the post while Bronk was head of the National Research Council. The relationship between Bronk and Richards, who had been close, long-time colleagues at the University of Pennsylvania, was such that Bronk's influence was strongly felt throughout Richards' term in office.

Most requests to the Academy for assistance at that time came from the executive branch of the government; usually from a department or agency, but occasionally directly from the White House. I noted only one explicit request from Congress since the Civil War. That request, in fact, dated back to the turn of the century, although the Academy had undoubtedly rendered the Congress much informal assistance over the years through Congressional hearings and by other means. With some exceptions, the executive agencies appreciated the advisory work of the Academy. It was relatively inexpensive, since the government only paid for the Academy's oper-

Figure 13.1. Detlev W. Bronk. The photograph was taken in 1948. (Courtesy of The Rockefeller Archive Center.)

ating costs. Advisors were reimbursed for travel and living expenses, but received no honoraria except when involved in summer studies that went on for extended periods. Occasionally some scientists at the National Science Foundation, who were outstanding in their own right, resented the work of the Academy, seeing it as competing with their own. This sentiment, however, was limited to a few individuals and did not affect other agencies.

During the 1960s, many academic scientists opposed U.S. involvement in the Vietnam War and therefore objected to the Academy having any relationship with the Department of Defense. The Department of Defense began to carry out some of its studies at other venues. For example, the Defense Science Board started using the Naval Graduate School at Monterey, California for many studies that might previously have been assigned to the Academy. The board also consulted a quasi-independent group of scientists, the Jason Committee, which interestingly enough included as members some of the Academy's war protestors.

The creation of, or rather the reorganization of, the President's Science Advisory Committee in 1957 opened up a new, important forum in which to discuss the myriad of policy matters affecting basic and applied science. Bronk was a member of the Committee from the start when James Killian was the president's science advisor. I, in turn, was invited to join in 1962, essentially as an ex officio member. When Jerome B. Wiesner was Science Advisor during the Kennedy administration, a number of studies carried out by the Academy were requested by the Advisory Committee. However most of the committee's research was handled by its own substantial and highly competent internal staff, which was based in the Executive Offices of the White House.

Since the centennial celebration of the Academy was to take place in October of 1963, a little more than a year after I took office, Bronk and I started developing our plans as soon as I arrived. One of our first steps was to try to have the event put on President Kennedy's calendar, in the hope that he would address the membership on a topic he considered appropriate to the occasion. We asked Wiesner to help. At first he encountered difficulties because the President was scheduled to be travelling in the South at that time. With the help of Arthur M. Schlesinger, however, Wiesner ultimately succeeded in altering the President's plans.

The principal ceremonial session (Figures 13.2 and 13.3) was held in Constitution Hall. Access to the speakers' platform was via two or three steps located at the rear of the platform, out of sight of the audience. I noticed that day that the President had great difficulty mounting the steps; he seemed to be tightly girdled in a way which inhibited free hip motion. When he was assassinated in Dallas a month later, I wondered whether he might have been prevented by such a brace from sliding to the floor of the car in which he was riding.

The Academy's Centennial was well-attended by members from home

Figure 13.2. The assemblage in Constitution Hall of Academy members commemorating the Centennial Anniversary in October, 1963. (Courtesy of the National Academy of Sciences.)

Figure 13.3. A portion of the platform committee at the Centennial Celebration. From left to right: Kenneth B. Raper, Harrison Brown, The Reverend Theodore M. Hesburgh, Detlev W. Bronk, President Kennedy, and me. Julius A. Stratton and Jerome B. Wiesner were also present, standing well to the right. (Courtesy of the National Academy of Sciences.)

and abroad, as well as by the presidents and other officers of a number of foreign academies. I had the privilege of meeting for the first time President Shi-chien Wang of the Academia Sinica in Taiwan. That meeting formed the basis of a long, friendly relationship with the scientific community and other groups in the Republic of China, such as those in the National Science Council.

It was a minor disappointment to me that no one from Stanford University attended the Academy's centennial meeting. I felt it was a strange omission, since Stanford was striving at that time, particularly through the work of Frederick E. Terman, to become regarded as the "Harvard of the West." However, a large delegation did attend from the University of California at Berkeley.

The gathering was marred by only one unhappy incident. We had many reporters attending our meeting to cover the festivities and to learn about the latest developments in various research programs. During one lunchtime, while most meeting participants were gathered at their table, an Academy member who enjoyed both controversy and publicity took a group of reporters aside for a "private" press conference. His purpose was to denounce the national space program and to urge its suspension. Other Academy members who were deeply committed to that program, and learned of the irregular meeting of the press through their friends in it, came to me and Bronk insisting that they be allowed to offer a rebuttal in the event the press conference went ahead. Tempers ran high and I found myself in the embarrassing position of having to push my way into the side room and cancel the interview. I remained on that particular member's black list thereafter. It was a distressing moment in an otherwise wonderful meeting.

I had the good fortune to inherit an excellent staff, too numerous to mention individually here, many of whom had been selected by Bronk. The Executive Officer was S. Douglas Cornell (Figure 13.4) with whom Bronk had worked when Cornell was on the staff of the Research and Development Board in the Pentagon. Cornell, whose aunt was the famous actress Katherine Cornell, was as close to an American intellectual aristocrat as one could find. He was well-informed on many issues, had a fine sense of language, knew the Washington area intimately, and was deeply devoted to the welfare of the Academy. His personality and exactitude reminded me a great deal of my high school physics instructor, Ralph Britton. Key members of the staff held him in high regard, and frequently sought his advice.

Cornell once told me that, when he first came to the Academy, he had had some difficulty with Bronk as the latter tended to be imperious with "staff." One day Cornell brought matters to a head by telling Bronk that he was prepared to leave the Academy unless Bronk modified his approach. Bronk heard him out and understood what was at stake. The two worked in harmony thereafter.

The head of the Physics Division, John S. Coleman (Figure 13.5), had

Figure 13.4. Dr. S. Douglas Cornell. (Courtesy of the National Academy of Sciences.)

Figure 13.5. John S. Coleman. (Courtesy of the National Academy of Sciences.)

spent World War II doing research in undersea warfare, and had joined the Academy staff soon after the war ended, working initially under the leadership of John Tate and Gaylord Harnwell. When Cornell eventually did leave the Academy in 1965 to found a special college on the upper Michigan peninsula, I appointed Coleman to succeed him as Executive Officer.

Julius Stratton (Figure 13.6), then President of M.I.T., was now Vice Chairman in 1962; he was succeeded in 1965 by George Kistiakowsky.

Another remarkable individual was R. Keith Cannan (Figure 13.7), a physiologist of English origin. Cannan had been on the faculty of the Medical College of New York University but had become involved in wartime medical activities in Washington and had elected to stay on there as head of the Division of Medicine of the National Research Council of the Academy. His knowledge, understanding, and capabilities were awesome to me.

When it was discovered that the drug Thalidomide could cause birth defects, the Food and Drug Administration decided to radically change its rules regarding the licensing of drugs. At that time, any drug could be put on the market if it seemed after minimal testing to be effective and safe; it could be withdrawn by request of the FDA only if it were demonstrated to be harmful or ineffective in the course of general use. Thalidomide babies brought home to the FDA how tragically backward this thinking was, and the new rules required that every drug be proven safe as well as effective *before* it could be licensed for sale to the public. The FDA called upon the Academy to carry out a study of the effectiveness and safety of nearly four thousand drugs already on the market. Cannan accepted the assignment, a profound responsibility, and carried it off from the start in a manner approaching perfection. Everyone affected felt that his conclusions were rea-

Figure 13.6. (Left) *Dr. Julius A. Stratton, the Vice-President of the Academy at the time of the Centennial. (Courtesy of the National Academy of Sciences.)* ***Figure 13.7. (Right)*** *Dr. R. Keith Cannan. (Courtesy of the National Academy of Sciences.)*

soned and fair. Approximately one-third of the already-licensed drugs were approved, another third were to be subject to further review, and the remainder were deemed ineffective and were removed from circulation.

As a diplomatic measure whose necessity he recognized, Cannan assembled a substantial group from the pharmaceutical industry to review some of the recommendations that were to be made. Those attending the conference were naturally apprehensive that they might have to pull from the market some of their most valuable products.

At one point, Cannan revealed that several physicians had recommended continued licensing for a very old antibiotic that had been around from the early days of sulfa-drugs. Despite the recommendation, he said his board was calling for the cancelling of the license. One of the pharmaceutical executives angrily jumped to his feet and demanded, "If those doctors think it is good enough, why take it away from them?" Then his voice trailed off and, turning to his colleagues, he said quizzically, "Who the devil would use that stuff these days?"

I later met Cannan's daughter, Cecily Selby Coles, in New York City. She showed me some poems her father had written while serving as an officer in the trenches during World War I. They displayed a somber sensitivity maintained in the midst of horror that was in keeping with his other remarkable attributes.

One day at lunch I recalled to Cannan that in the Crimean War, of the mid-nineteenth century, wounded officers were treated by medical staff, while privates were left to look after their own wounds. An observant medical officer noted to his chagrin that, on the whole, the privates fared somewhat better than the officers; this caused him to wonder about the status of medicine in his time. I asked Cannan when he thought it was that medicine had reached the point where it was more rather than less effective. He grimaced slightly and said, "Fred, I am not sure we are there yet."

My successor, Philip Handler, retained most of the staff who were in place when he took office in 1969. As a result I had the pleasure of meeting with many of my old colleagues during visits to the Academy over the next decade.

Hugh L. Dryden (Figure 13.8), then Deputy to James E. Webb at NASA, was the Home Secretary. He had spent many years at the National Bureau of Standards and had been named Administrator of the National Advisory Committee on Aeronautics in 1947. When that organization became NASA in 1958 following the Soviet launching of Sputnik, the congressional committee overseeing the transformation claimed, amid much noise and bluster, that he was too much of a traditional civil servant to serve as Administrator; in other words, he was not exciting enough to head up their glamorous new agency. President Eisenhower chose T. Keith Glennan, President of the Case Institute of Technology, as Administrator, with the understanding that Dryden would serve as Deputy.

Later, President Kennedy selected James Webb to head NASA. Fortunately, both Glennan and Webb recognized that Dryden was a truly excep-

Figure 13.8. Hugh L. Dryden. (Courtesy of the National Aeronautics and Space Agency, and the AIP Emilio Segrè Visual Archives.)

tional individual, absolutely committed to the advance of aerospace science and technology, and both gave him virtually unlimited authority. For example, Dryden was the author of much of the legislation creating NASA, and understood in intimate detail both the technical and human sides of the space program. If he ever resented the opinions expressed by the congressional committee, he kept those feelings completely to himself and remained thoroughly devoted to the mission of the agency.

At some point during the early 1960s, Dryden developed cancer; it was a type that responded to a degree to one of the chemotherapies then available. He carried on with his work as though the disease had absolutely no effect on either his body or his mind. Every few months, however, he would make a journey to the hospital, stay for a week or so, and return to work with his usual energy and dedication. Then, one day in 1965, he did not return.

Early in my term of office, Dryden noted my special interest in the history of the organization and gave me a book, the *History of the Academy for the First Fifty Years*, prepared by Frederick True as mentioned in the footnote on the first page of this chapter, which I have treasured and have referred to many times since.

I had a very special affection for Dryden; he was not made of ordinary clay and stood at the apex of our species in most of his virtues. His funeral was, in the main, a simple family affair. Betty and I felt privileged to be invited.

In those days, the President of the Academy was permitted to nominate an individual to replace an officer who was deceased, or who resigned prior to the end of his term. The nomination was of course subject to the approval of the Council. It was agreed that we should try to find a local resident to replace Dryden. There were two strong local candidates. One was Leonard Carmichael, previously Secretary of the Smithsonian Institution and now on the staff of the National Geographic Society. I had known him since Rochester days (Figure 6.5) and held him in very high regard. The other was Merle A. Tuve (Figure 13.9), who had just retired from the Carnegie Institution in Washington. I selected Tuve because I thought that he would benefit from a new interest, whereas Carmichael was fully and happily employed at National Geographic. Lloyd V. Berkner, our Treasurer, warned me that I might be getting more than I bargained for with Tuve, as turned out to be the case.

Berkner (Figure 13.10) was another highly exceptional individual with many accomplishments to his credit. He had been a Naval Reserve officer, serving in the Electronics Materiel Branch of the Bureau of Aeronautics during World War II; he had joined the Reserves soon after graduating from high school. After the war he became the Director of the Associated Universities, which operated Brookhaven National Laboratory. Among his many other activities, Berkner had fostered the creation of the Green Bank Radio

Figure 13.9. (Left) Dr. Merle A. Tuve, Dryden's successor as Home Secretary. (AIP Emilio Segrè Visual Archives.) Figure 13.10. (Right) Dr. Lloyd V. Berkner, a catalyst for many scientific activities and institutions. (AIP Niels Bohr Library.)

Observatory, and had played a prominent role in the establishment of the National Center for Atmospheric Research located at Boulder, Colorado. More recently he had started a graduate education and research institution in Dallas that eventually became incorporated into the University of Texas system as the University of Texas at Dallas. While serving as our Treasurer Berkner, working with Betty's brother Lauriston Marshall, developed a theory of the way in which the earth's atmosphere had been altered at the beginning of the Cambrian Period as a result of the production by algae of large quantities of oxygen. This atmospheric change in turn permitted land animals and plants to evolve. The basic concepts of this work have withstood the test of time and are now generally accepted. Sadly, Berkner died of a stroke while attending a meeting of the Council of the Academy in 1967. We appointed E. R. Piore (Figure 7.7) to be his successor.

Berkner's long service to the Navy entitled him to be buried in the National Cemetery at Arlington in Washington. A number of his close friends attended the ceremony along with his family. As we stood in formal rows around the grave site, I found myself next to an admiral who was weeping silently but copiously. When I spoke to him later, he said, "The only reason I am an Air Admiral is because of Lloyd's faith and persistence. We were high school students together in Minnesota when Lloyd noticed an announcement in a newspaper that the Navy was planning to take twenty air cadets from civilian life, train them to be Navy flyers, and have them join the reserve. He said, 'We're going to apply for that.' I said, 'Gee Lloyd, we don't stand a chance.' We applied and were turned down. I said, 'See, just what I told you.' Lloyd told me, 'Look, you and I are going to the place where they get their training. We're going to get up with the crowd and do everything they do. Sooner or later a few guys are going to drop out and we'll drop in. You watch!' We went to the training grounds. We got up at dawn and went through all the calisthenics and drill and went to all the meetings where they were discussing the theory of flight and what to expect. Then one day while we were standing off to one side following calisthenics, the officer in charge yelled at us, `Hey, you two, fall in line!' That was Lloyd all over."

H. P. Robertson (Figure 9.13) had been the Foreign Secretary of the Academy in the latter part of Bronk's term in office, but unfortunately he died following an automobile accident in 1961. Up to that time the office had not been an aggressively active one, its main business being the exchange of formal communications with foreign academies, and the creation, with the help of the membership, of a slate of foreign nominees for election to the Academy. The latter task, be it said, was by no means trivial, as it required minutely detailed research into the activities of prospective members. Nevertheless, Bronk felt that the role of foreign secretary could be a more forceful one, and he recommended Harrison Brown (Figure 10.6) to succeed Robertson.

I had known Brown since the days of the Manhattan District, when he worked at Oak Ridge in the field of radiochemistry. On returning to academic research he had carried out determinations of the age of meteorites by analyzing isotopic abundances. This work added significantly to our knowledge of the age of our solar system. During the same period, while at the University of Chicago, he became interested, with Szilard and others, in the possibility of achieving international control over nuclear energy. He was a major factor in the formation of organizations such as the Einstein Committee and the Pugwash group (see Chapters 10 and 11). This activism brought him to the attention of Bronk, who felt that Brown would give the position of Foreign Secretary a new and vigorous cast. In this Bronk was not disappointed.

Almost immediately, Brown began to establish foreign "desks" in his office to keep up with activities in various countries and regions. He had a strong personal attraction to Communist countries, and we soon had in place a system of exchange programs whereby U.S. scientists and scientists from behind the Iron Curtain would, in effect, change places for a time. This activity was funded through various sources such as private foundations in the United States, as well as the governments of participating foreign countries. Brown even developed a Mainland Chinese Committee in the mid-1960s long before it seemed at all likely that the doors to the Mainland would open to Western visitors. He also took the initiative, at my suggestion, to establish a program with Taiwan.

I never really understood Brown's special affinity for the Communist countries; whether it derived from an idealistic admiration for the concept of a "classless society," or whether he was attracted by the special privileges of the *nomenklatura*. Perhaps he did not know himself, although he did once comment to me, "There is no great difference between our societies. It is just that the leashes are a little shorter on their citizens than on ours." His perception of our freedoms was somewhat different from mine. In any event, his work opened many hitherto sealed doorways for the scientific community behind which were some fascinating opportunities. During periods of relatively cordial relations with the Soviet Union, such as occurred briefly just before the Cuban Missile crisis, Soviet scientists visiting the United States would drop into my office as freely as our own members. They were, of course, a highly select group in good standing with the Soviet authorities.

In the course of this work, Brown became deeply concerned about the lack of food in Third World countries and wrote extensively on the hazards associated with unlimited population growth. This work received much well-deserved public attention; unfortunately, the problem remains with us.

Sadly, Brown did not live to see the great political changes which have taken place in the world since the late 1980s; but they would have fascinated him. He developed lung cancer and suffered spinal damage as a consequence of the radiation therapy that was part of his treatment. One of his dreams had been to visit Ulan Bator in Mongolia, but fate ruled otherwise. His many friends and admirers published a small volume on issues of deep interest to him as a eulogy.[2]

The Council of the Academy consisted of the academy officers plus a group of elected councilors who were rotated every three years. On the whole, I found all the members to be highly supportive and I made some wonderful friendships, both professional and personal through those associations. One of the individuals whom I had known earlier as a result of his work in oceanography, Roger R.D. Revelle (Figure 13.11), proved to be particularly supportive. He was truly a Renaissance Man with an almost unlimited amount of physical and psychological energy.

In 1963, Douglas Cornell and I decided to examine the trends in membership of the Academy which was divided into fourteen "Sections" depending upon professional interest. We found that over time there had been a significant shift in its composition. For many years, approximately fifteen percent of the membership had had a professional engineering education, and about the same percentage was trained in medicine. The former group usually linked up with those sections of the Academy that were devoted to the physical sciences, and the latter with those concerned with the life sci-

---

[2] *Earth and the Human Future* (West-View Press, Boulder, Colorado, 1986).

Figure 13.11. Dr. Roger R.D. Revelle, another gigantic figure in American science. (Courtesy of the National Academy of Sciences.)

ences, until separate sections in medicine and engineering were created. With the rapid growth of basic science in the universities following World War II, both of those groups were being decreased percentage-wise in the course of normal elections in favor of members with Ph.D. degrees in the sciences. Since this would decrease the strength of the advisory services which the Academy could offer through its own membership in those fields, we brought the facts concerning the changing structure to the attention of the Council, pointing out that a large fraction of the advisory needs of the National Research Council related to engineering and medicine.

A committee was formed under the chairmanship of Herbert Carter to find ways to remedy this situation; it recommended the establishment of a system of membership quotas for the various fields of basic and applied sciences in the procedures for election. So far, that system has maintained a reasonable degree of balance within the membership. It remains to be seen, however, if this parity will be retained in the long run since there is a strong tendency among those in the basic sciences, and particularly among academics, to undervalue the importance of the applied fields.

A historic scientific development occurred early in my term: the confirmation of continental drift, which had been formally proposed by a distinguished geophysicist, Alfred Wegener, early in the century. It had been suggested previously, but on less tangible grounds. His idea had been ridiculed by most geologists, but the Princeton geophysicist, Harry H. Hess (Figure 13.12), had been pondering the issue in a constructively serious way for several decades. Hess noted that none of the basaltic rocks on the ocean

Figure 13.12. Dr. Harry H. Hess. He brought the study of Continental Drift into the heart of modern science, demonstrating the inevitable correctness of Alfred Wegener's original hypothesis. Additionally confirming research was provided by F.J. Vine, D.H. Matthews, and J.T. Wilson. (Courtesy of Princeton University.)

floor were more than about one hundred million years old, whereas continental strata had been dated back to the early history of the planet, several billion years ago. While in the Navy both during and after World War II, he also observed that ancient undersea volcanic islands, which he termed guyots, in honor of a famous geologist, were often submerged deeply in the ocean, as if they had been moved from an earlier, more elevated position by some movement of the ocean floor.

The climax came when an English geophysicist, Frederick J. Vine, working in collaboration with D. H. Mathews, demonstrated continental drift conclusively by studying the magnetic orientation of rocks on the ocean floor on either side of oceanic ridges. This study showed that new floor was being created continuously at those ridges as a result of upwelling of magma, which flowed symmetrically away from the ridge on both sides. Vine's measurements also demonstrated that the direction of the earth's magnetic field alternates on a time scale of tens of thousands of years.

The overall picture was consolidated further in the decade by the Canadian geophysicist, J. Tuzo Wilson (Figure 13.13), who introduced the concept of plate tectonics.

Bronk naturally wished to include this momentous new discovery in the Academy's centennial book[3] of essays. Hess, however, was much too busy consolidating the new work to be willing to take the time to write a review. Roger Revelle, known for his gallantry, wrote an essay on the topic, which appears in the book under Hess's name.

Hess was a versatile scientist who enjoyed a variety of relationships out-

---

[3] *The Scientific Endeavor* (The Rockefeller University Press, New York, 1965).

Figure 13.13. Professor John Tuzo Wilson, the Canadian geophysicist who developed the theory of plate tectonics soon after the pioneering work of Hess, Vine, and Matthews. (Courtesy of the department of physics of the University of Toronto with special thanks to Professor Jerry X. Mitrovica and Dr. Israel Halperin.)

side the academic, professional scene. He usually spent two weeks each year as a reserve officer aboard sea-going navy ships. He also told me that he greatly enjoyed associations with members of the indigenous population in the vicinity of Princeton University—the descendants of the original English settlers.

Shortly before the continental drift breakthrough, a group of American geophysicists, many of whom were associated with the Academy, launched a campaign to start a drilling program; the so-called Mohole Program. The drilling would pierce the earth's crust and reach well into the mantle. The project was eventually authorized and funded under the auspices of the National Science Foundation, and the Texas drilling company, Brown and Root, was selected to do the job. After much discussion, a site off the coast of Hawaii was chosen for the exploration.

After all the logistical questions had been settled and work was about to begin, the Academy committee overseeing the program was rotated per normal procedure, and came under the chairmanship of a very capable, sincere geophysicist who had previously worked in industry and was now a professor at Princeton. He thought he saw a better use for the drilling funds that would achieve the same scientific goal, not appreciating the fact that the program was now regarded as fixed so far as the government was concerned. Entirely on his own and without consulting his more experienced colleagues first, he approached Congress to propose an alternative program. As a result, the entire program was placed on hold. Seeing the unfortunate consequences of his actions, the chairman resigned. A small group including Harry Hess and W. Maurice Ewing, an oceanographer, held what amounted to a double wake in my office on the Saturday after President Kennedy was assassinated. The project was never revived in its original form, although similar exploratory drilling has since been carried out under other programs, some under the auspices of the National Academy of Sciences-National Research Council. In defense of the new chairman, who had inadvertently caused the cancellation, it should be said that his intentions were scientifically honest if politically naive. He had reason to believe that there were places on the globe where mantle rocks were exposed and could be studied easily. This has indeed turned out to be the case.

When I served on the Academy Council in the late 1950s, it was already apparent that housing the staff and committees of the NRC was becoming increasingly difficult. The Academy building itself had long since been outgrown and office space was being rented willy-nilly for each new project as it came along. Scheduling, budgets, and communications were a nightmare.

Just before I took office as President, Bronk and I agreed to attempt to rectify the situation. After talking with Donald Meid, our very able business manager, we approached the administration of George Washington University, which owned a great deal of real estate surrounding its campus that was mainly occupied by outmoded residential buildings and small businesses. It was agreed that the Academy would erect an office building, which it would then lease from the university at a nominal rent, at the southwest corner of Pennsylvania Avenue and 21st Street, N.W. After twenty years, the University would own the building free and clear and the Academy would have the option of negotiating a new lease for an additional ten years if it chose. The building, named after Joseph Henry, served the purposes of the Academy very well until the mid-1980s when the first lease ran out, at which time the Council of the Academy decided to select another option rather than renew the lease at a much higher rate for ten years.

A rather unpleasant and slightly bizarre incident occurred in connection with the final signing of the agreement with Washington University. I was in New England on business when Donald Meid phoned to tell me that the documents were ready and should be signed at once. Since the Home Secretary had signatory powers that could be exercised in the President's absence in matters such as this, I called Merle Tuve, who had recently taken over the post following the death of Hugh Dryden, and asked if he would sign the agreement in my stead. I thought that he would be pleased to have his name on an historically important document, but to my great surprise, he refused. I pressed him for the reason, believing that there might be some minor misunderstanding about the chain of authority that could be cleared up with a word or two. Instead, I found that he had a deep-seated fear of some unimaginable catastrophe occurring that would involve him in endless personal legal problems. I had no choice but to leave the meeting I was attending and rush to Washington in the middle of the night to meet the deadline. I had gained (at some cost) a special insight into the mind of a very brilliant scientist. I recalled that on several other occasions Tuve had seemed to become overwhelmed while trying to resolve issues that arose at Council meetings, and had had to be excused to compose himself. I presume that his great success as a scientist and leader depended in part upon this capacity to anticipate, and to be intensely concerned about all the possible hazards of a given situation.

Professor Howard W. Florey, who made the practical use of penicillin a reality during World War II, was elected President of the Royal Society of London in 1960, for a term of five years. He was a hands-on individual of Australian origin, and he wanted the Royal Society to become more active in matters of science policy at both the national and international levels. He came to see me during a visit to Washington and proposed a period of collaboration between our two academies during which he would attempt to introduce the changes he sought within the Royal Society to make it more responsive to external issues. We formed a joint committee which held several meetings in Washington and at Oxford University, Florey's home base.

We were not the first to attempt such a venture. While President of the National Academy of Sciences, Detlev Bronk, who was also a member of the Royal Society, had formed a joint committee of the two organizations, but for a somewhat different purpose; namely to discuss progress in the various fields of basic science. That committee was still functioning during my presidency and it served a useful purpose, but it was by no means as dynamic an entity as that envisaged by Florey. Since Bronk had retained membership on his own committee, I proposed that he be asked to join the new one. Florey demurred, however, presumably because he feared that Bronk's original goals would conflict with his own. Although we had many profitable discussions, Florey's five-year term proved too short to allow him to make substantial progress. He was succeeded by Patrick M. S. Blackett (Figure 13.14), who had further enhanced his considerable reputation by introducing operational analysis formally into the British military in World War II. While the committee continued to meet during Blackett's term, his approach to issues was more personal, and less global, than Florey's had been. I lost contact with the committee after leaving office in 1969.

I had been in office for eighteen months, commuting between Illinois and Washington, when it became apparent that the activities of the Academy were sufficiently myriad as to demand the attention of a full-time President. When I presented this conclusion at a council meeting, a committee was appointed to re-examine academy policy concerning the tenure of the President and it was recommended after only a brief period of reflection that the constitution of the Academy be revised to make the presidency a full-time position.

I was asked to continue on as President, now on a full-time basis, for a six-year term beginning in 1965. Betty and I weighed the offer for a number

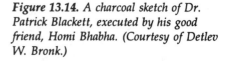

Figure 13.14. A charcoal sketch of Dr. Patrick Blackett, executed by his good friend, Homi Bhabha. (Courtesy of Detlev W. Bronk.)

of days before finally accepting. Our hesitation was caused by two factors. We loved our life at the University of Illinois and had made some long-term plans; we had purchased a piece of land near the campus,and had already acquired architectural drawings for a new atrium-style home. Moreover, I was keenly aware how quickly, and irrationally, the mood of the membership of an organization can change. I could become highly unpopular almost overnight because of some seemingly trivial issue. At that time, our country was involved in the war in Vietnam, and many members of the Academy were deeply hostile toward the administration of President Johnson (Figure 13.15) because of it, even though he had inherited the problem. Since it was inevitable that the Academy would be called upon to assist that administration in one way or another, it was not difficult to foresee troubled times ahead.

In 1964, the issue of creating a National Academy of Engineering comparable in size to the entire Academy of Sciences came to a head. The changes made within the latter by guaranteeing the election of a minimum number of engineers was deemed to be far from adequate by the engineering community. Several congressmen, such as George P. Miller of California, who had an engineering background, were understandably sympathetic to the engineers and listened seriously to the issues they raised regarding their status in the Academy. In principle they were capable, with the best of intentions, of introducing legislation which would create an academy of engineering, independent or otherwise. Those of us who knew how easily matters could get out of hand on the hill once they were brought to the attention

Figure 13.15. President Lyndon B. Johnson in a happy mood. He was greeting an admirer. (White House photograph.)

of Congress were very concerned that the Academy's Charter might be revised in ways that could damage the fundamental work of the organization.

After much deliberation in which Julius Stratton, President of M.I.T., and Eric A. Walker, President of Penn State, played a role, it was agreed by our Council that we would create a sister academy under the aegis of the original NAS Charter, and would share jurisdiction over the Governing Board of the National Research Council with the new academy. The President of the Academy of Sciences would serve as Chairman of that council. It was also decided that the various programs undertaken by the National Academy of Engineering would be subject to the approval of the Council of the Academy of Sciences. Most Academy Council Members were in favor of the plan; there was, however, an outspoken minority which strongly expressed its opposition before reluctantly accepting the new organization. In the main this group feared that the linkage with a large group of engineers would divert attention away from the basic sciences. I thought at the time that the National Academy of Engineering would detach itself from the Academy of Sciences within a decade or so, to act as a truly independent organization. It has not done so to date, presumably because the enormous effort required to make such a break is as yet too daunting a prospect.

The first President of the National Academy of Engineering was Augustus B. Kinzel, who, unlike most of the engineers, was also a member of the National Academy of Sciences. His approach was thoroughly constructive and cooperative, and he strove to ensure that matters affecting the two academies proceeded as smoothly as possible. He also raised a half-million dollars from his friend, Alfred P. Sloan, most of which went toward the construction of an auditorium to complete the building on Constitution Avenue.

Unfortunately, Kinzel was in office only one year. He was succeeded by Eric A. Walker, who became involved in other activities both before and after retiring as President of Pennsylvania State University, and found relatively little time for the Academy during his presidency. This greatly slowed not only the development of the National Academy of Engineering, but also the rate at which it established appropriately compatible relationships with the membership of the Academy. These developments became possible only later, after Robert C. Seamans became President of the National Academy of Engineering starting in 1973.

With the growing influence of medical schools, some members of the Academy who had an academic medical background felt that there should be a body within the Academy structure comparable to the National Academy of Engineering but devoted to medicine. It could, they argued, provide a voice on the issues that would complement that of the American Medical Association, which traditionally represents the views of private practitioners. The Academy members' efforts led to the creation in the mid-1960s, initially on a provisional basis, of the Institute of Medicine. The Robert Wood Johnson Foundation provided a substantial start-up grant to the Institute, on condition that the funds be devoted to a study of policies affecting patient care. The initial membership was selected to facilitate that study. Unfortunately, the Johnson Foundation did not renew its initial grant, and the Institute had a struggle for an extended period to keep its head above water. One of its Presidents (Samuel O. Thier (1985–91)), however, finally succeeded in raising a large endowment which has enabled the Institute to be a significant independent voice in the medical field.

Although I participated in the early meetings of the Institute in the mid-1960s, I note that 1970 is recorded as its founding date. Presumably that was the year that its basic charter was established in its final form.

Soon after taking office, President Kennedy appointed Glenn T. Seaborg Chairman of the Atomic Energy Commission. In order to acquire additional facilities for high energy physics, Seaborg obtained authorization and funding for an accelerator to be built at the Lawrence Berkeley Laboratory on Grizzly Peak, east of the Berkeley campus. As construction plans went forward, a significant and growing dissatisfaction became evident among the high energy physicists in other parts of the country, particularly along the eastern seaboard. The proposed Berkeley machine, they said, while undoubtedly functional, seemed to have an exceedingly rigid design that would allow for almost no appreciable future adjustments. If this turned out to be the last machine of its kind to be constructed in the United States, high energy research would be severely constrained by these design limitations.

Moreover, it was generally felt that the physicists at Berkeley, following a tradition established by E. O. Lawrence, had a tendency to make and push their own agenda. In-depth cooperation with other universities, while not entirely unknown, was relatively rare, and not easy. A number of physicists from various regions came to my office to discuss these concerns.

High energy physics was raising issues on other campuses beyond those concerning Berkeley. I spoke with several university presidents who were unhappy about the presence of informal high energy users groups at their institutions. These people tended to live in a world of their own, using university facilities, often without formal approval, and without any appropriate compensation to the school. The presidents hoped that something could be done to bring the situation more under control.

In response to the concerns expressed to me, I met informally with the governing board of the Brookhaven National Laboratory in the autumn of 1964 to discuss the national status of high energy physics. Brookhaven had been organized immediately after World War II to promote the development and use of nuclear reactors and to provide facilities for high energy research. Nine eastern universities had participated in the lab's creation, and managed it cooperatively through an entity known as Associated Universities Incorporated. I pointed out that a different approach might be needed in connection with the new accelerator if the national high energy community was to be accommodated and proposed that the Board of Governors of Brookhaven consider extending its membership more or less nationally. The new, nationally constituted organization could approach the AEC to reopen matters related to the new accelerator, particularly with regard to its design, management, and location. After some discussion, the Board rejected the suggestion, Rabi being the primary spokesman for the opposition. He emphasized that the nine Brookhaven universities formed a congenial and efficient group, and that they had no wish to complicate their lives by enlarging it.

After this unsuccessful meeting I convened a group of twenty five university presidents early in 1965 to review the situation and solicit their views. They recommended that a new national consortium of universities be created to manage the next accelerator. The consortium would include all professionally qualified and interested institutions. They also said that a reevaluation of the design was in order. I was asked to meet with Seaborg to convince him to accept these recommendations. Recognizing the militant mood of the physics community, Seaborg agreed, although he undoubtedly found the situation most trying, as he had been at Berkeley since the start of his professional career. As far as I know, his colleagues at Berkeley never faulted him personally for the turn of events.

In retrospect, the meeting of university presidents that I convened was a high point in the successful collective influence of such presidents during that period. Soon thereafter, most of those who had attended the meeting were engulfed in the almost overwhelming disturbances that swept over campuses across the country.

There was some objection in Congress to the new proposal, but it died out rapidly once it became clear that the plan had nationwide support. The new organization was called the Universities Research Association; the Academy served as its temporary home until it acquired its own base in Washington, Leonard Lee Bacon, who was on the Academy staff, serving as midwife to the infant association. A committee under the chairmanship of J. C. Warner developed a structure for the Board of Directors that would allow for regional representation rather than representation by each participating university. It proved to be very successful. Norman Ramsey (Figure 12.1) of Harvard University was the first President. The board, in turn, selected Robert R. Wilson of Cornell University to be the first Director. In the meantime, Brookhaven Laboratory has retained separate identity and management.

The responsibility for selecting a site for the new laboratory and accelerator fell to an Academy committee selected from a nationally distributed group of scientists and chaired by E. R. Piore. A list of requirements for the new site was drawn up including size, accessibility, location relative to electric power stations, and water supply. All interested states were provided with the requirements and invited to submit proposals. President Johnson asked the Academy to give him a "short list" of the six most promising sites, from which he himself would make the final choice. While this of course led to claims that his selection was unduly influenced by politics, I personally regarded the President's request as a godsend since it removed the burden of final choice from the Academy. There has been much speculation as to the President's reasons for settling on Illinois. On the whole, give or take a grumble or two from a few individuals from other regions, the site has worked out well since it is accessible to a major airport and has proven to be an acceptable place to carry on research by both the national and international scientific communities.

One of Bronk's many memorable achievements as Academy President was his very successful fundraising effort to complete the two wings to the west and east of the main Academy building, which had been built just after World War I. Wallace K. Harrison, who as a young architect had worked under Bertram Goodhue, the designer of the original building, was commissioned to plan the new wings. Bronk had used Harrison's services for a number of buildings on The Rockefeller University campus in New York City. In the process, the two became good friends.

Soon after I took on the full-time Presidency in 1965, John Coleman, now the Executive Officer, and I decided to try to raise funds to construct the auditorium that would complete the academy quadrangle at 2101 Constitution Avenue. We had help in this effort from many good friends, especially friends of the late Hugh Dryden in the then affluent aerospace industry.

Dryden had been much admired in the aerospace field for his ability, wisdom, and dedication. James E. Webb, the NASA Administrator and head of the Apollo program, wrote numerous letters on our behalf and, in Dryden's honor, donated his own lecture fees to the building fund. Other important benefactors included William O. Baker (Figure 13.16), who persuaded his friend Paul Mellon to make a major donation through the Mellon Foundation, Leonard Carmichael of the National Geographic Society, and the National Institutes of Health. Just prior to my departure from the Academy in 1969, Coleman and I had managed to raise the funds we needed. Ready to commission an architect, Coleman and I naturally thought of Wallace Harrison first, but when we mentioned the idea to Bronk, we found that he had had a falling-out with Harrison and had limited enthusiasm. A difference of opinion over the failure of a decorative sheath of external ornamental mosaics on The Rockefeller University campus had apparently escalated into a serious rift. The two were no longer on speaking terms, and I was given to understand that Harrison, whom I had never met, was impossible to work with. Nevertheless, as Harrison had a history with the Academy and had, I knew, previously made some preliminary plans for the auditorium out of personal interest, I decided to meet with him anyway, and at least solicit his opinion about the project. Perhaps he extended himself in an attempt to atone for his row with Bronk; whatever the reason, we hit it off well from the start, and he accepted the commission. Plans for the new building went ahead rapidly and smoothly. Bronk later expressed satisfaction with the result.

As Coleman and I had foreseen, building costs began to escalate at a frightening pace in 1970, as the national inflation rate began a rapid expan-

Figure 13.16. Dr. William O. Baker, one of the foremost scientific statesmen of our time. (Courtesy of the National Academy of Sciences.)

sion that would continue for the next decade. Had either our fund-raising or the architect's planning been delayed for even a year, we would have found ourselves chasing rainbows, and it would have been left to a later generation to complete the building.

From time to time members have recommended that the auditorium be named for an individual whom, for one reason or another, they hold in high esteem. It is my unqualified opinion that, as most of the money was raised in Hugh Dryden's name, it should be named for him if for anyone. A commemorative bust of Dryden stands in the upper display room of the auditorium.

In the course of seeking funds for the construction of the auditorium, I began a parallel program to increase the endowment funds of the Academy so as to give it more freedom for selecting study programs of its own choosing. Frank Press continued this activity with noteworthy success during his twelve years in office (1981–93).

A position I assumed *ex officio* as President of the Academy was membership in the Defense Science Board (see the previous chapter), as stated in the Board's original charter. It soon became evident that the Board had a serious problem. Harold Brown had been chosen for the important post of Deputy Director for Research and Development (DDRD) soon after Robert S. McNamara was made Secretary of Defense in 1961. Brown selected an engineer friend to act as his principal aide, and assigned to him the overseeing of some activities which fell under Brown's jurisdiction, including the Defense Science Board. The aide regarded the Board with great suspicion, perhaps fearing that it could inhibit the freedom of Brown's office by attempting to exert control over his technical decisions. As a result, board meetings became rather perfunctory exercises.

Secretary McNamara also brought into the Pentagon a group of young operations analysts upon whom he depended heavily for advice, and who hence tended to dominate the whole advisory structure. This group soon became known as the "whiz kids" and were the object of much scrutiny and commentary in Washington during that period.

The Chairman of the Defense Science Board was my longtime friend, Clifford C. Furnas, a sturdy, intelligent and experienced Scot, who was also President of what was then the University of Buffalo. He came to my office one day in 1963 to say that the Chairmanship was proving to be a waste of his time, and to ask if I would take over the position since I was in Washington anyway. To this he added one of his trademark witticisms: "Remember Fred, you can't fight City Hall!" We agreed that the Board's activities were pretty *pro forma* at the moment, given the attitude of Brown's aide and the whiz kids, but that the Board should be kept alive until the climate at the Pentagon improved. I accepted the Chairmanship, working first with Lawson McKenzie and then with William W. Hammerschmidt as Executive Officers. In 1966, when John Foster became the Deputy Director for Research and Development, the atmosphere changed completely; he made the Board a centerpiece of his program. I remained Chairman until 1968 when other activities demanded more of my attention.

Patrick E. Haggerty (Figure 14.2) became Vice Chairman when Foster took over and brought to the Board a welcome new spirit. I had known him slightly as the Chief Executive Officer of Texas Instruments Incorporated and as one of the individuals who helped develop the semiconductor industry. Haggerty was one of those rarely gifted engineers who, in addition to having special entrepreneurial skills, could look beyond the immediate horizon and foresee significant future developments. When he joined what is now Texas Instruments, he was willing to stake the company's future — and his own — on the promises of the newly invented transistor. He sponsored both the invention of and the early work on the integrated circuit because he appreciated its potential to create a revolution in electronics and in the broad field of communications, even though he knew that that revolution would be a long while in the making. Haggerty's life was cut short by cancer in 1980. None of the technological developments which have occurred since would have surprised him.

Harold Brown became Secretary of Defense in the Carter Administration and brought his former aide with him again. This time the aide realized that, if he himself were to serve as Chairman of the Defense Science Board, he would acquire that much more authority. He then remained on the board as Vice Chairman for another ten years.

The mood of the scientific community became engulfed in conflict as the war in Vietnam progressed — a situation that reflected the tensions brewing in the population as a whole. These tensions were, of course, exacerbated by the university students and the media, and the whole contributed to a great unrest within the Academy. Several newly elected Academy members resigned because of the academy's traditional links with the military. One member from Harvard announced his intention to resign in a fiery speech at the April general meeting of the Academy in the late 1960s. He displayed such eloquence that he was often interrupted by thunderous applause. Official acceptance of his resignation occurred in 1971.

A significant number of Academy members became active as advisors in the United States Arms Control and Disarmament Agency, forming something of a private group within the academy opposing President Johnson's war policies.

Early in the Vietnam War, George Kistiakowsky tried to help Secretary McNamara deal with the conflict by offering a tactical solution to the infiltration of South Vietnam by Vietcong from the North. His solution was to establish fixed lines with the use of electronic and other alarming devices for purposes of detection and reprisal. The military, however, were mindful of the heavy losses of life and material that were sustained during the siege of Petersburg near the end of the Civil War, as well as during the trench warfare of World War I. They were not sympathetic to what became designated the "McNamara Line," and the plan went sour. Kistiakowsky washed his hands of the whole affair and became an active member of the Academy's anti-war, disarmament group. Council meetings became fraught with the stress engendered by the war.

I had first known Kistiakowsky by reputation during my Princeton student days, as he had spent some time there before moving on to Harvard. He was rightly regarded as both brilliantly creative and, occasionally, somewhat disturbingly temperamental. It was a privilege to work with him during his creative periods, particularly when he was at the Bruceton Laboratory (see Chapter 9). During his service as Vice President of the Academy, however, I found his temperamental side to be increasingly in evidence.

In the mid-1960s, James A. Shannon (Figure 13.17), who had done so much as head of the National Institutes of Health to make that body a principal supporter of forward-looking biomedical research in our country, began to encounter difficulties with Congress. Up to that time, he had had strong support from powerful leaders in both the House and Senate. One of them, Representative John E. Fogarty, died, and the other, Senator Lister Hill, decided to give up his post. A new chairman of the committee overseeing the NIH in the House took exception to many of Shannon's activities, and began to make life very difficult for him. As a result, he resigned his position and I was pleased to offer him an appointment as Resident Scholar of the National Academy of Sciences, where he proved an exceedingly valuable asset.

This appointment caused Shannon's adversary in Congress to do some rather senseless things. First, he attempted to cut off *all* funds for Academy

Figure 13.17. Dr. James A. Shannon, the father of the modern grants program of the National Institutes of Health. (Courtesy of the National Academy of Sciences. Photograph by Edward A. Hubbard.)

studies, an effort that was obviously doomed to failure. Next, his staff noted that the Academy was holding records of studies of the effects on patients of various pharmaceutical agents. These records had been accumulated over a long period, and were extremely useful to the regulatory agencies. The representative's staff arbitrarily decided to subpoena them, claiming that he wanted to see if they were being used "properly." The privacy of reports on individual patients was compromised, and we had no choice but to discontinue a worthy program since there would be no way to avoid "leaks" of the information, accurate or otherwise, to the discomfiture of physicians and patients. The irresponsible way in which confidential medical records were subpoenaed by the politically motivated staff makes one wonder how secure genome records could be maintained if the Human Genome Program proceeds under government auspices.

Shannon was delighted to join the staff of The Rockefeller University when I moved there and remained with us for several years as a most effective member of the university.

I was associated with the President's Scientific Advisory Committee (PSAC) for most of its life in one way or another; first when I worked for NATO and reported to the Committee each time I returned to Washington; then during my service to the NAS; and finally as a member of a long-term study panel when Edward E. David was Science Advisor during the second two years of President Nixon's first term in office.

Taken as a whole, the members of the Committee were a dedicated and highly efficient group. The Committee had an excellent staff and had little difficulty in enlisting the services, as needed, of panelists from the highest levels of various professional groups. Its studies were carried to completion and usually had significant results to report. And yet, in a sense, the Committee failed, for one crucial reason. A very small number of its members saw themselves as being in service not to the President but to some private cause of their own, or some constituency elsewhere, thereby in a sense undermining the dedicated work of the steadfast members, such as Harvey Brooks and Charles Slichter, as well as Donald Hornig, who served as Chairman of the Committee and Science Advisor to the President during the Johnson Administration. I could see this defect clearly, and it must have been far more evident to the politicians whose careers depend on such perceptions.

I have often been asked, "Just what did all those people associated with the PSAC manage to do in the White House over the years?" The answer is both simple and complex. The President's Science Advisor was always on call to help the President deal with any special issue that might arise in the Oval Office involving science and technology. The matter might be a highly privileged one, or one that could be referred to the Advisory Committee and its panels.

Beyond this, the Committee advised the White House on countless matters related to the activities of the various executive departments and agencies. Some questions might pertain to details of their operating budgets; others to determining the feasibility of an entirely new technical project. Essentially the Committee functioned as both facilitator and watchdog between the departments and agencies and the group in the White House that managed the national budget. The Advisory Committee and its panels did their best to place their own stamp on all matters that passed through their hands. This meant that, depending on one's agenda, its work was either a great help or a great hindrance.

The President's Science Advisor also served as Chairman of a committee known as the Federal Science Council, composed of representatives of the different departments and agencies, which, at least in principle, was designed to improve communications and coordination.

Each President, naturally, interacted with PSAC in his own way. President Kennedy had inherited the Committee from President Eisenhower, who undoubtedly had created it in the hope of placating the public after the embarrassing surprise of Sputnik. While there is no doubt that Kennedy was initially enthusiastic about the Committee, I often wondered if that enthusiasm was dimmed after an incident which took place in 1963, at the George C. Marshall Space Center in Alabama, where Wernher von Braun and his group were situated. In pursuit of Kennedy's Apollo space program, plans had been made at von Braun's recommendation to launch a manned lunar landing vehicle from lunar orbit. Earlier, a PSAC panel had advised that the lunar vehicle should be launched from earth orbit, but von Braun's views prevailed. On his visit to the space center, the President was accompanied by a leading member of PSAC; television cameras were also on hand to follow the president's tour. On live television, the PSAC member suddenly chose to enter into a heated debate with von Braun concerning the plans for the Apollo program. The President was evidently surprised and, as I read it, not a little dismayed at this unexpected addition to his agenda.

During both the Kennedy and Johnson eras, a great deal of energy was devoted in one way or another to matters concerning Vietnam. It was clear to me that, as time went on, President Johnson became more and more dubious about the value of the help he received on this issue from PSAC. And, since President Johnson had reservations about the Committee for a number of other reasons as well, I believe, from fairly direct evidence, that he would have disbanded the Committee had he been elected to another term in 1968.

President Nixon's attitude toward the intellectual community is demonstrated by two incidents which occurred in January of 1969, after his election. His staff had asked a group of individuals, mainly from the academic world, to develop "transition plans" in the weeks following the 1968 election that might assist Nixon in settling into his new administration. As a gesture of his gratitude, Nixon hosted a reception and lunch for the group at the Hotel Pierre in New York City. He spoke briefly at the reception preceding the lunch, telling those present, "I know that none of you voted for me, but thank you for your work."

At about this same time, Nixon asked Melvin R. Laird, an important member of the House of Representatives, to be his Secretary of Defense. When they discussed the invitation, Laird said that he would accept the job if, and only if, PSAC were henceforth debarred from reviewing the affairs of the Department of Defense. President Nixon agreed without public comment.

Laird's animosity toward the PSAC was, I believe, a reaction to the behavior of some members of the Committee who did their best to demean any military officer, and on occasion, even civilian members of the Department of Defense who appeared before them for a presentation. News of this treatment of Defense staff had obviously reached Laird and, equally obviously, the situation was one that he was not prepared to tolerate as Secretary. Laird's request was one more nail in the coffin of the advisory committee in the form developed in the wake of Sputnik.

I was not surprised when President Nixon eliminated PSAC after his reelection in 1972, even though Lee DuBridge, Edward David, and the entire PSAC staff had done their best to serve him well during his first term. To maintain some connection with science, the president appointed H. Guyford Stever, the Director of the National Science Foundation, as his Science Advisor. He was available on request.

In mid-1963, President Kennedy decided, quite properly, that Robert Oppenheimer should receive the Fermi Award — the highest national award in the field of nuclear science and technology — for his outstanding work during World War II. The date previously fixed for the award ceremony, December 2nd, occurred just after Kennedy's assassination. President Johnson, mindful of the state of public mourning, presented the award privately at the White House, and a more public reception was later held the National Academy of Sciences for a few select guests, including previous recipients of the award.

Betty and I hosted the Academy reception, and had a small receiving line to greet our guests. Many reporters and photographers were present for the occasion, and were keeping an alert eye on the proceedings. When Edward Teller, a previous Fermi recipient, arrived in the reception line he reached out to shake hands with Oppenheimer and the photographers promptly surged forward to capture the picture. Betty happened to be standing between Teller and Oppenheimer at the critical moment, so that she became an inextricable part of the tableau. With traditional journalistic gallantry, one of the photographers shouted, "Get that dame out of the way!" But it was too late. Betty was highly amused at the situation and said, "I guess I am part of history." The unmodified photograph appeared in certain issues of *Life* magazine in various areas of the country, and Betty received a number of "fan" letters from delighted friends who saw it. Unfortunately, the East Coast editions of the magazine carried a version of the picture that had been subjected to an airbrush, and Betty was reduced to a very faint shadow, barely distinguishable from the background (Figure 13.18).

One of the many rewards of our years in Washington was the privilege of making friends among those who worked on the Hill and in various other governmental agencies. Prominent among these relationships was the renewal of our rewarding friendship with James Webb (Figure 11.20), now the Administrator of NASA. Webb consulted with me frequently, and he and his wonderful wife Patsy invited Betty and me to several Apollo launchings as well as other exciting events.

Webb was very close to President Johnson, and even succeeded in having Betty and me included in a number of gatherings at the White House (Figure 13.19). On one memorable occasion, we were there to see the first video returns from the cameras that had landed on the planet Mars. They made it clear that, whatever might have been the case in past eons, Mars was now a dead planet.

Betty and I were also invited to a very special farewell party for the President and his family at the White House in January of 1969. It was attended by many close friends of the Johnsons, and I remember it as being a relaxed, family affair. One of the highlights of the evening was a presentation of Haydn's one-act opera "Il Mondo della Luna" depicting a fanciful journey to the moon.

Again through my connection with Webb, I was invited to travel with

Figure 13.18. Edward Teller greeting J. Robert Oppenheimer at the reception at the National Academy of Sciences following the presentation of the Fermi Award to Oppenheimer at the White House by President Lyndon Johnson on December 2, 1963. We were all still deeply in a state of shock and mourning as a result of the assassination of President Kennedy ten days earlier. (Ralph Morse, LIFE Magazine © 1963 Time, Inc.)

Figure 13.19. Betty in mid-day regalia being greeted by the First Lady, Ladybird Johnson. (White House photograph.)

the President's party aboard Air Force One when Johnson went to Houston to speak at the dedication of the Lunar and Planetary Institute of the Universities Space Research Association in the old Mansion of the West family bordering the NASA complex. The President occupied a private forward compartment on the plane while Webb and I were in the rear with reporters and White House staff. Halfway through our journey the President sent for Webb and me to come forward. He cocked his head to one side and gave me a quizzical look. Then with much anger in his voice he said, "Yesterday I spoke with President Eisenhower. He told me that he had received an urgent call from Kistiakowsky, who apparently believes that I plan to use the atomic bomb in Vietnam; he was calling to ask Eisenhower to intervene. Eisenhower asked me if there was any truth to Kistiakowsky's statement. I assured him there was not." President Johnson shook his head wrathfully and added, "Don't you scientists have the sense to know that I could not afford to take any such step without consulting Eisenhower? We talk on the phone for at least an hour each week!" His anger subsided slowly as Webb hastily steered the conversation to other matters with my full cooperation. I have no idea from what source, good or bad, Kistiakowsky might have obtained such a notion but it was clear that the credibility of the Advisory Committee had suffered another blow.

During the second half of the 1950s, some members of the scientific community became very interested in the prospect of sending scientific research equipment into space. For example, Herbert Friedman[4] and others were using any high-altitude rocket available to carry equipment that would search

---

[4] See H. Friedman, *The Astronomer's Universe* (Norton, New York, 1990).

out x-ray emitting stars. In 1957, Detlev Bronk created the Space Science Board within the Academy. Under the leadership of Lloyd Berkner, the board's task was to obtain a consensus of opinion among members regarding research in space, and to provide guidance as to its feasibility. While the initial group of members was primarily interested in sending scientific instruments in space, they were open-minded about manned flights in view of the fact that the Soviet Union made a major point of such missions.

Early in his term of office, beginning in 1961, Webb encouraged the Academy to continue the work of the Space Science Board along broad lines. With the passage of time and the inevitable turnover of membership, however, the Board gradually became more heavily weighted with scientists who preferred to emphasize instrumented scientific missions in space rather than manned missions, a subject on which the opinion of the scientific community became divided once President Kennedy initiated the Apollo Program—the manned mission to the moon. While the latter might be more exciting to the public, they argued, they were not the most cost-effective. Webb was deeply distressed to see support for "his" program turn to opposition; unfortunately, there was little he could do, as the Board, although financed by NASA, selected its own membership and chairmanship in accordance with the Academy policies.[5] Any attempt to change this policy in a given case would be certain to create a row.

Hubert H. Humphrey (Figure 13.20) served as Vice-President from 1964 to 1968. Following an emerging trend, he took an interest in scientific affairs. He called me early on to ask if he could visit the Academy occasionally to be briefed on new and ongoing research. He came by about once a month and usually spent three or four hours. The vice-president's visits always included a leisurely lunch; since his arrival time was unpredictable, the luncheons were planned around food that could hold up well on the back of the stove. We all became very fond of Humphrey in the course of these visits, during which we covered many topics of special interest to him, ranging from space research to medicine.

He had a fascinating dietary practice. During his barnstorming days as a young politician, he was obliged to eat whatever food was offered to him. He had apparently discovered that he could make any food palatable by liberally dousing it with ketchup. This had long since become an unbreakable habit, and we saw him devour fine lobster and delectable filet mignon floating in a sea of the sauce.

It was a great sadness to Humphrey that many of his liberal supporters declined to vote for him for President in 1968. They did not vote for Richard Nixon, but simply deserted the polls. He regained some of his aplomb when he returned to the Senate in 1970.

---

[5] I gained the impression at the time of the first moon walk that the educated European community was more deeply awed by the accomplishment than the corresponding community in the U.S. which apparently had come to take its technological prowess for granted.

James Webb was bewildered when he saw so many liberal Democrats deserting Hubert Humphrey. Once, when he and Patsy were visiting, he expressed his concern—how in the world could this happen? I went to the bookshelf and got down my copy of Mark Twain's *Huckleberry Finn*, and had him read the passages where Huck's father explains how he will get even with society. He will simply not vote. After he had read it, I said, "Well, Jim, there are certain types of individuals who feel that way about the power of the vote." Huckleberry Finn had not been part of Webb's early reading. He simply shook his head, still baffled.

Upon leaving his post as Administrator of NASA at the end of the Johnson administration, Webb (Figure 13.21) began to play a broad, private, non-partisan role in Washington with the desire of increasing unity of purpose and cooperation within government circles, along with the hope of countering the disruptions of the 1960s. He continued to play this role in a most remarkably determined way even after he contracted a serious case of Parkinson's disease in the mid-1970s—an enervating illness which he fought to over-ride with great stamina and courage until his death in 1992. I saw him frequently in New York where he was treated initially by Dr. George C. Cotzias.

Two House members whose friendship I enjoyed were George P. Miller, a fellow native Californian, representing Alameda, and Emilio Q. Daddario from the Hartford, Connecticut area. They established a study committee within the Academy that led to the creation of the Office of Technology Assessment, which was originally meant to speculate on what important benefits as well as problems might emerge from new forms of technology. In keeping with the prevailing tenor of the 1960s, however, the OTA man-

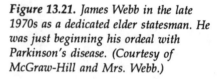

Figure 13.21. James Webb in the late 1970s as a dedicated elder statesman. He was just beginning his ordeal with Parkinson's disease. (Courtesy of McGraw-Hill and Mrs. Webb.)

aged to devote most of its energy to finding real or imagined negative effects of technology. In the 1980s I served briefly on the committee assessing the Strategic Defense Initiative. I concluded that the administrative staff of the office had intentionally selected a committee that would return with a negative recommendation, even though its chairman, Guyford Stever, did his best to be neutral.

Miller and Daddario were initially responsible for the appearance in many Congressional bills of a statement to the effect that the Academy's advice should be sought concerning the implementation of any scientific or technological facet of the bill. This led to a flood of study requests from Congress that Handler and Press had to deal with in subsequent periods. As mentioned earlier, in my readings of the history of the Academy I found only one such request from Congress in the entire period from the Civil War to the 1960s.

The year 1966 was a special one for the Académie des Sciences in France — its three hundredth anniversary. I was invited to attend the anniversary festivities on behalf of our own Academy, and had the distinction of being the only native-born United States citizen in the assembly. The opening event, held in the Académie headquarters in Paris, was spectacular. President De Gaulle officiated, seated in an elevated, throne-like chair. A magnificent military band accompanied the formal arrival of the members of the Academy, who were arrayed in resplendent uniforms with swords and cocked hats. The military tempo was not precisely suited to the gait of many of the older members, but was essential for the proper rendering of the "Marseillaise." A more appropriate string ensemble replaced the band in later sessions.

Most of the meeting was given over to formal addresses related to the remarkable achievements of the Académie since its founding by Louis XIV in 1666, aided by the great Dutch physicist, Christiaan Huygens. The final elegant reception and dinner were held at the richly renovated Petit Trianon palace, originally built for Marie Antoinette.

Relations between France and the United States were somewhat strained at that time, but all animosity was suspended for the occasion. And I was personally gratified as the event gave me the opportunity to revisit the Passy district, where Betty and I had spent so many exciting months earlier in the decade.

In 1966, reported observations of unidentified flying objects became sufficiently numerous that the United States Air Force called on the Academy for help in carrying out a serious scientific study of the phenomena. I knew that Edward Condon, now at the University of Colorado, would relish leading such a study, but I also knew as a result of my many years of valued, friendly association with him, that he would carry it through with such forthright investigative vigor that he would soon become involved in altercations with the dedicated group of individuals who believed that our planet is under regular surveillance by beings from outer space. I tried in vain to find a suitable alternate to Condon; the Air Force finally made a contract with the University of Colorado at Boulder in order to obtain his services. It was agreed that Condon's final report would be subject to review by a committee of the National Academy of Sciences before being made public.

The Condon study, which had aroused much public interest, was ultimately published in book form by Bantam Books with the sponsorship of *The New York Times* (*Scientific Study of Unidentified Flying Objects*, Bantam Books, New York, 1969). This version catches some of the flavor of the controversies that surrounded the study, engendered in part by Condon's enjoyment of such rows. At one point in the process, a reporter who was interviewing him said, "Dr. Condon, you seem to have been the center of controversy fairly frequently." He cheerfully responded, "It is not very hard."

The President of the University of Colorado at that time was Dr. Joseph Smiley, an old friend of mine who had succeeded Louis Ridenour as Acting Dean of Graduate Studies at the University of Illinois. Once when a group of the true believers held a press conference to accuse Condon of falsifying evidence, Smiley commented to Condon, "I see you are not getting a very good press these days, Professor."

Perhaps the most important item in the report is the brilliant essay, "Conclusions and Recommendations," written by Condon and describing the types of UFO-related phenomena which can and cannot be made the object of valid scientific study. Only one thing is certain: This subject will not go away.

In 1967 I received a request from Dr. Jaime Benitez, the President of the University of Puerto Rico, to serve on the advisory committee of the university's Puerto Rico Nuclear Center. The Center was funded by the U.S. Atomic Energy Commission; its director was Dr. Henry J. Gomberg, an experienced nuclear engineer. The committee was composed of well-known scientists and engineers selected from various areas on the mainland, including Latin America. William O. Baker, the Director of Bell Laboratories, was a member. While the center was modest compared to the mainland's great national laboratories, it provided a useful link with those laboratories for students and professional scientists on the island who were interested in atomic and nuclear physics as well as engineering. Those of us on the committee also had an opportunity to gain a better understanding of life on the island. Unfortunately, Benitez's position as University President was regarded as a political appointment by the parties on the island. After the next election, he was replaced by an individual who preferred to have a more restricted group of advisors, and our appointments were canceled in 1970. Dr. Benitez became the Representative of Puerto Rico in the U.S. House of Representatives and Gomberg accepted a position at a research laboratory in Michigan.

In 1967, Daddario in the House and Edward M. Kennedy in the Senate decided to revise the charter of the National Science Foundation and asked the Academy to endorse their plan. The version of the bill that I saw emphasized the desirability of including some support for applied work by the National Science Foundation. This seemed reasonable to me, provided the work was appropriately controlled and funded. The bill that finally passed, however, permitted the Congress itself to introduce line items for projects into the annual appropriation. Previously, the staff and the Science Board of the National Science Foundation had had exclusive authority to determine the way in which the budget was spent; now, it seemed, Congress was to have its many fingers in the pie. In the absence of concrete evidence, I assume that Kennedy's staff was responsible for the change and that the House team went along.

Daddario was a loyal party member and was drafted to run for governor of Connecticut in 1968 — unsuccessfully as it turned out — otherwise he might have stayed in Congress indefinitely. He joined a Washington law firm after the election; and became a good friend to the scientific community, serving for a term as the President of the American Association for the Advancement of Science.

Just before President Johnson left office, the members of his Scientific

Advisory Committee met to speculate on the inevitable changes that were to come in the White House. We also wanted to decide how best to show our respect for the departing President and to thank him for the privilege of serving his office. It was finally decided that we would request a meeting with the President and that Charles H. Townes, who could speak with a distinguished southern accent when he wanted to, would say a few well-chosen words expressing our gratitude.

The meeting was arranged to be held a day or two prior to the President's departure. We assembled in the conference room adjoining the Oval Office. While standing by the door to the Oval Office, it became evident to me that the President had instructed his aide to call him out in ten or fifteen minutes for a hypothetical important call so that he could diplomatically terminate the meeting. When Townes began speaking, however, he was so eloquent that, much to my surprise and gratification, the President relaxed in his chair, and adopted a contemplative air, as if listening to fine music. When Townes had finished and a few other comments had been made by committee members, the President thanked him and the group, and then began a soliloquy that went on for nearly thirty minutes. The aide appeared with the prepared message, but was waved off.

In brief, Johnson told us that he had tried to do his best for our country through the creation of the program that had come to be called The Great Society, and that he had hoped to help preserve democracy in the world by continuing the struggle in Asia that he had inherited. He trusted that he had done some good. He knew that he had made a number of mistakes, some of which were well-known to the public and some of which might only become evident much later. He said that his greatest mistake had been to place more confidence in what he termed "experts" who had the distinguished educational credentials that had been denied him, rather than in his experienced political friends and his own instincts. It was a most remarkable declaration from a very remarkable man at a major turning point in his life.[6]

The Academy years gave Betty and me a fine opportunity to travel abroad on special missions. There was a meeting of the Council of the International Union of Pure and Applied Physics, on which I served as national representative, in Bombay in January of 1963. Betty accompanied me on a journey that involved visits to scientists in Japan, Hong Kong, India, and Egypt. The meeting in Bombay made it possible to renew our acquaintance with the

---

[6] Although the student activism which occurred in the 1960s and 1970s is normally ascribed to the onset of the war in Vietnam, the historian, A. Hunter Dupree, who was at the University of California at Berkeley between 1956 and 1966, noted the rise of significant levels of activism there much earlier, associated with concern about voter registration in the southern states (see footnote 5, Chapter 16). The war in Vietnam apparently gave additional focus to an underlying crusading spirit in the generation that emerged after World War II, demanding social and political reform, which was bound to be expressed in one way or another.

brilliant scientist, Homi J. Bhabha (Figure 13.22), our host and a newly elected member of our Academy. Our friendly relationship lasted until his untimely death at the age of 57, in a plane crash at Mont Blanc in 1966 — a great tragedy for India. Bhabha was developing plans to establish a version of our Academy in India; his death unfortunately put an end to those plans.

Bhabha was involved in many other aspects of national planning as well. For example, he played a leading role in the creation of the nuclear power system that is so important to India's industry today.

I experienced one embarrassing moment in Japan on this trip. Since Professor Yukawa (Figure 11.14), who had won the Nobel Prize for his prediction of the particle now known as the pion, was a member of our Academy, I decided to pay a courtesy call on him and arranged an appointment at his institute in Kyoto. After greeting me, he ushered me into a conference room containing six or seven of his colleagues. He then said, somewhat severely, "We Japanese scientists are greatly disturbed by the fact that your country interferes with our desire to form close relationships with our Mainland Chinese colleagues." I thought quickly how Rabi would have handled the situation, and said in an earnest, sincere voice, "I fully sympathize with you. As you undoubtedly know, we had a much closer relationship with China than with Japan before the recent war." The atmosphere in the room changed at once for the better. After a brief discussion of scientific work, Professor Yukawa offered Betty and me his car and driver for a special tour of nearby Nara, the original capital and cultural center.

Betty and I also visited the Atomic Bomb Casualty Commission Center in Hiroshima while in Japan. (See Chapter 15.)

Figure 13.22. Dr. Homi Bhabha. (AIP Emilio Segrè Visual Archives.)

In October of 1963, I made an official visit to the Soviet Academy of Sciences to meet President Mstislav V. Keldysh and his staff. I spent an interesting day with the physical chemist, Nikolai N. Semenov, who took me mushroom hunting, an autumn tradition. We ended up with a small group at a mansion on a large pre-revolutionary estate, where the conversation turned eventually to politics. Semenov expressed the view that Western Europe would never again be militant and that the United States and the Soviet Union should work together to maintain world peace – a proposal which may come to pass in another form in accordance with recent events.

On my way to a meeting in Bombay in 1964, I stopped off in Beirut, Lebanon to meet with my old friend, Jack Clark, who was then serving as United States Science Attache to the Arab nations in the Near East. Lebanon was deceptively calm. While wandering through the Moslem quarter of Beirut with Clark, I saw a picturesque scene and took out my camera to photograph it. Within seconds I heard violent shouting nearby. A passerby urgently grabbed my arm. "I sincerely hope you understand French," he said in that language. "You are in very great danger. Get out of here immediately." We did. A month later a tourist was killed in the same area under identical circumstances.

Clark's base at this time actually was the United States Embassy in Cairo. I once visited him there for a few days in 1965 after a left-wing, anti-American group had burned down the cultural, educational, and science library building that the U. S. government had built for the use of the Egyptian public.

Also during this period, a Pugwash meeting was held in Udaipur in the Rajasthan Province of India (see Chapter 11), where we met Indira Gandhi – a proud, humorless woman – just before she was made Prime Minister. The province was suffering from one of its worst droughts in recent history. It was heart-breaking to see the people driving their cattle along the dusty roads, desperately seeking non-existent water.

While in Bombay for another special meeting in 1966, I had the privilege of attending several informal gatherings at Homi Bhabha's home with him and Patrick Blackett (Figure 13.14), who was then President of the Royal Society. These two had been friends since their student days, and enjoyed bantering, argumentative discussions. Most of the talk was about the best way to bring India into the modern world and to tap into the tremendous latent talent in the people of that remarkable land. Bhabha, of course, felt a deep sense of responsibility to his country and was doing everything in his power to improve education, not least in the basic and applied sciences. He was also working to improve the energy supply to help promote the development of industry. Blackett, in his turn, while highly sympathetic to Bhabha's endeavors, felt that India's primary strength lay in the long-term development of agricultural resources.

It is only now, as this book is being written, that India is beginning to take advantage of the opportunities that come with foreign investment in a country's industrial centers—investments that were crucial to the initial development of countries such as Korea, Taiwan, and Singapore. Not surprisingly, some foreign countries have found India a good place wherein to establish facilities for the production of software—mathematical, logical traditions have deep roots in that country.

In 1965 I had the extraordinary experience of taking an extended tour through Africa south of the Sahara through the courtesy of George Harrar (Figure 13.23), President of the Rockefeller Foundation, on whose Board I served, and the Wildlife Foundation. The wildlife portion of the trip was arranged by Harold J. Coolidge, a very valuable member of the Academy's staff and by Judge Russell E. Train. I was fortunate in visiting Nigeria before its oil boom of the 1970s, since I was able to take a serene journey by automobile to Ibadan. The Rockefeller Foundation had a research center there, and a colleague was carrying out research. I spent a night or two on the university campus, in a small chalet usually used by one of the governmental officials. Tragically, he was assassinated a few weeks after I returned home in one of the many coups that periodically racked that country.

The Wildlife Foundation made it possible for me to visit, by private plane, several parks in Kenya and Tanzania, including a stop at a temperate-climate plantation halfway up Mt. Kilmanjaro. I also had the privilege of spending part of a day with Louis Leakey at Olduvai Gorge. He had just discovered some of the remains of the hominid he eventually named Homo Habilis, considered one of the first tool makers. Leakey's fame was mounting at this period; while his various pronouncements sparked a great deal of controversy in the field, most have now been accepted in principle.

Figure 13.23. Dr. J. George Harrar, the founder of the Green Revolution and the president of the Rockefeller Foundation during most of my period on the board. (Courtesy of the National Academy of Sciences.)

After this visit to Africa in 1965, I flew to Melbourne, Australia to attend a meeting of the International Union of Pure and Applied Physics. I learned a great deal about the wonders of that continent and renewed a number of old friendships. The Australians were just beginning then to appreciate the remarkable qualities of the Aborigines, to whom the early settlers, particularly the farmers and ranchers, had been quite hostile. I attended an exhibition of Aborigine art created by a newly founded society which was cultivating the friendship and the well-being of the original natives. I acquired several remarkable bark paintings.

I also enjoyed the frank openness of the people who, as others have observed, are somewhat like Texans — although, with quite a different background, as many Australians had ancestors who came from urban London.

On an outing to the Blue Hills near Sydney, the members of my group delighted in stopping along the way to show me abandoned jails where some of the early settlers, their ancestors, had been incarcerated for causing various troubles.

Betty and I also visited Hungary and Rumania, and were greeted in both places by officials of the local academies. The scientific community owes a great debt to Harrison Brown for opening the doors to these countries although I must admit that I always breathed a sigh of relief on returning to the free world.

On arriving in Budapest, I thought it wise to pay a courtesy call on the U.S. Embassy. When I arrived there, I thought I would simply leave my card and go. The receptionist, however, said: "Let me call the Ambassador." She turned the phone over to me and Ambassador Elim O'Shaughnessy said, "If you have the time, come on up." We spent an hour or so discussing Hungarian affairs. At that time he had Cardinal Josef Mindszenti as his house guest. The cardinal had been charged with treason by the Communist government and had sought sanctuary in our embassy. He was eventually released and allowed to go to Rome. As the person whose job it was to cultivate cordial relations with the local government, the Ambassador found that he had a great deal of spare time on his hands.

Much the same situation prevailed at our Embassy in Bucharest. In fact, Ambassador Richard H. Davis used our visit as an excuse to hold a reception, to which he was able to invite many individuals who ordinarily would have felt compelled to send their regrets. The invitation list was cleared by the Rumanian Government.

As part of our visit to Rumania encompassed a weekend, President Nicolescu of the Academy suggested that Betty and I take the opportunity to visit either the Carpathians or the Black Sea. We chose the Black Sea since I knew that there were remains of Fifth Century B.C. Greek communities on the coast, and I was anxious to visit them. We were driven to the sea in a large Mercedes sedan at a speed of one hundred and ten miles per hour over roads which, if I had been driving myself, I would have taken at a maximum of fifty. In addition to visiting the Greek ruins, we managed to consume a large fraction of the caviar available in the town of Mangalia, the resort community of the Black Sea.

There was an interesting meeting on space science in Vienna in August of 1968. James Webb invited me to be part of his delegation, which included his wife and daughter. It was a most rewarding experience in one of my favorite cities. I took the opportunity to visit President Frantisek Sorm of the Czechoslovakian Academy in Prague. It was the height of the so-called "Prague Spring;" that wonderful country was in a state of exaltation at the prospect of being free again. When I left, Sorm escorted me to the airport; by the time I arrived in Frankfurt to change planes, the Soviet tanks were already rolling into Prague. Sorm died soon after, but President Dubcek lived to see his country freed again in 1990, and to participate in the new era about which he had dreamt.

In the same year, Harrison Brown arranged a very large conference on Third World agriculture in Jakarta, Indonesia. That country had just escaped a Communist takeover, thanks to a complex plot involving the complicity of Premier Sukarno. Many prominent Indonesian scientists, who had attempted to play a neutral role in the previous struggle between Communist and Democratic forces, found to their horror that they had been placed on the Communist death list. Brown rode out all the turmoil with a characteristic air of serenity, his eye on higher goals.

Several moments during that visit to Indonesia remain particularly memorable. One evening at a garden reception, I asked our host, "Isn't Krakatoa near here?" Krakatoa was the volcano that exploded in 1887, darkening the local skies for days and spreading dust throughout the northern hemisphere. All other conversations stopped as the Indonesian guests took turns in reciting the family tales they had heard regarding the momentous event. I was irresistibly reminded of my father's accounts of the San Francisco earthquake of 1906.

Also during that visit. I had an opportunity to make a tour of the tropical gardens which had been developed by the Dutch when Indonesia was a colony. It was there that they took the rubber plants that had been spirited out of Brazil, and started the great rubber plantations that would completely change the nature of the international rubber trade.

In 1977, Detlev Bronk decided to retire as President of Rockefeller University at the end of the academic year. The institution had recently acquired its new name after a long period when it was called The Rockefeller Institute for Medical Research, and a shorter one when it was called The Rockefeller Institute. I had been on the University's Board since 1964 through the courtesy of its Chairman, David Rockefeller, and was asked to serve as an advisor in the search for Bronk's replacement. Ultimately the Chairman and other members of the Board asked me to be the next President, undoubtedly as a result of Bronk's strong recommendation. Needless to say, I accepted the post.

Philip Handler (Figure 13.24), a biochemist from Duke University, was elected as my successor at the Academy. Unfortunately, his wife was an invalid, and he decided he needed an additional year before taking office. Consequently, I spent the academic year 1968–69 commuting between New York and Washington, doing my best for the two institutions. I think it is safe to say that Handler came to the post with many attitudes about the virtues and vices of our country that are fairly typical in academic circles. Along the way, however, his views changed. He became a heroic supporter of human rights, and used the freedoms our country offered, which he appreciated in full measure, to act as an eloquent spokesman for those rights both at home and in many international meetings.

I had recommended that Handler be appointed to the Board of Rockefeller University. He served well both as Academy president and Board member until a few months after the end of his term with the academy when he died of cancer.

Commentary

During the more than sixty years that I have been intimately familiar with Washington, D. C., it has been transformed from a community akin to a provincial city, with a somewhat "southern" aspect, to a major metropolis with qualities of a world city. Unlike London or Paris, however, which are also capitals, it is neither the cultural nor the entrepreneurial center of the country. Instead, it is the center of money transfer (tax money) and deeply steeped in politics. The greatest danger the National Academy of Sciences

Figure 13.24. Dr. Philip Handler, my successor as president of the Academy. (Courtesy of the National Academy of Sciences.)

faces as Washington gains more and more control over the operation of the country as a whole (so far with voter consent) is that the group of scientists who are in charge of the Academy and its operations at any given time will focus much more attention on the political rather than the scientific mission with which it is charged—and on a continuing basis. Such a focus would subvert its main value both to the scientific community and to society as a whole.

Generally speaking, the well-developed scientific mind is quite different from the typically political one. In fact, the two are poles apart. When involved in basic or applied research, the actions of the scientist must conform to tightly controlled principles of logic while trying to uncover the rulings of nature; the politician is concerned with influencing constituents and usually employs a different form of logic, one in which facts may be less important than perceptions, and there is willingness to compromise on principles to achieve agreements. There is a strong movement in the scientific community at present to attempt to become "close" to Washington with the hope of benefitting the professions by trying to learn and play the politicians' game. It is possible that instead the scientists will earn contempt.

Since federal funding of research is here to stay, it is clear that the scientific community must become conversant on a selective basis with the key individuals in the agencies as well as the formal structure of the government and the rules by which the agencies and legislators must operate. However, it must not forget that in the past the scientific community gained much of its respect from the fact that it appeared to represent a form of priesthood which served the public interest by carrying out its main investigative functions in accordance with its own well-established rules at whatever level it worked. The deeper it becomes embroiled in the political process, the less likely it is to help the cause of good science.

Chapter Fourteen
Rockefeller University,[1] *1968-1978*

Although Rockefeller (Figure 14.1) is by no means a typical university, perhaps being best defined by its original classification as a research institute, I welcomed the opportunity to return to a more nearly academic environment. I was not a complete stranger to the institution, having served on the Board for four years before becoming President. At the time of writing, I have had the privilege of close association with this special place and the remarkable individuals in it, both administration and faculty, for thirty years. David Rockefeller was the Chairman of the Board when I joined. He was succeeded first by Patrick Haggerty (1975–1980) who was, in turn, followed by William O. Baker (1980–1990), and Richard M. Furlaud (1990–). David Rockefeller has continued as Chairman of the Executive Committee. Along with service on the Board came a number of new and interesting friends such as Ralph Ablon (see next chapter), Mrs. Vincent (Brooke) Astor, and J. Richardson Dillworth. Detlev Bronk served as President between 1952–1968. My own tenure succeeded his, and was followed by the terms of Joshua Lederberg, David Baltimore, both biochemists, and Torsten N. Wiesel (1993–__), a neurobiologist. I am indebted to these and other leaders for generous and friendly support both in and out of office.

The institution had been founded in 1901 by John D. Rockefeller, Sr. as the Rockefeller Institute for Medical Research. He realized that the United States was far behind Europe with respect to biomedical research and hoped that such an institution would help accelerate the national development, both through its own research and through the training of research-oriented physicians. In the process he was wisely advised by a long-term colleague, Frederick L. Gates. With customary wisdom, he selected William H. Welch, who had been one of the founders of the Johns Hopkins Medical School, as the Chairman of the Advisory Committee for the new enterprise. Welch, in turn, recommended that one of his former students, Simon Flexner, then at the University of Pennsylvania, become the Director. Flexner believed that it was very important to have a research clinic included in the Institute – a clinic that began to operate in 1910. The overall productivity of the institu-

---

[1] For a fifty year history of the institution see *A History of The Rockefeller Institute (1901–1953)* by George W. Corner (Rockefeller Institute Press, New York, 1964).

Figure 14.1. *The entrance gate to The Rockefeller University. Founder's Hall, the original building, is in the background.*

tion far exceeded Rockefeller's initial dreams. For about a half-century it helped set the standards for biomedical research in the United States and continues to provide unique opportunities for a highly committed group of scientists.

One of the benefits of my association with Rockefeller University and the larger medical community on the east side of New York City was the privilege of working with a number of remarkable individuals outside the scientific community, particularly David and Laurance Rockefeller (Figures 14.2 and 14.3). As mentioned, David was the Chairman of the Board of Trustees at Rockefeller at the time I joined it, while his brother Laurance was Chairman at Memorial Sloan-Kettering Cancer Center just across York Avenue. David had been on the Board since 1940 and had taken over as Chairman in 1950 when his father, John D. Rockefeller, II retired. Laurance had similarly followed in their father's footsteps at Sloan-Kettering. Both David and Laurance were deeply committed to philanthropic works and public service, as their father and grandfather had been before them.

It is commonly held, of course, that it was Henry Ford who put the United States on wheels. We all tend to forget that this achievement would not have been possible without the previous entrepreneurial work of individuals such as the senior John D. Rockefeller, who made available large quantities of inexpensive petroleum products and steel.

Alfred P. Sloan had been the principal benefactor in the creation of the

Figure 14.2. (Left) David Rockefeller, right, with Patrick E. Haggerty. The latter succeeded Rockefeller as chairman of our board. Founder's Hall is on the left; the university hospital, the first research clinic in the United States, started in 1910, is in the background behind David Rockefeller. (Courtesy of The Rockefeller University.)
Figure 14.3. (Right) Laurance S. Rockefeller, a major humanitarian philanthropist in our time. (Photograph by Yousuf Karsh.)

Sloan-Kettering Cancer Research Institute, which was associated with Memorial Hospital. Sloan insisted that there be a definite separation as well as a definite link between the hospital and the research institute. His concern was that the research program might become too closely tied to, and therefore unduly directed by the immediate practical aspects of the hospital's treatment programs.

Another impressive entrepreneur who gave his time and energy selflessly to Memorial Sloan-Kettering is Benno Schmidt, the Chief Executive Officer of J. H. Whitney and Company. He played an important role in formulating the legislation that led to the expansion of the National Cancer Institute in the 1970s.

During his tenure as President, Detlev Bronk had made a number of significant changes in the University with the concurrence of the Board. He founded a small graduate school with about one hundred carefully selected students, and he added small faculty groups in physics (Figure 14.4), mathematics, and philosophy. The additions to the physics department included George E. Uhlenbeck, Abraham Pais, and E.G.D. Cohen. The "house" mathematician was Marc Kac. In addition, he encouraged the senior staff to seek

Figure 14.4. Professor George C. Uhlenbeck, one of the stellar physicists Bronk persuaded to join the faculty. (Courtesy of The Rockefeller Archive Center.)

grants from federal agencies to increase the resources available to them. Until the 1950s, most activities on the campus had been paid for by the endowment, which was proving increasingly inadequate.

Perhaps most importantly, Bronk gave the institution a psychological lift by introducing a number of amenities, including a new faculty center, Abby Aldrich Rockefeller Hall, a gift of the Rockefeller brothers, and a fine auditorium, with funds donated by Mr. Alfred H. Caspary, a friend of Board Member George Murnane. Bronk also added a student dormitory and a small on-campus apartment house, the latter with funds provided by a well-wisher. He built a new laboratory which faces northward toward New York Hospital; this building now bears his name, by faculty and board acclaim.

Finally, he had one of the most handsome residences in New York, to be used both for receptions and as a possible residence for the President, constructed at a lovely spot on campus overlooking the East River. The atrium-style mansion was designed by Wallace Harrison, who was also responsible for all the other new buildings up to that time. A highly talented landscape expert, Daniel Kiley, was engaged to beautify and maintain the grounds.

Bronk did leave me with a few problems, among which was a growing deficit. His style of management was quite different from my own. He and his able associate, Mabel Bright, employed a procedure whereby all of the senior staff reported directly to them, with few formal connections at other levels. Most of the staff were veterans of the institution and carried on magnificently under this form of administration. The Controller was Roger C. Elliot, the Manager of Purchasing was Anthony J. Campo, and the Supervisor of Buildings and Grounds was Bernard Lupinek. Lupinek had a capable deputy, Paul R. Penndorf, who took his place when he retired, and was

followed in turn by Thomas P. McGinnity. All of these individuals transferred their loyalty to me as soon as I arrived, although I knew their primary loyalty, of course, was to the institution. Some of them had been at
Rockefeller almost from the time the Institute first began operating. They
regarded their work almost as a sacred trust, an attitude which I believe
had played an important role in elevating the institution to prominence.

Much as I admired the administrative achievements of Bronk and Bright,
I had my preferred form of executive organization. I first brought in Albert
Gold, with the title of Vice President for Academic Resources. He was a
former postdoctoral fellow under me at the University of Illinois, had been
for some time on the staff of the Dean of Graduate Studies at the University
of Rochester, and welcomed a new administrative post. One of his first challenges was to organize better lines of communication with the postdoctoral
population. It was clear that they were not being advised adequately, since
a large fraction of them, about three hundred, expected to remain at
Rockefeller as tenured faculty, which was simply not possible. Gold was
invaluable in clarifying the situation. Working with the faculty, we established a ranking system for all postdoctoral positions, with a clearly defined appointment period, assigned to terms of appointment, associated
with each rank.

Gold also helped me find someone to computerize the institution's operating budget, so that we could have virtually instant updates on where our
various budgets stood, as well as gain the economies of automation. The
individual he recommended was David J. Lyons (Figure 14.5), a colleague
of his from Rochester. Lyons was not only highly proficient at his vital computer work, but also proved to be a warm and sympathetic person to whom

*Figure 14.5. Dr. David J. Lyons, Vice-
President for Business and Finance as well
as Treasurer of the University. A wise and
sympathetic counselor to the entire staff.
(Courtesy of The Rockefeller Archive
Center.)*

all members of the campus, regardless of rank, turned for personal help. His response to any such appeal was unfailingly generous and full-hearted.

The University had constructed most of its buildings with funds derived from its endowment or, in some cases, from special private gifts. While at the University of Rochester, Gold had learned that under certain conditions it was possible for an institution like ours to secure long-term, low-interest loans from the State Government, using the so-called Dormitory Authority. He made a detailed study of this possibility, venturing deep into the labyrinth of the State Government at Albany in the process. Thanks to Gold's efforts, we were able to construct two much-needed buildings on campus using funds borrowed in this way. One, an apartment house for the junior faculty, was built at a time when such housing was in very short supply. The other was an animal research facility (about which more later). Gold, in addition to discharging his other duties, managed to keep a weather eye on the planning and development of both projects; the number of change-orders during construction was negligible.

In his later career Gold became Associate Director for Administration of the Division of Applied Sciences at Harvard University.

Another happy addition to our staff was Eugene Sunderlin, an old friend and Rhodes Scholar, with a chemical background, whom I had known since Rochester days, and who had worked closely with me at the National Academy of Sciences as a special assistant. I asked Sunderlin if he would be willing to serve as Vice-President and Secretary to our Board. Fortunately he agreed, and was of enormous help, not only with activities directly related to the Board, but with many other matters that deserved special attention. I had previously asked Mabel Bright to remain on as Board Secretary, but she preferred to continue working with Bronk on the numerous projects he carried out after retirement.

Finally, I brought in Rodney W. Nichols to help with a variety of policy issues both internal and external affecting the institution. He held a Bachelor's degree in physics from Harvard, and thus was no stranger to the scientific world. We had met in Washington earlier in the 1960s when he was on the staff of the Deputy Director for Defense Research and Engineering in the office of the Secretary of Defense. The National Academy of Sciences had been asked by the Royal Society of London, with which it had long-standing links, to intervene in a plan being developed by the British Air Force to turn the Aldabra Island complex in the Seychelles off the east coast of Africa into a base for observational aircraft to survey the African continent. The Nature Conservancy and the Smithsonian, which had scientific interest in preserving the flora and fauna of the islands in as near their natural state as possible, hoped that this plan could be forestalled. Among other things, Aldabra was an important bird sanctuary. We discussed the matter with Secretary McNamara, since the U.S. worked closely with the U.K. on some defense issues. He agreed to find a suitable alternative, and assigned the task to Nichols. Survey satellites were just coming into general use, and they seemed a superior alternative means of surveillance. Another possibility was to establish the proposed base on the island of Diego Garcia which was not so important to the area's wildlife. In any event, Aldabra was spared. Nichols was a great help both to me and to my successor, Joshua Lederberg, at the University. In 1993 he became Chief Executive Officer of the New York Academy of Sciences.

In addressing the various problems that came our way, my staff and I relied on the help of Rockefeller's Faculty Council. The Council, equivalent to the executive committee of a university senate, was created by Bronk in the later stages of his term in office and proved to be a very useful group. It not only saw me through my acclimatization, but was invaluable in securing the cooperation of the academic staff on many problems.

The first graduate students brought to the University in the mid-1950s proved to be so attractively stimulating that there was a tendency to allow their number to grow more or less unchecked. It became evident to Frank Brink, the Graduate Dean, and to me, that as the size of the group grew, so did the percentage of students who were really not sufficiently self-motivated to take full advantage of the special opportunities available at the institution. Brink and I consulted, and we agreed to limit the total number of students to about one hundred starting about 1970.

Soon after this decision was made, the program was further modified. Some students were given the opportunity to carry out medical studies at the Cornell University Medical College, just north of our campus, while simultaneously pursuing research programs at Rockefeller with the ultimate goal of obtaining joint M.D.–Ph.D. degrees from the two institutions. The program was supported initially by both the Commonwealth Fund and the National Institutes of Health, and did exceedingly well for the next two decades. Those faculty members who were interested in medical education cooperated in the initiation of the program. James G. Hirsch, who succeeded as Dean when Frank Brink retired, encouraged the further development of the cooperative enterprise.

With the rapid evolution of cellular and molecular biology in the 1950s and 1960s, some of those working in the field came to believe that their research would bring about such a great revolution in our knowledge that the practice of medicine would change fundamentally. In the future, the patient would turn to the biochemist rather than to the traditional practitioner, specialist, or otherwise. While there was unquestionably a substantial element of truth in this view, as the advances in some major aspects of medicine since 1970 have demonstrated, it must be remembered that in addition to scientific expertise, there is a human art, based on wisdom, skill, and experience, that is indispensable to the practice of medicine. It is vitally important to maintain links between the research scientist and the practitioner, and the University Hospital at Rockefeller had performed that function with distinction since 1910 when it was created by Simon Flexner and led by Rufus Cole, the first head of the hospital.

Bronk was apparently ambivalent towards the hospital, perhaps because he saw a great revolution in biomedical research in the offing and did not want it to be hindered by previous traditions; in consequence he did not expand its activities appreciably during his period in office. Fortunately, he wisely appointed Maclyn McCarty[2] (Figure 14.6), a brilliantly creative scientist with an excellent medical background, the Physician-in-Chief, with the understanding that the high standards that had been set would be maintained. In my turn, I did my best to further the interests of the hospital and appointed Attallah Kappas (Figure 14.7), another active research investigator and clinician, as McCarty's successor on the latter's retirement. In the meantime, he has been succeeded, first by Dr. Jan L. Breslow and then by Dr. Jules Hirsch.

Figure 14.6. (Left) Dr. Maclyn McCarty, a key member of the triumvirate involving Oswald T. Avery and Colin M. McLeod which demonstrated that DNA, rather than a protein, carries the genetic message – one of the monumental discoveries of the century that has been strangely ignored by the Nobel Committee. (Courtesy of The Rockefeller Archive Center. Photograph by Ingbert Grüttner.) Figure 14.7. (Right) Dr. Attallah Kappas, McCarty's successor and deeply valued friend. (Courtesy of The Rockefeller Archive Center.)

---

[2] Maclyn McCarty's book, *The Transforming Principle* (Norton, New York, 1985), gives a detailed account of the discovery that DNA, rather than amino acids, provides the basis for the gene.

I find it distressing that many molecular biologists, including some who should know better, continue to denigrate hands-on clinical work. A similar attitude is found among some physical scientists and mathematicians who believe that good engineering begins and ends with the development of a computer program.

Although Bronk had encouraged the staff to augment University funds by applying for government grants, he did not establish a formal program to seek private funds in any systematic way. In the meantime, our endowment income was proving increasingly inadequate to sustain the expanded activities of the institution. The Board of Trustees therefore agreed to set up a formal fund-raising program, the first in the University's history, once I was in a position to describe our growing plight. Several Board Members took the initiative to get the program moving, with David Rockefeller making a very generous start-up gift. Our plan was developed with the help of Marts and Lundy, an independent organization created after World War II, with the encouragement of private foundations, to provide fund-raising guidance to non-profit organizations. The staff of the University, scientific and otherwise, joined in this effort in highly effective ways.

My experiences with fund-raising at both the National Academy of Sciences and at The Rockefeller University taught me that the exercise has rewards far beyond the monetary. If those from whom one seeks support are not drawn to the cause, the association is usually brief, and one hopes to, and generally does, part on good terms. On the other hand, in those who do help financially, one frequently finds a common spirit that provides both inspiration and encouragement, and that can be the basis of a rewarding long-term friendship.

One of the practices our executive group developed as part of our fund-raising efforts was to have a potential donor meet over lunch with one or more members of the scientific staff who would describe their research programs in more or less popular language. The faculty members were exceedingly cooperative in participating in these discussions, and they proved to be stimulating for all concerned. Many close links with townspeople and others were formed in this way.

It is perhaps worth mentioning that the University followed a policy of paying its senior faculty ample salaries with University funds, thereby providing them with a strong reason to identify their own well-being with that of the institution. I have long felt that the widespread practice of compelling tenured faculty to raise a substantial portion of their personal income through outside grants weakens an academic institution. It causes even more stress at times like the present, when federal budgets are lean and there is fierce competition for grants among even the best scientists.

In order to stay in touch with the "outside world," and to keep our work in perspective, we felt we would be wise to seek advice from a broad range of individuals, not only those in our immediate community. We therefore created an Advisory Council, comprised of one hundred or so members from the private and philanthropic world, who met on campus several times a year to review university activities. James A. Linen, a close friend of David Rockefeller and a key figure in the Time-Life organization, chaired the Council and kept it vigorously active, to the benefit of both the University and the participants. The members proffered much advice and helped extend our relationship with new donors; they in turn gained fresh new insights concerning problems and advances in science, particularly in areas of biomedical research.

Once the development program got underway, David Rockefeller, who was deeply involved with the Chase Manhattan Bank, asked Patrick Haggerty, Chief Executive Officer of Texas Instruments and one of the bank's advisors, to assume the Chairmanship of the University Board so that Rockefeller could spend more of his time fund-raising. Haggerty took a keen interest in all University affairs. Not the least of his contributions was his analysis of our financial situation, conducted in cooperation with our staff, which enabled us to bring our budget closer to a state of balance.

In his turn, Haggerty had asked me to serve on the Board of Texas Instruments, which I looked upon as a special privilege. My twelve years on the board corresponded to the time when the potential of the integrated circuit came to be realized — a development that is still revolutionizing world-wide communications and information processing, almost on a daily basis. The Board contained some of the original founders and developers of this dynamic company who were, with their spouses, very generous philanthropists. In addition to the Haggertys were Cecil and Ida Green, the S.T. Harrises, Erik and Margaret Jonsson, and Eugene and Margaret McDermott. Eugene Fubini and Paul W. McCracken respectively added much technical and economic wisdom.

The heads of the research laboratory at Texas Instruments during most of this period were old and valued friends, first Dr. J. Ross Macdonald and then Dr. Norman Einspruch. The first subsequently joined the faculty of the University of North Carolina and the second became Dean of Engineering at the University of Miami. In addition, I made many new friends in the community.

While at the Academy I had spoken frequently with Donald Griffin, the biologist who first confirmed that bats actually use sonar in their insect-hunting forays, and whom I had first met as a young scholar during my Schenectady days (Chapter 6). Griffin had been at Rockefeller since 1965; one of the pleasures of my years at the University was the increased opportunity for detailed discussions of his work. His principal interest during this period was the attempt to discover the means by which birds navigate when their sight is obscured by fog or clouds. He needed access, he told me, to a reasonably convenient field station where he could use radar to continuously track bird migration. Such a station, we thought, would also be useful for his colleagues, Peter Marler and Fernando Nottebohm, who were studying, both in the wild and in the laboratory, the ways in which birds acquire their ability to sing. Some learn by imitation, much the way humans learn to speak, and exhibit regional "dialects."

We instituted a search for a suitable site in the nearby countryside, and soon discovered that at her request the Dutchess County estate of Mary Flagler Cary, the philanthropic daughter of the railroad magnate Henry Flagler, was to be donated to an appropriate not-for-profit institution by the Foundation bearing her name. We made an application for the estate, which was located about eighty five miles north of our New York City campus. Our bid for Mrs. Cary's estate was not successful; however, the Foundation instead gave us the funds required to purchase a large piece of the nearby Beck estate, which proved eminently suited to our purpose.

While Griffin never did find a complete solution to the problem of bird navigation, other remarkable discoveries were made. For example, Nottebohm learned that the number of brain cells in adult birds of some species actually wax as the season for birdsong approaches, and wane at the season's end. Studies continue of the agents which stimulate such growth. In connection with other investigations, a student discovered that some birds acquire a small ferromagnetic element in their breast which may react with the earth's magnetic field, and hence may play a role in bird navigation.

One of the most difficult duties I had to undertake as President of Rockefeller was the termination of the philosophy program that had been initiated by Bronk. He had started it in the 1950s when he brought to the campus from The Johns Hopkins University Ludwig Edelstein, a versatile philosopher with much interest in science, who enjoyed the special features of life on our campus and spent many hours discussing research and other activities with the scientific staff. The great value of the addition of such a wise and experienced generalist was his willingness to provoke interdisciplinary discussions within a staff whose members had a strong individual tendency to work in relative isolation.

When Edelstein died, Bronk attempted to fill the gap he left with Ernest Nagel, a philosopher from Columbia University who had interests in science. Nagel agreed to take the post on condition that he not come alone as Edelstein had done but be allowed to build up a senior core staff involved in other areas of philosophy. Bronk agreed.

Unfortunately, Nagel returned to Columbia within a year, presumably because he missed his work and students there. He left behind him a group of four tenured philosophers some of whom had little personal or professional interest in affairs at Rockefeller—two of them, in fact, spent most of their time on other campuses. We resolved the situation in 1977 by offering them each a lump sum equivalent to three years' salary if they chose to leave. If they elected to stay, however, they were assured continued tenured appointments at the University, without penalty. All of those who left immediately obtained excellent positions elsewhere. Their students followed them with continued support. One member of the group, who was mathematically oriented, did remain and served out his years as a highly respected and dedicated member of the University.

A much distorted version of these events was reported in the *New York Times*. Apparently some of the reporters on that newspaper had combined philosophy and journalism in their education and took a strong adverse view of our decision. As it happened, the story broke when I was attending a meeting in Europe. I returned to find some members of the faculty on campus in an uproar in the mistaken belief that the code of academic tenure was being violated. Fortunately, tensions eased as soon as the real details of the transaction became known. That was perhaps my darkest moment at Rockefeller University.

I had quickly learned that, when one is attempting to serve the needs of a university as its president, the unusual becomes the rule rather than the exception. It is worth mentioning a few of the challenges that made up our daily routine.

University fellows program. It is important to give young postdoctoral fellows who have unusual talents and goals as much leeway and support as feasible. With the help of the A. W. Mellon Foundation, we were able, early in the 1970s, to create a University Fellows Program that permitted us to offer special opportunities with regard to acquisition of equipment and choice of research topic to a select group of young investigators.

New animal research facility. I discovered soon after my arrival that the facilities for maintaining the animals used in research (which dated back to the early history of the institution) were not only well below national standards, but were regularly condemned by city authorities, who warned that they would shortly insist on their being closed down permanently. In response, we revived a series of as-yet unresolved studies which had been started by Detlev Bronk; we then moved ahead vigorously to construct a thoroughly modern facility, using funds borrowed from the state. Professor Richard M. Krause and Albert Gold took the lead in this endeavor, making a detailed investigation of the best national facilities, and following advice provided by the National Institutes of Health.

Staff committees. With the help of two of our trustees, William O. Baker and Walter N. Rothschild, Jr., we created two committees to deal with special concerns expressed by the research staff and the students. Protests at the verbal intellectual level were common, but there was no significant violence. Baker became Chairman of the Standing Committee on Scientific Affairs, which oversaw the programs and needs of the research investigators, while Rothschild met with groups of students, postdoctoral fellows, and alumni to discuss a wide range of policy issues, including social and political problems, which loomed large in those days. Questions were raised, for example, about some of the companies having branches in South Africa in which the University endowment funds were invested. Also there was a desire to have student and junior staff representation on the Board of Trustees and on selection committees previously reserved for senior faculty. Mutually acceptable compromises were found.

Inflation. As the decade of the 1970s progressed, inflation became a major problem, not least after the sudden jump in energy costs that accompanied the oil embargo of 1973 by some of the major oil-producing countries. Careful attention had to be given to fund-raising and to conserving resources. As mentioned earlier, we had a great deal of help in this work from our new Board Chairman, Patrick Haggerty, whose diligence was rewarded by success. We managed to bring the university budget into balance near the end of the decade.

Trust and estate program. Once fund-raising became a major endeavor, we followed the advice of one of our trustees, E. Whitney DeBevoise, and established an advisory committee on estate planning for members of the Board and others. Mr. Alexander D. Forger, later a trustee, was of special help in the process. With the passing of time and the continued help of the trustees, this has turned out to be a valuable source of income.

With all due respect to many other attractions, it is my opinion that the single most impressive feature of The Rockefeller University has always been the exceptional quality and character of its senior staff, and of the young scientists who work with them. F. Peyton Rous, who joined the staff in 1909 and who, in 1911, first demonstrated the existence of virally induced cancers, once said to me, "It is not a place where anyone feels compelled to do anything trivial." The fact the Rockefeller faculty boasts a number of Nobel laureates is partly accounted for by this truism, but the full explanation goes much deeper. For senior staff, routine work is minimal. There are no formal classes and few committees. In fact, the most onerous task they currently face is probably the need to prepare competitive proposals for federal agencies—a fact of life in a democratic society. No private endowment is so great these days that one can ignore federal funds. To ignore their availability would be to abuse the special flexibility provided by limited private funds, which usually have fewer strings attached to them at present.

A number of the eighty-odd senior members of the staff who were at Rockefeller when I assumed the Presidency had joined the institution prior to Bronk's arrival—some, indeed, had been there since before World War II. These included, in addition to Rous, Armin C. Braun, Merrill W. Chase, Lyman C. Craig, Vincent P. Dole, Rene J. Dubos, Walther F. Goebel, Samuel Granick, Rollin D. Hotchkiss, Henry G. Kunkel, Lewis G. Longsworth, Rafael

Lorente de No, Maclyn McCarty, Philip D. McMaster, Alfred E. Mirsky, Stanford Moore, John B. Nelson, Alexandre Rothen, Theodore Shedlovsky, William H. Stein, Norman R. Stoll, Igor Tamm, and William Trager. The great majority of the staff, however, had been hired during the Bronk era, and were comparably illustrious. One, Purnell W. Choppin, a brilliant virologist and good family friend, eventually became President of a greatly expanded Howard Hughes Medical Institute and served the cause of biomedical science very broadly.

I had known Herbert S. Gasser, Rockefeller's second President (or Director, as the post was then called), as a member of the Academy of Sciences, and had actually heard a lecture by Abraham Flexner, the younger brother of the first Director, Simon Flexner, while I was a student at Princeton. Abraham was a professional educator, deeply involved in the evolution of methods of education in the U.S., and was intrigued with his brother's new institute when the latter was selected to head it in 1901. It led Abraham to make a study, under the auspices of the Carnegie Corporation, of medical schools in the United States, and then to recommend in 1910 a radical revision in medical education. He proposed that medical schools should be linked to universities in order to benefit from exposure to their scholastic standards and involvement in research. At the time, many such schools were independent and of low quality. His idea received strong support from the Rockefeller Foundation's General Education Board, and had a profound influence on the subsequent development of medicine in United States. For several decades thereafter, most heads of university-associated medical schools, before assuming those directorships, spent several years at our Institute becoming familiar with medical research.

Abraham Flexner's lecture at Princeton dealt with this development and with the founding prior to it of the Medical School of The Johns Hopkins University by William H. Welch. As mentioned earlier, Welch had been Simon Flexner's teacher and mentor and had been chosen by John D. Rockefeller Senior to head the Advisory Committee for the new institute in New York. Abraham Flexner was both courtly and imperiously arrogant, if my own brief experience with him is any indication.

Soon after I arrived in Washington in 1962, I was invited to be a regular honorary dais guest at a series of fund-raising dinners for the Senator Robert F. Wagner Fellowship Program, held in New York City. This program for indigent New York State college students was started after World War II, and was named in honor of Robert Wagner, a much-admired political figure. Senator Wagner served for twenty two years in the U.S. Senate, specializing in labor relations. Since I had reason to visit the city occasionally, in any case, I made it a practice to attend the dinners, at which the Governor of the state served as host.

Initially, the gatherings were held at one or another of the city's major hotels, in a gigantic ballroom-dining hall with between fifty and one hundred tables occupied by ten or so individuals each. Most of those attending were businessmen who had achieved their success through hard work, having taken full advantage of the opportunities offered by public education, as Wagner had done. They were often accompanied by their families. It was not uncommon, at some point in the proceedings, for one or more guests to rise at their places and announce a special contribution to the fund beyond the donation already given to acquire a seat or a table. Several million dollars were raised each year in this way.

Once the student riots that had started in California spread east, attendance at the dinners dropped dramatically. The last dinner, which took place soon after I moved to New York City at the end of the decade, was held in a relatively small dining room and barely made its expenses. The program was dropped. I was deeply depressed at the time by this sharp decline in public support for the fellowship program. To me, it indicated the emergence of a generational cleavage in attitudes toward education. The older generation had looked upon education as providing an opportunity to rise in the world and add to the strength of the commonwealth; the younger generation regarded education as a means of gaining access to a vantage point from which to propose radical social reforms. It was evident that a significant change in the national climate was underway and that the change would have far-reaching effects.

The country's universities were deeply affected by these upheavals. Administrations became defensive as new centers of power emerged in both faculty and student groups. Ideology, which previously had played a relatively minor role in matters affecting instruction and faculty appointments, became almost crucially important. A new cycle in university life had begun.

Six individuals from Rockefeller received the Nobel Prize during my decade in office. Two, Stanford Moore and William Stein, were awarded the prize in chemistry in 1972; Gerald M. Edelman, one of our former students and now a faculty member, shared the prize in medicine that same year. Two years later, Albert Claude (Figure 14.8), Christian de Duve, and George E. Palade were awarded the prize in medicine. All three, with the exception of de Duve, who was now a faculty member, had carried out their basic work in our institution.

Albert Claude, who in close cooperation with Palade and Keith R. Porter had founded modern cell biology early in the 1940s using the osterizer, the centrifuge, and the electron microscope, had retired and was living in Belgium. After receiving the prize, Claude returned to Rockefeller University frequently; his periodic visits added up to many months and we became good friends.

Stanford Moore invited Betty (Figure 14.9) and me to be his guests at the

Figure 14.8. (Left) Dr. Albert Claude, the father of modern cell biology who carried out his basic work at The Rockefeller Institute in the 1940s. (Courtesy of The Rockefeller Archive Center.) Figure 14.9. (Right) Betty as the President's wife in our grand home on campus designed by Wallace Harrison. (Courtesy of The Rockefeller University.)

1972 award ceremony in Stockholm. Unfortunately we were just starting our fund drive and were not able to leave New York. Dr. Stein had to accept the award in a wheelchair. He had contracted an undiagnosed infection, possibly polio, during a visit to Europe soon after I arrived at Rockefeller. Evidently he had received the appropriate available vaccines before his trip, but had then dosed himself heavily with immunosuppressives in an attempt to combat a family of bothersome allergies.

In the same period, Vincent Dole and his wife, Dr. Marie Nyswander, were establishing clinics in New York City to start an unusual drug treatment program. They planned to use methadone as a substitute for heroin for the relief of heroin addicts. Needless to say, the program received a great deal of publicity and generated considerable controversy. Methadone, a synthetic material developed abroad during World War II to alleviate pain in the war wounded, is addictive. Its advantages over heroin are that it need be taken much less frequently, does not generate the same extreme highs and lows as heroin, and permits the recovering addict to function at work or school while receiving counselling and rehabilitation.

The methadone program proved to be successful in New York City. The rise of Acquired Immune Deficiency Syndrome (AIDS), however, deflected medical attention to that affliction. Actually, there appears to be a strong correlation between AIDS and the misuse of drugs, both because of the careless use of needles and the lowering of immune resistance, which is currently being recognized. For this and other reasons linked to AIDS, the methadone program is receiving renewed attention at the federal level. The research program has been continued under the leadership of Professor Mary Jeanne Kreek.

While making the transition between the National Academy of Sciences and The Rockefeller University, I was elected to a four-year term on the Princeton Board of Trustees as an alumnus of the Graduate School. It proved to be an exciting appointment because campus tensions that had been building up everywhere during the 1960s were reaching their peaks. I was deeply impressed by the steadfast loyalty of the members of the Board of the University, both to the institution and to President Robert Goheen who was at the center of the local hurricane. It was very rare during that intense time for any Board Member to miss a meeting without a very good reason. Betty's family, incidentally, had been close friends of the Goheens, Robert's father having been a medical missionary in India where Robert was born.

At the start of my term on the Board, two burning issues came before us: the admission of women undergraduate students and co-habitation in the dormitories. Princeton, with the backing of the Board, followed the national trend and became a coeducational institution in that period in which dormitories were shared by sexes. New dormitories were built; old ones were renovated; the Princeton Inn became a dormitory. The transition went ahead fairly smoothly although some of the students did display meanness in their approach to the administration.

One characteristic of the Board Members that impressed me was that none, including the very successful businessmen, looked upon the job of administering the University as trivial when compared with their own responsibilities, as is sometimes the case. Perhaps they paid particular attention at that time because of the uncommon turbulence on campus, but all Board Members appreciated that the demands of university administration, although different, were fully as great as any they faced in their other affairs.

I noted that the Graduate School had long since lost the special features which had made it such a wonderful refuge for me in the 1930s. It was now co-educational and there were nearly two thousand graduate students, so that the combination of privacy and intimacy which I had treasured were gone. Taken as a whole, however, the environment was still remarkable compared to that on most other campuses.

One incident remains particularly fixed in my mind. There was some student agitation to drop military training. The board agreed to review the matter at a meeting in January of 1971. I was "kidnapped" on my way to Nassau Hall by a group of student activists and held with two other Board

Members just outside the hall within a circle of protestors. Their purpose apparently was to prevent the Board from having a quorum until the activist students could have the assurance their views were represented. I noted that our captors took most of their instructions from a rather scruffy middle-aged man standing near the front entrance to Nassau Hall. He looked like neither student nor faculty, and I assume he was a member of some national left-wing organization that had succeeded in forming a cell within the student group. The great mass of students on campus were going about their own affairs as if our circle did not exist.

The morning was bitter cold and a very distinguished fellow captive, Livingston T. Merchant, who had had open heart surgery during the previous autumn, began to show signs of acute suffering after fifteen minutes or so. He said to the group, "I am suffering from the cold as a result of recent surgery. I beg you as Princeton ladies and gentlemen to allow me to go inside where it is warm." The male captors seemed willing to relent in his case, but the women objected, making their comments in gutter language that indicated they really were not ladies. Fortunately, a young, newly elected member of the board, then a graduate at Yale Law School, came out of Nassau Hall and assured the group that their cause would get a fair hearing. With that the protestors allowed us to go to the meeting, clearing the move with their colleague on the front steps.

After an hour or two of discussion, the Board agreed to retain military training on a continuing optional basis. When the news was broken to the student activists, they gathered at a window and screamed invectives. Female voices were prominent.

The picketing and posting of handbills that I witnessed on The Rockefeller University campus at the same time seemed mild in comparison.

One of the consequences of my membership on the Princeton Board was a request to serve as Chairman of the Board of the Woodrow Wilson National Fellowship Foundation, a private, not-for-profit foundation, headquartered in Princeton, which provides fellowships to advanced students with funds derived from philanthropic sources. As mentioned earlier, public concern about some of the violence associated with student activism made that a particularly difficult period in which to obtain such funds, although we did manage to sustain the organization for the arrival of the better times which lay ahead.

The Board of the fellowship foundation is composed of a mixture of individuals from both the academic and business worlds. At one of the meetings, a discussion of the ethical practices of private business organizations emerged — a common topic at the time. One of the business members raised, as a question, which features of private business were looked upon as the most questionable. An academic member promptly stated: "The profit motive!" Since the individual who responded came from a private university which thrives on the profits of business in one way or another, it became clear that naivete can intrude on academic wisdom.

Shortly after I moved to New York City, the head of the American Institute of Physics, William H. Koch, asked if I could take the time to give him some help. He was having trouble with the heads of some of the member societies and thought it could be useful if I were to meet with them. As a past Chairman of the Governing Board, I might have enough influence to reassure them of the great advantages of membership in the Institute. These were, of course, times that tried men's souls — everyone seemed to be looking for something to complain about.

I was delighted to accept and met with the society representatives several times, but only seemed to make matters worse. Richard Crane, who was then Chairman of the Governing Board, attended one of these meetings and resolved the problem in a most remarkable way, demonstrating his genius once again in the process. My strategy of describing the privileges the group enjoyed produced no positive effect. Crane then rose and said clearly (with tongue in cheek I am sure), that he and the staff had been guilty of gross incompetence and mismanagement, and that he was grateful to those at the meeting for making their point so clearly. To my surprise, the bickering stopped and the group got on at once with some constructive work. The staff of the Institute, who had been doing an excellent job all along, carried on exactly as it had in the past without further problems. I still am not certain what I should have learned from this experience. I knew the value of true confession, but it took great genius to realize that false confession also has its place. Perhaps Crane took the stance that one might in a poker game and "called their bluff."

In 1971, the Tinker Foundation, headed by Martha Muse, decided to sponsor a series of meetings in Latin American countries so that university presidents from north and south could discuss common problems. Wives were also invited to participate. Grayson L. Kirk, the President of Columbia University, was Chairman of the Board of the Tinker Foundation and assisted in the planning, which was directed by Kenneth Holland, the President of the Institute of International Education in New York City. Betty and I attended meetings in Peru, Brazil, and Guatemala. I ceased to be invited to the meetings after retiring as President of the University in 1978.

Although Betty and I had previously been to meetings in Mexico City, this deeper experience in Latin America opened up entirely new vistas for us. It was noteworthy that while the Latin American countries all had significant bilateral relationships with other countries, particularly the United States and France, their academic leaders rarely met together in Latin America itself. In this respect, the work of the Tinker Foundation was significantly catalytic. In the 1980s, a French-speaking Canadian group developed a somewhat similar program with its own funding and invited Betty and me to attend one of its meetings in Argentina. It has continued over a longer period and has been very successful.

In Peru, we were graciously received by the Beltran family, which had lived in the country since colonial days. The family had previously owned *La Prensa*, Lima's principal newspaper, but it had recently been confiscated by the new, dictatorial government. David Rockefeller was a good friend of the Beltrans and provided a letter of invitation which made it possible for us to see their magnificent colonial-period home in downtown Lima. Mrs. Beltran had been born in the United States and had attended Stanford University. Mr. Beltran commented, with what I was inclined to regard as questionable pride, that it had been a tradition in the Beltran family to speak only French when at home.

The Beltrans were forced to leave the country as the revolution became more violent. The revolutionary government eventually bulldozed their home as an "obstacle to progress," instead of treating it as the national treasure it was.

In 1972, President Nixon signed the National Cancer Act, which had been promoted by a group of private citizens and public figures led by Benno C. Schmidt, who was very active in supporting cancer research at Memorial Sloan-Kettering Cancer Center in New York City. The bill, in addition to providing funds for cancer research and treatment, created both a Cancer Advisory Board and a Cancer Panel. The board was part of the structure of the National Institutes of Health while the panel reported directly to the President. I served on the Advisory Committee for ten years with one brief hiatus. Benno Schmidt was the first Chairman of the Board and provided remarkable leadership during his period of service. Schmidt proposed to President Reagan that I be elevated to serve as Chairman during the early 1980s; however, the appointment went to Armand Hammer, who had been a friend of the President for many years.

The cancer program was extraordinarily successful in that it provided funding and expanded facilities for both research and treatment. More recently, however, the emergence of the AIDS epidemic has shifted public attention to problems associated with that disease, even though cancer remains much more prevalent among the general population.

I had been a member of the Board of the Rockefeller Foundation since the mid-1960s and especially appreciated the privilege of working on that board with J. George Harrar (Figure 13.22), the President. During World War II, he had initiated in Mexico what came to be called the Green Revolution; he was a highly intelligent, hands-on scientist. Our relationship became especially close after I moved to New York. John D. Rockefeller III was the Chairman of the Board until his tragic death in an automobile accident in 1978.

George Harrar felt strongly that it would be highly desirable to create an archive center near Pocantico, the Rockefeller family estate that would be open to the public at some time in the future. The goal was not to glorify the family, but to preserve important records and make them available to scholars. There was another estate, which had belonged to the late second wife of

John D. Rockefeller II, and which was located across the road from the Pocantico Estate, that was selected as the ideal location for the archives. Harrar felt that the center should be managed by the University as part of its permanent structure, but should have its own board of trustees, suitably chosen to serve the specific needs of an archive center.

The plan faced a few hurdles at first, as some family members were concerned that they might be subject to undue public comment when it was announced. Fortunately the matter was resolved, after suitable discussion, and the center has fulfilled its purpose well since it was opened in 1973. I served on its Board until my own retirement as President of the University in 1978.

In 1973, the Shah of Iran visited our campus at the invitation of David Rockefeller. The shah was fascinated, and asked if our staff would consider establishing a companion institution in Shiraz, Iran, in honor of his wife. That city was free of the turmoil of Tehran, and already had a good medical school. The price of petroleum had just been increased by a large factor and his country was feeling relatively wealthy at the time.

I had had the privilege of meeting the shah before, in the mid-1960s. During a visit to India, I was invited to stop over in Tehran by Dr. Saleh, the shah's personal physician, who was planning to establish an academy of science in Iran. At the time of that earlier visit, the country was colorful but backward. The best hotel in Tehran was a pre-war structure, reminiscent of buildings found in small U.S. cities in the 1920s. Even then, however, a new, modern hotel was under construction two miles or so away across the arid, desert-like landscape I could see from my hotel window. By 1974, when we started our planning, the new hotel was crowded with visitors and the arid landscape I had seen was brimming with new buildings. The traffic jams in the streets were monumental.

A group of colleagues from Rockefeller University spent a very exciting two-year period visiting a number of cities in Iran, and made many friends, both professional and personal. Shiraz was a particularly attractive community, somewhat off the beaten track but close to Persepolis where the great Persian ceremonies had been held dating from the time of Cyrus and his followers. Persepolis had been revived as a modern cultural and ceremonial center by the new Shah. In the course of our travels, we met with the shah several times to review our progress. The principal responsibility for providing us with guidance and help was assigned to Minister of the Interior, Jamshid Amouzegar, an exceptionally capable and cooperative individual who also managed many other major national programs.

During the course of our studies in Iran, we had a privileged session with Prime Minister Amir-Abbas Hoveida, the powerful right-hand to the Shah. For reasons of his own, he seemed somewhat dubious about our mission but was nevertheless quite prepared to carry out the Shah's instructions. Hoveida came to an unfortunate and untimely end during the revolution; he was executed by revolutionaries after a sham trial. Amouzegar was spared and moved to the United States.

We also had the pleasure of meeting David E. Lilienthal, who had been the first commissioner of the U.S. Atomic Energy Commission, and before that, the head of the Tennessee Valley Authority. He had established an international consulting service and was very busy in Tehran helping to build up the infrastructure of the country, particularly the hydrodams and power stations. In the United States, the liberal press was doing everything it could to paint the Shah and his colleagues as villains, but the activities we witnessed were all highly constructive and progressive. James Linen, who was a good friend of the Shah, was making frequent visits to Iran at this time. He was most helpful in establishing our credibility.

I recall one unforgettable tour over the Elburz Mountains to the north of Tehran and down to the shores of the Caspian Sea, well below sea level. At that time, the Soviet naval fleet did not allow the Iranians to use the inland sea. Fortunately, the roe-bearing sturgeon spawned in the rivers of the Elburz, so the supply of Iranian caviar was not obstructed by the Soviet action.

Incidentally, in ancient times the Elburz mountains were the site of the legendary Citadel of the Assassins — a fortress from which operated a mafia-like organization of murderers-for-hire. The killers were highly trained professionals who were specially recruited for their work. They were usually paid in part with hashish — in fact, the word "assassin" apparently has its roots in the word "hashish." The final part of the legend has it that in the thirteenth century Genghis Khan, fearing for his own life, ordered his army to destroy the citadel and its inhabitants, bringing its activities to an end.

We prepared and submitted a detailed report of our experiences in Iran and our recommendations for the new institute. The report is now in the Archives of Rockefeller University. Among our other recommendations we had nominated Dr. Faramarz Ismail-Beigi, a young medical-biochemist at the Medical School at Shiraz as director. As of this writing, he is at Columbia University in New York City.

Of course, all of our plans collapsed almost overnight as the Shah's fortunes turned and a reactionary revolution swept the country. The revolutionaries opened the jails, which were assumed to be full of political prisoners, but found only a few criminals. Most of the charges of human-rights violations that had been made against the Shah had been manufactured, although the myth haunted him thereafter and made his life especially difficult when be became very ill while in exile. The students at the University of Tehran, which we visited quietly several times, were led by a revolutionary group whose members appeared regularly on television during the uprising. One wonders how many of them survived the later war with Iraq.

Having a long-standing interest in Persian rugs, Betty and I had become acquainted with some of the dealers in Tehran. We were pleased and relieved to meet some of them several years later in New York where they fled for safety and reopened their businesses.

In 1974, when Gerald Ford took over as President after Richard Nixon's resignation, and Nelson A. Rockefeller had been named Vice-President, a new Presidential science advisory structure was created. It consisted of two committees, one devoted to basic and the other to applied research. William O. Baker chaired the former committee; Simon Ramo the latter. The atmosphere was quite different from that which had prevailed in the previous era. The meetings were open to the public, and the press was made welcome, unless matters related to a proposed executive budget were under discussion. I served on the Baker committee, but the two committees met together frequently. We saw President Ford only rarely, but Vice-President Rockefeller attended regularly, particularly when major decisions were being made or reviewed. He seemed to prefer getting his information from verbal discussions rather than from written reports; apparently, he suffered from a form of dyslexia that made reading something of a trial.

When the Carter Administration came to the White House, an Office of Science and Technology Policy was created by a Congressional act and the President's Science Advisor became available to Congress for discussions of activities that did not infringe upon privileged work for the President. Frank Press (Figure 14.10) became President Carter's Science Advisor.

When the drive on the east side of Manhattan Island named in honor of Franklin Delano Roosevelt was built in the 1930s, we agreed to allow the University-owned strip of land along the river to become public domain.

Figure 14.10. Frank Press, the distinguished geophysicist who served as Science Advisor to President Carter (1977–81) and as President of the National Academy of Sciences (1981–93). (Courtesy National Academy of Sciences and Ivan Massar of Black Star.)

As we began to plan for the possibility of erecting new buildings on campus, it became evident that we would benefit substantially if we could gain access to air rights over the drive. We applied for those rights to the New York City Council, following city real estate practices, and with the advice and support of our Board of Trustees, particularly Thomas G. Cousins. The New York Hospital and the Hospital for Special Surgery, our neighbors to the north, joined us in order to share in the air rights on the land along the river next to their property.

While we anticipated the inevitable delays associated with any such request, we were somewhat surprised at the amount of random opposition we encountered from groups of private citizens, many from areas well outside our district. It seems that many citizens in a large city have little to do but express some form of indignation about anything that appears on the horizon. Their reaction is probably born out of a combination of boredom and frustration. Their complaints become magnified by the local media, which delight in publicizing their causes and inevitably bring in more opponents of the same type.

Fortunately the City Council saw the benefits to be derived from granting our request, since it would obviously add to overall institutional strength. As a result they approved our application in 1973, with the proviso that any buildings constructed in that air-space should be erected within twenty years. In the interval Rockefeller University constructed two buildings, one an apartment house and the other a research laboratory, the latter supported in part by the Howard Hughes Medical Institute. New York Hospital also made use of the rights.

The University celebrated its seventy fifth anniversary in 1976, the same year that marked the bicentennial of the Declaration of Independence. Dinners and symposia were held, and the University staff was joined by many guests, including a number of scientists and physicians who had spent a period at the University.

In the early 1920s, John D. Rockefeller II, who was then chairman of our board, purchased a copy of a painting by the distinguished French artist, David, showing Lavoisier and his wife in their laboratory. The painting hung in our library and attracted many visitors. It was lent to the National Gallery in Washington to be part of an exhibit of artwork and artifacts associated with the theme, "The Age of Jefferson." Because of this loan, Betty and I were invited to be honored guests at all the functions and festivities during the bicentennial celebrations in Washington. This experience left us with a lifetime of rich memories. After the revels had concluded, the University's Board of Trustees agreed to sell the David painting, one of his masterpieces, to the Metropolitan Museum of Art. Once again, it occupies a special setting, and it is now both more secure and more easily available to the public. The proceeds of the sale were used to endow two professorships bearing John D. Rockefeller, Jr.'s name.

During the autumn of the same year, a Japanese society named after a distinguished microbiologist, Hideyo Noguchi,[3] who had spent his years of creative research at the Rockefeller Institute between its early days and his death in 1928, invited Betty and me for a ten-day celebration of the centennial of his birth. Among other things, he was the first investigator to demonstrate the existence of live spirochetes in the brain tissue of victims of a late-stage syphilis infection. The celebration took place partly in Tokyo and partly in Noguchi's family home at Inawashiro, north of Nikko. The importance of the celebration was indicated by the presence of members of the royal family at the ceremonies in Tokyo.

In the mid-1970s, the Department of State asked if I would be willing to serve as the United States Representative on a United Nations Committee for Science and Technology for Development. Many less-developed countries hoped that the United Nations would create a large fund, on the order of two billion dollars or more, derived from the wealthier nations, to help the so-called Third World countries accelerate their rate of social, economic, and industrial development. It was anticipated that the work of our committee would eventually lead to a global U.N. conference devoted to this subject. I agreed with the stipulation that Rodney Nichols be allowed to serve as my alternate in the event that my schedule at the University prevented me from participating in a given meeting or event.

As anticipated, the Committee soon focussed on planning a great meeting of representatives of member nations in Vienna in mid-summer of 1979. Special preparatory meetings were held in many countries.

As the year of the meeting approached, President Carter, who had taken office in 1977, decided, quite appropriately, that he preferred to have an individual who had previously worked with him head the U.S. Delegation to Vienna. He selected the Reverend Theodore M. Hesburgh (Figure 14.11), whom I had known since my days at the University of Illinois when we served together on a number of review committees, including one for Argonne National Laboratory. In fact, he had given the invocation at the ceremony commemorating the centennial of the National Academy of Sciences (Figure 13.3), as well as at my installation as President at Rockefeller University. I remained a member of the large U.S. delegation. Thomas R. Pickering, then the Assistant Secretary for the Bureau of Oceans and International Environment and Scientific Affairs of the State Department, was designated Co-Chairman.

Once I arrived in Vienna I spent most of my working time in the conference hall listening to the presentations of various national groups; I was mainly interested in the overall purpose of the meeting. There were many

---

[3] A detailed biography of Noguchi was written by Ms. Isabel R. Plesset (Farleigh Dickenson Press, Rutherford, Madison, Teaneck, 1980).

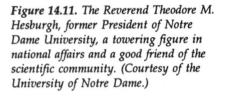

Figure 14.11. The Reverend Theodore M. Hesburgh, former President of Notre Dame University, a towering figure in national affairs and a good friend of the scientific community. (Courtesy of the University of Notre Dame.)

other events that would have been diverting, such as press conferences and sub-committee meetings on special topics, but I had relatively little involvement in them. It soon became obvious that science, technology, and development issues had become eclipsed by more political matters. In brief, the international political corps of the various nations took advantage of the gathering, since travel funds were readily available, to raise many other questions, and they succeeded in pushing science and technology into a corner. To the best of my memory, only one nation, South Korea, devoted its presentation specifically to the ways in which science and technology might be used to promote development, citing its own significant experience. Taiwan could have provided a similar chapter from its own recent history, but it was no longer a member of the United Nations, having been lately ejected in favor of the mainland.

The Conference ended with the recommendation that a multi-billion dollar fund be created by developed countries to help promote economic growth through the application of science and technology in the developing nations. The money actually offered by the wealthier countries was far less than the suggested amount, and the subject remained confused for the next decade.

The "politization" of the meeting came as no great surprise to those of us who had observed the evolution of the proceedings over several years. We became convinced that much more direct interaction between scientists and technologists in the developed and the developing countries was needed. With the help of Rodney Nichols and a few individuals such as Roger Revelle and Carl Djerassi, we formed a private group, which met under the informal sponsorship of the National Academy of Sciences, to draft our own proposal. We recommended that President Carter create a new agency which would provide help to developing nations on a bilateral basis. One developed country would be paired with a developing nation, and the program would emphasize personal interactions between scientists and engineers in the two countries involved. The new U.S. agency would have a staff comprised principally of experts in specific areas of technical development who were familiar with the problems of developing countries. As we envisioned it, it would not interfere with the work of the Agency for International Development, which has broader goals and provides funding at the upper levels of government for use in governmental programs.

Frank Press (Figure 14.10), then the President's Science Advisor, was enthusiastic about the proposal and enlisted the cooperation of President Carter, who gave a speech endorsing the plan while visiting Venezuela. Unfortunately a bill that originated in the White House proposing the creation of the agency was defeated by the Congress. Subsequently, the President prepared an order requesting that the sum of ten million dollars be awarded to the Agency for International Development (AID), an already existing agency, for use by the National Academy of Sciences in support of our program.

Initially, the staff of AID was enthusiastic about the opportunity to cooperate with the scientists, engineers, and other individuals that the Academy could provide, and the program made a fine start, working with carefully selected countries in Latin America, Asia, and Africa. Staff changes at AID, however, brought in new individuals who resented the transfer of what they came to regard as their own funds. The program was dropped in 1988 after demonstrating that the basic concept was sound. I served as the chairman of the advisory board during the life of the program. The Executive Officer for the committee, Dr. Michael Greene, a member of the Academy's staff, was a superb leader.

During interludes in which the United States and the Soviet Union had good relations, Rockefeller University provided a fine haven for delegations from the Soviet Academy of Sciences. The visitors enjoyed being in New York City, and found our campus one where they could relax in peace without having to endure the noise and heckling that was apt to occur on some university campuses.

Professor Peter Kapitsa (Figure 14.12), who was kept a virtual prisoner in the Soviet Union, was allowed a trip to the United States in 1969 and made our campus his home for part of his visit. Betty and I became good friends with both Professor and Mrs. Kapitsa, an engaging, friendly woman with whom we exchanged Christmas cards for many years thereafter. We took them to the Metropolitan Opera and had several enlightening discussions about Russian opera. Through the Kapitsas we became familiar with Mussorgsky's "Khovanshchina" which I had not previously known. They

Figure 14.12. Dr. Peter L. Kapitsa, a great scientist and patriotic standard bearer of humane principles in the Soviet family of scientists. This photograph was taken during his sojourn at Cambridge University in England in the 1920s and 1930s. (AIP Niels Bohr Library. Gift of E.T.S. Walton.)

regarded it as the greatest of Russian operas—an honor I had always assigned to "Boris Godunov." Khovanshchina has since become a fairly standard offering at the Met.

Somewhat later, President M. V. Keldysh and a group of his colleagues from the Soviet Academy used our campus as their base while visiting the eastern United States. We had a number of interesting meetings with them and earned their special thanks by taking them to a good fish restaurant instead of to the type of steak roast to which they were treated endlessly in other parts of the country. A piquant element was added to their journey when one of the KGB agents who tagged along to keep them honest vanished for a few days, having met an interesting young lady along the way.

Still later, Professor N. N. Bogoliubov and his wife spent several days with us while preparing for an extended visit to a number of university campuses. Unfortunately their trip came at a time when their government was bearing down strongly on dissidents. This action stirred up political and social groups on many U.S. university campuses, and the Bogoliubovs became targets of criticism and pressure. As a result, they cut short their other visits and returned to the friendlier atmosphere of our campus, where they remained quietly for a month. Bogoliubov said to me somewhat wryly, "If I went back home and said to my government all of the things that the people at the other universities asked me to say, I would promptly be thrown in jail."

The Rockefeller Foundation, assisted by the University staff, had created what was the Peiping Union Medical School in Beijing during the World War I era. Once mainland China opened up after President Nixon's visit there in 1972, there was considerable interest on our campus in visiting the mainland and seeing what had transpired in the past fifty-odd years. Betty, having spent most of her childhood and youth in China, was particularly anxious to go—it would be her first visit there since she left to come to the United States for college.

Rodney Nichols and I visited the Mainland Embassy in Washington several times and managed to arrange a trip for ten faculty members and their wives in May–June of 1977 with the help of a special study grant from the Milbank Memorial Fund in New York. The V.I.P. visit, which was hosted by the distinguished physicist Chou Pei-yuan of the University of Beijing, and which extended over three weeks, included stops in Beijing, Nanking, Suchou, Shanghai, and Guangzhou. We were greeted everywhere by scientific and political officials who made it clear that the winds were changing; with the death of Chairman Mao in the previous year, the "Cultural Revolution" which had wreaked so much havoc on the educational system was at an end. A report is in the University Archives.

In Nanking we spent half a day with guides admiring the great bridge across the Yangtze River. The span was to have been built by the Russians using a special alloy. The Soviet Union abandoned the plan, however, during a period of tension between the two countries. The Chinese picked up the project, redeveloped the alloy, and completed the bridge, very proud of their accomplishment.

The scientists we met along the way were friendly but apprehensive, having recently endured an extremely difficult period of repression engineered by the "Gang of Four," who in turn were led by Chairman Mao's wife. I had the privilege of seeing my former student, Y.Y. Li, who was at the impoverished Physics Institute of the Academia Sinica in Peking.

As was expected, we were taken to many hospitals and observed a great deal of surgery, some of which employed acupuncture. In Shanghai, we also met Dr. Chen-yu Yang, now in the U.S., the biochemist sister of C.N. Yang, the distinguished physicist and Rockefeller University Trustee.

We had been travelling primarily by air, but the last stage of our journey took us from Guangzhou to Hong Kong by rail, through the lands Betty had known as a child. Her father had been buried in what was then called Canton, but the graves had been desecrated, and we made no attempt to find the site of his former cemetery. Most of the hotels we stayed in were of an older vintage but more than adequate for our needs. Since that time China has greatly expanded its hotel system.

The friendly hospitality we received each day had a remarkable effect; on leaving China, we felt none of the sense of release from constraint that one normally experienced leaving the Communist countries of eastern Europe.

There is one incident that bears repeating. The Chinese had established what were called "friendship stores," where one could purchase special items for hard currency that were not available to the general public. We visited these stores in each city, in part just to see what was available. The service at friendship stores tended to be rather poor. The staff received no special benefits from their sales; as a result, they preferred to spend their time in conversation. In fact, they behaved like a privileged group of party members.

On reaching the top floor of the friendship store in Shanghai just ahead of Betty, I saw several glass cases containing attractive objects; all of which were small and elegant. As I started to browse in one of the cases, a hand reached in, took out a tray, placed it before me and pointed out several items, while a voice observed, "These would make fine gifts for you." Betty quickly joined me. It turned out that the man behind the counter had owned the principal jewelry store in Shanghai before the revolution and had stayed on because his children thought there was great promise in the Communist system. He had been sent to an agricultural commune to do manual labor but was called back when the government decided to establish the friendship stores early in the 1970s and requested his help setting them up. He had carried out his task very effectively and was then permitted to serve as a salesman at some of the counters on the top floor. His attitude was a refreshing change from that of his colleagues, as he actually preferred selling to chatting with other employees.

He gave us directions for finding Betty's old prep school. We found on visiting that it was now a military office.

At the start of my period as President, the then Board of Trustees had decided to establish a retirement age of 65 for the office with the understanding that the actual date would be determined by their success in finding a replacement. Actually, I stayed on for a full ten years until 1978 when Dr. Joshua Lederberg accepted the appointment. Through the generosity of the Board and the special kindness of Dr. Attallah Kappas and his successors, I was provided with a wonderful haven in the University Hospital which made it possible to continue a productive professional life among much admired friends.

Chapter Fifteen
1978 and Beyond

Joseph B. Platt, an old friend who had been a student at the University of Rochester and was now President of the Claremont University Center in Southern California, had invited me to become a resident faculty member upon my retirement from Rockefeller University. The prospect was very attractive, not least because it would have allowed me to renew my association with Vladimir Rojansky. I soon found, however, that many friends on the East Coast to whom I was indebted for various kinds of help were now asking in turn for my help in ways which made it much more sensible to remain professionally on the East Coast and near New York City for an extended period. The university generously allowed us to occupy one of the University-owned apartments as required and Dr. Attallah Kappas provided me with access to a fine suite of offices in the University Hospital, complete with part-time use of a conference room. The following years were full and satisfying beyond anything we imagined. I was soon involved in a number of activities and as busy as ever. Furthermore, the National Academy's Grants Program continued well into the next decade and provided an initial means of establishing an ongoing relationship with that institution; others soon developed.

It was my good fortune, at this major transition point in my life, to receive a request from Ralph E. Ablon (Figure 15.1), a friendly and dedicated member of the Rockefeller Board of Trustees. He asked if I would be willing to join the Board of the Ogden Corporation, of which he was then the Chief Executive Officer (since succeeded by his son Richard Ablon), and help him select other "outside" directors as recommended by the Securities Exchange Commission. The Corporation had been created in the early 1960s through the consolidation of a number of previously quasi-independent companies. The time had come to reorganize the Board, and Ablon wanted to include a number of outside directors.

This offer, which I was pleased to accept, opened new vistas to me by introducing me to many aspects of American business that were relatively unfamiliar to me. It also brought me an interesting new circle of friends. At the time I joined the Board, most of the Corporation's enterprises were closely linked to manufacturing of one kind or another. Mr. Ablon, with characteristic foresight, saw the difficulties manufacturing industries would face in the years ahead and skillfully shifted most of Ogden's enterprises to areas of the service sector which remained relatively lively.

345

Figure 15.1. Ralph E. Ablon, a strong supporter of the University and a valued personal friend. (Courtesy of the Ogden Corporation.)

An important addition to the business of the company came with the licensing of the Martin technology, a waste-to-energy system that had been developed in Europe and was widely employed there and elsewhere. In the following years, the licensing led to the creation of a useful and profitable subsidiary. Having become familiar with the various advantages of such systems, I continue to be shocked at how the City of New York still piles its garbage on Staten Island or elsewhere when it could generate a significant fraction of its electrical power through safe and efficient burning. It is very difficult to believe that the population could imagine that accumulating a rotting pile of garbage with all its effluents is in any way healthier than having it burned in a sanitary unit that is carefully designed to produce safe, well-controlled emissions, in addition to useful steam and electricity.

Although Philip Handler had his own personal agenda for the National Academy when he first took over as President, his mood changed as time went on and we became fairly close friends during the second half of his twelve years in office. Many factors affected our relationship, but I believe the most important one was that his realization that we both had worked toward common goals at the Academy.

He had served on the board of a small foundation, the Lounsbery Foundation, for several years and had taken much pleasure in the work. When a vacancy opened on the board, he recommended that I be asked to fill it. This led to a long, interesting association with the Foundation's President, Alan F. McHenry (Figure 15.2). I enjoyed working with him on many issues of mutual concern. The other members of the small board were William J. McGill, a past president of Columbia University; Lewis Thomas, the fa-

Figure 15.2. Alan F. McHenry, the president of the Lounsbery Foundation. A wise and experienced individual.

mous biomedical investigator, statesman, and author; and Benjamin F. Borden, a lawyer and good friend of Mr. McHenry.

In the wake of the suppression of the Boxer Rebellion of 1900 in Peking in which the Chinese attempted to rid themselves of the oppressive burden of colonial powers, the latter exacted an indemnity of about five hundred million U.S. dollars from China as restitution for damage caused by the uprising. Most of the countries in question used that money for their own purposes. After some inevitable debate, the U. S. government, however, returned its share to China following World War I, with the understanding that it would be used for educational purposes. The China Foundation was established; its expenditures were to be approved by a Board of Trustees composed of both Chinese and United States citizens. Until World War II it played a significant role in permitting Chinese students to go abroad for advanced education.

Most of the endowment was unfortunately invested in Chinese government bonds, and was eventually lost in the inflation that accompanied the Japanese War. A fraction of the money, however, was invested in the United States and was for the most part preserved, so that the foundation, based in Taiwan, is still able to serve its purposes.

In the late 1970s I was asked to join the Board and was delighted to accept, partly to serve a useful cause and partly to maintain close relationships with a dedicated group of Chinese and Americans, including Joseph Platt mentioned earlier, who had become admired friends. When I first joined the Board, the chairman was Dr. Shi-liang Chien, an organic chemist who had studied under Roger Adams at the University of Illinois. He had served as President of the National Taiwan University before becoming President of the Academia Sinica and Chairman of the Board of the China Foundation. One of his sons is Frederick Chien, who has served as Minister for Foreign Affairs of the Republic of China.

In his youth, the senior Dr. Chien was nicknamed "the little Confucius" because of his strict adherence to formalities. I experienced his exact sense of discipline firsthand on more than one occasion during our meetings when I proposed slight deviations from the approved agenda.

At this writing, the Chairman of the Board of the China Foundation is the physicist, Dr. Ta-you Wu (Figures 15.3 and 15.4), who served as President of the Academia Sinica for ten years after the death of Dr. Chien in 1983. He completed his graduate work in physics at the University of Michigan and is known as a distinguished professor who inspired many of his students, including T. D. Lee and C. N. Yang, who jointly won the Nobel Prize in 1957. Above all, Dr. Wu has a clear understanding of the value of science for its own sake—a concept that may be at risk in Taiwan if its political and industrial leaders look to Japan as their role model in deciding what kinds of research to support.

When President Carter decided to shift U.S. recognition from Taiwan to the Mainland in 1979, the move generated considerable apprehension in Taiwan since there was understandable fear that our country might desert the Republic of China completely. The earlier failed attempt of the Communists to take over Taiwan by military force made it clear that great danger remained. By that time, I had made many visits to Taiwan, both as a representative of the National Academy of Sciences, and on behalf of Texas In-

Figure 15.3. Dr. Ta-you Wu, the president of the Academia Sinica and the China Foundation. A leading figure in Chinese science. He was also an inspired teacher of several generations of Chinese physicists including Tsung-Tao Lee and Chen-Ning Yang. (Courtesy of the Academia Sinica of the Republic of China. Photograph by Ten Hui-En.)

Figure 15.4. *Betty in Chinese dress with Dr. Wu at a reception in Taipei. The photograph was probably taken in 1988. (Courtesy of the Academia Sinica of the Republic of China.)*

struments, on whose Board I had served since 1970. The company had an excellent manufacturing and assembly plant in Taipei which I had visited every other year or so since joining the Board.

In the course of establishing relationships in Taiwan, Texas Instrument's Chief Executive Officer Patrick Haggerty had come to know the leaders of the government on a first-name basis. He had particularly close relations with Premier Y. S. Sun (Figure 15.5), who was an engineer by profession, and Minister K. T. Li (Figure 15.6), a physicist who had worked with Rutherford in England and had held many important posts in the government of President Chiang Kai-shek, including that of Minister of Economic Affairs. Premier Sun and Minister Li had done much to help the island republic in its nearly miraculous economic and industrial development.

Immediately after President Carter's decision, Haggerty came to me with the suggestion that we form an advisory committee of a dozen or so members from the United States, to continue to provide whatever advisory help we could, and to make it clear that many individuals in our country supported the Republic of China. He proposed that he serve as Chairman of the advisory group and that I should be Vice-Chairman. Minister Li was enthusiastic about the plan and, with Premier Sun's help, created the Science and Technology Advisory Group (STAG) in Taipei, attached to the Premier's office, to serve as a link between the advisory committee and the Premier.

Figure 15.5. K. T. Li and me during a visit to former Premier Y. S. Sun's home during his convalescence from a stroke. The photograph was taken about 1985. (Courtesy of the Republic of China.)

Figure 15.6. Dr. K. T. Li, one of the principal figures in the evolution of the Taiwan miracle. He spent several years in the early 1930s with Ernest Rutherford but left to serve his country in many capacities. (Courtesy of the Republic of China.)

The committee was then named the Science and Technology Advisory Board. Others who joined the Board at that time were Thomas L. Martin, then President of the Illinois Institute of Technology; Ivan L. Bennett, the Dean of the Medical School at New York University; Carl A. Gerstacker of the Dow Chemical Company; B. O. Evans, initially of IBM; and Chauncey Starr of the Energy Producers Research Institute. Our first meeting was held in Taiwan in January of 1980. Many Taiwanese officials, business leaders and academic scientists was present at the plenary sessions, chaired by Premier Sun.

The creation of the Board proved to be much more than an act of friendship. It generated useful advice on many government decisions in the following decade or so and fully justified the inspiration to create it. Perhaps its greatest single contribution came through the work of Ivan Bennett (Figure 15.7), who introduced a program to vaccinate all infants against hepatitis B. The plan caused a great deal of public concern when it was first proposed, but the anxieties subsided once it was understood that vaccination would eventually eliminate the disease, which is widely prevalent in Asia.

In the course of time the Advisory Board became more international with the addition of French, German, Italian, and Japanese, as well as other U.S. members, such as Kenneth G. McKay, formerly of the Bell Laboratories, and Robert L. Sproull, former President of the University of Rochester. Dr. K.T. Li's role as Minister and Convenor of STAG was eventually taken over, in succession, by Drs. Nan H. Kuo and Han M. Hsia. Both had extensive experience in academic and governmental service.

Most unfortunately, Haggerty died of pancreatic cancer in October of

Figure 15.7. Dr. Ivan L. Bennett, a member of our advisory committee. He introduced hepatitis B vaccine to Taiwan and ultimately to the whole of Asia. A very great, persistent biomedical scientist, The Emperor of Japan honored him with one of the highest national awards. (Courtesy of Mrs. Martha Bennett.)

1980, after a very brief illness. His passing was an enormous loss to the organizations with which he was associated. I was asked to take over as Chairman of the Board, and we soldiered on. We continued to follow the initial pattern of semi-annual meetings for over a decade, becoming involved in almost every facet of the scientific, technical, and industrial life of the country, and being aided along the way by excellent members of the STAG staff such as Drs. Chi-wu Wang, Patzen Wu, and Sung-mao Wang. Over the years, Taiwan's economy has expanded and now approaches the level of the most affluent, advanced countries.

In the meantime, Taiwan has been the site of one of the most remarkable political and social shifts in modern times, having made the transition from a tightly controlled dictatorship fending off threats from the Communist Mainland to a prosperous, open democratic society. The key figure in the transition was President Chiang Ching-kuo (Figure 15.8), the politically oriented son of President Chiang Kai-shek. Ching-kuo, following the example of George Washington, decided to terminate the period of rigid martial law required during the early, dangerous phase, and open the doors to a multi-party, democratic system. His initiatives were carried to fulfillment by his successor, President Lee Teng-hui, a native Taiwanese and descendent of some of the earliest Chinese settlers from the Mainland, the Hakka.

Figure 15.8. *A first meeting with President Chiang Ching-kuo, soon after he became President of the Republic of China in 1978, following the footsteps of his father. This remarkable man preferred the virtues of democracy to marshall law and instituted the reforms which have led to a multi-party system in Taiwan. Dr. K. T. Li is in the center. (Courtesy of the Republic of China.)*

Although the dramatic Communist victory in Mainland China in 1949 and the retreat of the Nationalist government to Taiwan appeared at that time to represent a massive defeat for the free-enterprise system in the Chinese world, the reverse may, paradoxically, turn out to be the case. Taiwan, with an initial population of nine million (now about twenty million through natural growth) provided a society of manageable size which, with good leadership and some initial outside help, could move rapidly up the industrial, economic, and social ladder. Within a generation, it provided the leaders in the mainland with a model of success which they could not ignore and which has unquestionably caused them to introduce far more liberal policies. Today Taiwan businessmen are among the major investors in the rapidly growing free enterprise movement on the mainland.

In 1993, K. T. Li, who by this time had gained world-wide fame, was invited by the World Bank to participate in Beijing in a series of general discussions of international business and finance but with special emphasis on the economic situation in Mainland China. The Mainland authorities treated him with the deepest admiration and respect, knowing of the major role he had played in the remarkable series of developments that had taken place in Taiwan. One of his many recommendations was that the Mainland establish a central bank to exert some control over both the banks and the flow of currency—a recommendation which apparently was adopted soon after his visit.

In the final year of my term at Rockefeller University, J. Paul Sticht, a friend of David Rockefeller and the Chief Executive Officer of what was then the R. J. Reynolds Corporation came to visit me with his colleague H. C. Roemer, the company's Legal Advisor. Sticht had decided that his corporation should support a coordinated medical research program in various universities to the tune of several million dollars per year. Certain individual gifts had already been made, but it was thought desirable to establish a special advisory committee to help organize the large program. Sticht asked if I would put together the Advisory Committee which would plan the system of grants.

This stimulating program lasted for nearly ten years, until Sticht's retirement and the financial disturbance caused by a leveraged buyout of the company after his departure brought it to an end. Serving on the committee were Maclyn McCarty, James A. Shannon, Edward J. Jacob, a New York lawyer with an extraordinary understanding of medicine, A. Wallace Hayes, the director of research of the Tobacco Division of the company, H. C. Roemer and several other company officials who served at review meetings in Winston-Salem.

Some of the programs we supported had been initiated somewhat earlier at the suggestion of Jacob. These older projects included studies of the effect of renin on blood pressure, which were carried out at Harvard under the direction of A. Clifford Barger and Edgar Haber; studies of the factors which sustain the normal development of cells, which G. Barry Pierce was directing at the Medical School of the University of Colorado; and Russell Ross's studies of the factors leading to arterial sclerosis at the University of Washington.

Perhaps the most successful of our programs, in the sense of its being seminal work, was the research conducted under the direction of Stanley B. Prusiner (Figure 15.9) at the University of California at San Francisco, which we funded from a very early stage of its development. Prusiner had been attempting to determine the cause of "scrapie," a neurological disease that occurred in sheep. The agent was commonly believed to be an as-yet unidentified virus.

At the time we met Prusiner, he had begun to accumulate evidence suggesting that the agent might be a relatively pure protein, or at least an entity having very little DNA or RNA. Because of its apparently singular nature, Prusiner termed it a Prion. The suggestion that a protein could be an infectious agent ran so counter to the conventional wisdom of the time that Prusiner would have had great difficulty getting adequate support for his research through the usual channels of the National Institutes of Health. The Reynolds grants, and subsequent support through the Sherman Fairchild Foundation, were essential in enabling him to carry his work through to the point where its extraordinary features and results could be appreciated.

The board spent two years observing a program supported by the Reynolds Grants that examined the possible influence of psychological stress upon the activity of the immune system, an issue which has since been the

Figure 15.9. Dr. Stanley B. Prusiner, a highly productive recipient of R.J. Reynolds Grants.

object of frequent discussion. A great deal of the work was done with mice and involved examinations of immune system cells. By the time the Grant Program ended and the work was terminated, its results were still inconclusive, showing a wide statistical spread.

While at the National Academy of Sciences, and later at Rockefeller University, I followed with great interest the course of biological research, and was impressed with the remarkable ingenuity of the investigators exploring the inner workings of the biological cell. Such research makes it evident that these systems are enormously complex; moreover, the level of complexity has tended to increase rather than decrease during successive stages of investigation. For example, soon after the basic genetic code was first discovered, in all its elegance, it became clear that the form in which important genetic messages are coded by the nuclear DNA is quite intricate; and its seemingly redundant sequences and other features are in apparent violation of any accepted rules of economy of structure. Similarly, the facts disclosed regarding the constitution and operation of various components of the biological system, such as those that control our immune response, or the growth of cells, are found to be more and more complex as research proceeds.

This evidence of almost unlimited complexity leaves one wondering if the process of reductionism which has played such an important role in the physical and engineering sciences can ever be applied successfully to gain anything approaching integrated, or holistic, understanding of biological systems.

Once having relinquished my duties as President of Rockefeller, I thought I would attempt to gain a better understanding of this matter and began reading everything I could find on the subject. In the course of my researches, I encountered the papers of Professor Walter M. Elsasser (Figure 15.10), then at the University of Maryland. He had given considerable thought to the nature of biological systems and to the significance of their inherent complexity. I had first met him soon after World War II, when he was involved in research on the source of the earth's magnetism.

While engaged in geophysical studies, Elsasser was inspired to speculate on the place of biological systems in the physical world, and he drew several major conclusions: The immense complexity found in living systems must be seen as fundamental to their existence. At the same time, he pointed out, this organization of complexity occurs under highly improbable circumstances. The properties of individuality and reproducibility that we associate with systems are also intimately linked to this complexity. The structure of such systems is not to be regarded as any more mysterious than that of any other part of the natural world, but it has its own unique integrity. Underlying all of this is the knowledge that we cannot hope to understand biological systems in the holistic sense using the methods of analysis nor-

Figure 15.10. Dr. Walter M. Elsasser, a remarkably wide-ranging physicist. He had the courage to ponder the status of living systems in the physical world. (AIP Emilio Segrè Visual Archives.)

mally applied to physical systems. Their complexity, improbability, and capacity to incorporate individuality preclude the use of normal methods of analysis.

Elsasser and I began corresponding, inspired by our mutual interest in this topic. His work in the field had been ignored to such a degree that, when we first started writing, I found him quite discouraged about carrying it any further. His interest was soon rekindled, however, as a result of our exchange and he went on to produce several eloquently written booklets that employ modes of thought and examples from his own experience. Since my own background was somewhat different, I decided to write an essay on the topic from my personal viewpoint. This piece appears in a small book, *The Science Matrix*, under the title "A Physicist's View of Living Systems" (See footnote to Chapter 16).

My friendship with Elsasser, which developed out of this association, proved to be a very rewarding one. He was one of the early pioneers in the field of wave mechanics, having studied at Munich with Arnold Sommerfeld and then at Goettingen with Max Born. He left Germany for France in 1933 and came to the United States in 1936. An excellent theoretical physicist, he also had interests in a wide range of other subjects. He suffered, however, from deep, complex inner tensions. As a result, he became something of a roving investigator, holding some ten different scientific posts in the course of his American career. I was particularly honored when he asked me to give the commendatory speech at a dinner given for him in Baltimore on the occasion of his eightieth birthday in 1984. In one of his last letters to me, he expressed his gratitude for the recognition I gave to his work in biology in "A Physicist's View of Living Systems."

In the early 1960s, while serving at the National Academy of Sciences, I encountered a remarkable French medical scientist, Professor Maurice Marois (Figure 15.11), a histologist and member of the College de France VI in Paris. He was very concerned by the fact that many scientists, disturbed by the potential horrors of nuclear weapons, were focussing their attention on matters related to war and arms control. He felt that this concern should be counterbalanced by paying comparable attention to the precious features of life in all its forms, focussing on its preservation and enhancement, the prevention and alleviation of human suffering, and the expansion of human knowledge and wisdom through science. To that end, Marois had founded an organization, the Institut de la Vie, which sponsored many highly constructive meetings on such subjects. In the process he gained a large and enthusiastic following within the scientific community, not a few of his supporters coming from biology and medicine.

In the early days of the Institute, Marois invited me to cooperate with him as an advisor or member of one of his boards. Unfortunately I was much too deeply involved, first with the Academy and then with Rockefeller University, to accept his invitation, in spite of my sympathy for his cause. I did, however, follow his work as an admiring observer. His conferences, all well-organized and well-attended, dealt with such diverse topics as the treatment and prevention of genetic defects; the preservation of endangered species; the links between physics and biology; the care of the autistic child; scientific approaches to the world-wide epidemic of AIDS; the affects of diet on health; and the interrelation of man and the biosphere.

I was able to cooperate with Marois once I was freed from university administration, and did my best to support his mission in the United States

Figure 15.11. Professor Maurice Marois, the founder of the Institut de la Vie. (Photograph by his son, Jean-Pierre Marois.)

and abroad. After some thirty years of devoted service, one of his fondest desires has been that the work of the Institute should continue beyond his own life and career, focussing its activities upon the very broad topic of "Human Destiny." To do this would require the creation of a study and conference center, preferably in the vicinity of a university, possible in the United States, with a reasonably stable financial base. In spite of the outstanding quality of the Institute, the support needed to achieve this longer-range goal still remains to be found.

Early in the decade, President Margaret Mahoney of the Commonwealth Fund provided support for a series of books which would introduce the interested reading public to areas of frontier science. Drs. Lewis Thomas and Alexander G. Bearn directed the program with the help of an advisory committee on which I had the privilege of serving. The book on astronomy by Herbert Friedman and that on DNA by Maclyn McCarty listed as footnotes in Chapters 13 and 14, respectively, were published in the series.

Also during this period, Betty and I had the pleasure of participating in several summer sessions at the Ettore Majorana Centre for Scientific Culture at Erice in Sicily, organized under the imaginative directorship of Antonino Zichichi and his able staff. The meetings were much enriched by the presence of individuals such as Dirac, Teller, Wigner, and others, including colleagues from the Soviet Union. As was mentioned in Chapter 5, Peter Kapitsa was permitted to leave the Soviet Union to participate in one of the sessions. I returned there in May of 1993 for a special conference, led by Pope John Paul II, giving recognition to the importance of science for human progress.

Soon after the end of World War II, the United States government established research centers in both Hiroshima and Nagasaki, to study the effects of atomic radiation upon the survivors in both cities. The twin goal was to provide medical advice and treatment to those exposed and to gather information on the long-range effects of irradiation. Creation of the Centers came about as a result of the recommendation of a group of scientists, partly medical, of which Detlev Bronk was a member. They were placed under the direction of the National Academy of Sciences and were financed by the Atomic Energy Commission. The guiding organization became known as the Atomic Bomb Casualty Commission (ABCC). The lands surrounding the epicenter of the explosions were divided into concentric circular zones of increasing distances from the center. Several hundred thousand people were included in the overall survey.

I had first visited the Center at Hiroshima with Betty late in 1962 when I was President of the National Academy of Sciences. At that time, Dr. George B. Darling, a physician and administrator of great charm as well as great ability, was directing the program with the help of a dedicated staff from the United States and Japan. It was evident at that time that the principal short-range pathological effect of radiation was an approximately six-fold increase in the occurrence rate of leukemia over what had been the norm. It was subsequently found that, as radiation victims age, they suffer from other forms of cancer at a somewhat higher than normal rate of occurrence as well.

The most important finding of all, however, is that no observable genetic effects appear to have been passed on to the children and grandchildren of the survivors. Descendants appear to be completely normal. Some children who were exposed *in utero*, however, do exhibit brain damage and other defects.

While in Tokyo during one of my early journeys to Japan, I visited the United States Embassy and had the privilege of meeting with Ambassador Edwin O. Reischauer, who was on leave from Harvard University. The Ambassador told me of the strong criticism that was being leveled at the United States by various Japanese protest groups over the establishment of the Atomic Bomb Casualty Commission and its laboratories. The protestors claimed that the United States government was treating the bombing victims like so many laboratory animals — mere experimental subjects. Reischauer felt deeply embarrassed by these attacks, and wondered if the creation of the clinics really was in the best interests of our country; he said he intended to raise the issue with the Secretary in Washington. I did my best to assure him that the clinics were providing quality medical services with the health and well-being of the participants well in mind, and that extremely valuable scientific information was being accumulated through the studies. I am not sure how persuasive my arguments were; however, I never heard subsequently of any attempt by Washington to terminate the clinics as a result of the Ambassador's remonstrations. In the meantime, criticism has subsided and the clinics have amply demonstrated their worth.

My own first visits to the clinics were prompted by the onset of a long period of labor agitation at the centers. The Japanese economy, which was in a greatly straitened condition when the Centers were started immediately after the war, began picking up and was soon growing at a greater rate than that of the United States, mainly because of the deep trough from which it had started. This relatively rapid growth became reflected in the budget requests which the Academy presented to the Atomic Energy Commission and produced financial stresses in the entire chain from the Washington agency to the Japanese Centers. By 1975, the situation had reached a point at which it was agreed that the Japanese government should take principal responsibility for the future of the Centers and operate them with the assistance of a bi-national advisory committee. The main danger they face is that the United States will withdraw its share of financial support as a result of short-sightedness regarding the value of some aspects of science that is prevalent in the 1990s. The name of the organization was changed to the Radiation Effects Research Foundation (RERF).

Initial estimates of the radiation levels to which individuals in the various zones had been exposed were based on the best measurements that could be obtained during and immediately after World War II, using then-current computing techniques. By the 1980s much more sophisticated techniques were available. It was concluded that those who had been exposed at Hiroshima had received a much lower neutron dose than had been originally supposed. As a result, the Department of Energy, which had taken over from the defunct Atomic Energy Commission, decided to carry out a completely new and much more thorough study, both of the radiation produced by the bombs and of the exposures experienced by individuals at various sites about the epicenters. The National Academy of Sciences staff members who guided the programs were Seymour Jablon, a veteran of the Atomic Bomb Casualty Commission, and William H. Ellett, a highly experienced analyst and administrator. They were joined by Dr. Takashi Maruyama, an equally experienced member of the Radiation Effects Research Foundation.

In 1982 I was asked to join Dr. Eizo Tajima, Chairman of the Senior Dosimetry Committee, in leading the Advisory Committee for the new study. Professor Robert F. Christy of Caltech agreed to direct an in-depth study of various aspects of the problem with the aid of several groups of specialists. Along with studies of residual radioactivity by other investigators, James H. Schulman of the Naval Research Laboratory aided the work with novel determinations of the radiation dosage received by various artifacts using a technique based on the thermal release of luminescence from radiation-induced defects. Whereas the bomb used at Nagasaki was one of a group which had been extensively tested, the bomb dropped at Hiroshima was one of a kind. In order to obtain much clearer information concerning the radiation produced by the Hiroshima bomb, a full-scale model was constructed and tested at Los Alamos at a low, experimental level. As anticipated, the group concluded that the neutron dosage at Hiroshima was indeed very low, so that, in fact the effects observed in both cities were mainly the result of gamma radiation. Moreover, the levels of effective exposures proved to be about half the previously estimated figures. The far greater capacity of modern computer systems made it possible to estimate exposure incidence for a relatively large number of individuals in different situations.

The Director of the Radiation Effects Research Foundation during that period was Dr. Itsuzo Shigematsu (Figure 15.12), an energetic individual who had extensive experience as a research investigator in public medicine and public health. He held degrees from both the University of Tokyo and Harvard University and had been made a Fellow of the Royal College of Physicians of London. He was both a fine leader and generous host. In 1985, Betty and I had the pleasure of participating in the celebration commemorating the tenth anniversary of the centers in their newer incarnation. The

Figure 15.12. Dr. Itsuzo Shigematsu, the director of the Radiation Effects Research Foundation in Japan. (Courtesy of the Japanese Radiation Effects Research Foundation.)

celebrants enjoyed the full range of hospitality the Japanese can offer on such occasions—needless to say, a stimulating time was had by all.

In 1983 the United States government obtained evidence, through intelligence sources and by satellite, that the Soviet Union was employing, perhaps on an experimental basis, highly toxic agents of biological origin in both Vietnam and Afghanistan. The material appeared to be dropped by air in Vietnam in a form the natives referred to as "yellow rain." The technology used in Afghanistan, however, appeared to be quite different and was never as well understood. A few American scientists stated publicly that the yellow rain was a result of bee droppings. The basis for their claim was not clear, but it received backing nonetheless from some parts of the media in the United States, as well as the East European press.

The U.S. government asked if I would be willing to spend several weeks visiting the NATO nations to see what the consensus of opinion was regarding the situation. That tour proved to be very fruitful. I found that the military toxicology departments in several of the countries were very much aware of the potential problem, and were already working to develop countermeasures.

The available evidence indicated that the compounds being used were most likely special derivatives of a group of toxic organic materials known as the "tricothecenes." Some forms of these toxins can occur naturally in stored grain, particularly if it becomes moist. In fact, grain is normally tested before being used as food for either humans or animals to be certain that any amount of the compounds that may have developed is well below the hazardous level. In high concentrations, the toxins apparently can cause serious nerve damage, and can act as catalysts for significant degradation of proteins, including the walls of blood vessels. Severe hemorrhaging is a common effect of tricothecenes.

There is evidence that several decades ago the Soviet Union had a catastrophe in the Ukraine when both humans and animals consumed improperly stored grain containing high levels of tricothecenes, with resulting serious illness and fatalities. A group of biochemists studied the disaster, ascertained its cause, and began experimenting with the compounds and their derivatives in the hope of finding a practical degradable pesticide. The all-pervasive Soviet army learned of this work, and promptly began developing its own program to find possible military uses for the toxins.

One NATO official, who had extensive experience in the area of toxicology, informed me that there are derivative compounds of the family of tricothecenes that are exceedingly deadly. He said that the Soviet agricultural chemists, and presumably the military ones as well, had become highly expert in this field.

The news of the probable nature of yellow rain galvanized agricultural chemists in the United States who immediately stepped up their own research. I participated in several open meetings of toxicologists which focussed on this field of research, learning much that was relatively commonplace to them but new to me.

In June of 1991 I attended a conference at an American Chemical Society center near Baltimore, where I met two Russian scientists who were close to President Yeltsin. One of them was preparing a document for publication that openly described the catastrophic dumping into the local river of radioactive waste from a plutonium plant east of the Ural mountains. Many individuals who worked in the compound died as a result of gross exposure. The Yeltsin supporters were anxious to publicize such disasters in Russia to counter the influence of those there who would plead for a return to the "good old days."

I brought up to the Russian group the question of the Soviet army's use of toxic chemicals in Vietnam and Afghanistan. They told me that Andrei Sakharov had uncovered some of the details before he died, but that he met great resistance from the army when he tried to probe more deeply. Further information may come to light with the passage of time.

The facilities used for materials research in the United States had been improving at a reasonably steady pace in the 1950s and 1960s, but a period of stagnation set in during the 1970s. In the meantime, European governments had moved ahead more systematically and began to install new, large-scale research equipment, such as neutron and x-ray sources, which were unmatched in the United States. To remedy the situation, the Department of Energy and the National Science Foundation asked the National Academy of Sciences in 1983 to carry out a study of the steps that would be required to update experimental facilities in the United States. I was asked to co-chair the study with Dr. Dean E. Eastman, a brilliant young physicist from the IBM Laboratories in Westchester, New York, with a broad knowledge of the most advanced areas of research in condensed matter physics. The Executive Officer selected was Dr. William Spindel of the National Research Council; he had the versatile editorial support of Dr. Norman Metzger. The study used the services of a broad-based Advisory Committee as well as a number of consultants from the United States and Europe.

After an extended period of deliberation, the Committee recommended both the construction of several major new pieces of equipment and the upgrading of existing experimental facilities. The new equipment included two synchrotron radiation accelerators, an advanced steady-state neutron reactor, and a high-intensity pulsed neutron facility. Facilities expansion included the development of centers for cold neutron research, insertion devices on existing synchrotron accelerators to generate x-rays, and the extension of experimental facilities at Los Alamos National Laboratory in connection with a pulsed neutron source there.

The Committee also recommended a broad-based review of the facilities available in the United States to conduct research with high magnetic fields. While we knew that very useful facilities existed at the Francis Bitter National Magnet Laboratory at M.I.T., our discussions revealed that both Japan and the European countries were developing more advanced equipment.

Essentially all the Committee's recommendations were accepted, although the advanced steady-state neutron facility is still under consideration because of its high cost. In particular, a separate study was made of magnetic-field facilities in 1987. I served as co-chairman of this study along with Professor Robert C. Richardson of Cornell University. After learning the results of the Committee's work, the National Science Foundation requested bids from academic institutions nation-wide for a new magnetic research laboratory. Proposals were received from three strong institutions, including M.I.T., but that from the State University of Florida at Tallahassee was deemed the most favorable, partly because of the extent to which the state was willing to contribute buildings, staff, and funds. A system of cooperative efforts has been established between the Bitter National Laboratory and the State University of Florida with the understanding that the main center will evolve in Florida later in the decade.

In December of 1983 Secretary of State George Shultz notified the Director-General of the United Nations Educational, Scientific, and Cultural Organization (UNESCO) of the intention of the United States to withdraw from that Organization effective 31 December, 1984. Shultz said that, although the decision of the United States government was definite, he would appoint a committee of U.S. citizens who were conversant with UNESCO's work to review the matter in the intervening year to determine whether there was any significant reason to reverse the decision. I was asked to serve on a monitoring panel for the committee under the chairmanship of Ambassador Leonard Marks, then a partner in a Washington law firm.

UNESCO had been created in 1946 to advance educational, scientific, and cultural relations among the nations of the world on a non-partisan basis. By 1983, it had begun to seem as though UNESCO was developing a strong anti-Western stance. The director-general during this period was Dr. Amadou M. M'Bow, a distinguished scholar and diplomat from Senegal, who had appeared to be an ideal candidate at the time of his election and had been strongly supported by the U.S. government. The three main constitutional organs in UNESCO are the General Conference, consisting of representatives of all member states (one hundred and sixty one at the time), an Executive Board of about fifty members elected by the Conference, and the Secretariat headed by the Director-General who is appointed by the Conference on the nomination of the Executive Board.

Two issues that had caused special concern to the United States government had been the recommendation by the UNESCO Conference that Israel be expelled from UNESCO, and a further recommendation that the United Nations endorse what the U.S. government regarded as a left-wing proposal to establish economic balance among the various nations. This proposal went under the designation, The New Economic Order.

Our committee held a number of meetings both in the United States and in Europe during 1984 and, with some reluctance, ultimately agreed that we could not provide strong reasons for reversing the U.S. decision to withdraw from UNESCO. Whenever I focussed my attention on the scientific activities of the organization, I found hard-working, dedicated individuals who would have been assets to any agency. There was no doubt, however, that the general political climate of UNESCO was anti-Western. There were essentially no U.S. citizens in key positions in the organization. During a meeting of the executive committee which I was permitted to attend, several members made it clear that they would regard the absence of the United States as a benefit. This was probably a minority view, but still only served to reinforce the original U.S. position.

During the course of our work, our panel received a substantial amount of correspondence from various groups in the United States. It became apparent that many of the outstanding international science programs had been taken over by bodies associated with the International Scientific Unions, so that support from the scientific community for remaining in UNESCO was not strong. In contrast, a few groups of social and political scientists who had found useful service on some UNESCO committees hoped for a reversal of the decision.

The United Kingdom decided to follow the lead of the United States and also withdrew from UNESCO at the same time.

In 1987, Dr. Federico Mayor Zaragoza followed Dr. M'Bow as Director-General and brought an entirely new attitude with him. It would now seem appropriate for all the members of the United Nations, including the United States and the United Kingdom, to review the status of UNESCO and their own membership in it.

In 1984, a year after President Reagan announced his plan to support the Strategic Defense Initiative—the program referred to by the press as his Star Wars Initiative—General James A. Abrahamson (Figure 15.13), who was placed in charge, decided to create an advisory committee which would undertake critical reviews of the program and its development.

George A. Keyworth, the President's Science Advisor, asked if I would be willing to chair such a committee and help select the members. I accepted, and helped the General in establishing a two-sided structure for the Committee. On one side were individuals from institutions which were not under contract to the Office of Strategic Defense, and on the other were scientists and engineers from industrial organizations that might be directly or indirectly involved with SDI. The first group was the official Executive Committee, but both groups met together for general reviews and discussion. During the Bush administration, the leadership of the office was taken over by Ambassador H. Cooper and our committee served under the chairmanship of William R. Graham who had followed Keyworth as advisor to President Reagan.

The program was remarkably successful in the sense that it demonstrated, in its first few years, that it is possible to have one missile track-and-destroy another in space and on a global scale. Generally speaking, the research and

Figure 15.13. General James A. Abrahamson, the first director of the Strategic Defense Initiative. An inspired and flexible leader. (Courtesy of the Department of Defense.)

related studies demonstrated that the most effective way to defeat an armed offensive missile, which may be carrying multiple warheads, is to destroy it in the boost phase before it can deploy its contents. This would require a space-deployed defensive system involving both observational and conventionally armed counteractive units. The initiative encountered much opposition, some of which was purely political and arose from a desire on the part of some Senators and Representatives to weaken the authority of the President. This led to severe budget cuts and other restrictions. Another source of opposition, widely publicized by segments of the media, arose from academics and related groups led by a core body that was concerned with arms control and disarmament. Some objectors claimed that the program could not succeed, and hence was a waste of money, others that success would only escalate the arms race. Early on, some proponents of the program spoke of developing an "impenetrable shield" which clearly would be very difficult to maintain if hundreds of attacking missiles had to be dealt with in a short period of time. The more sober and realistic view taken by our advisory group was that we could hope to achieve a sufficiently high level of defense as to severely limit the effectiveness of a first strike by the Soviets while making U.S. retaliation a virtual certainty.

Our group concluded that the program would benefit from the creation of a free-standing, not-for-profit, federally funded research organization that could work in tandem with General Abrahamson's office to expedite aspects of planning and subcontracting. Unfortunately, Congress denied him this freedom. Senator Carl Levin of Michigan led the opposition to the plan.

The changes that took place in the Soviet Union in the late 1980s and early 1990s have, of course, radically altered this picture. The goal of the program was modified so as to aim for a higher degree of effectiveness against fewer missiles. This kind of program seems quite feasible if adequate support is given to it—support which is by no means extravagant relative to the overall scale of our defense budget, particularly if we consider the human cost that will result if such missiles and bombs fall into the wrong hands and are launched in a wanton burst of hate. At this writing, the program seems to have been side-tracked by what are regarded to be more immediately pressing issues.

In the autumn of 1987, I learned that Rabi was under treatment for cancer at Memorial Hospital in New York City. I spent a number of hours with him at the hospital, spread over several visits, discussing a wide array of topics of mutual interest, some relating to science, others not. He was in good spirits during these visits, although he was undoubtedly suffering.

Having met him in 1932, when he was in his mid-thirties and at his most creative scientifically, I had the privilege of observing him over more than five decades, and through several stages of activity and influence. One of the most brilliant and influential American scientists of the past century, as well as one of the most active, he always knew how to have fun, too, and enjoyed life to the fullest. He was always in good humor, at least in public, and he had a high-pitched, infectious laugh that was one of his trademarks. He enjoyed his work. He enjoyed enlisting others in his activities when it made sense to do so. He enjoyed good arguments and had developed singular techniques for winning debates. Above all, he enjoyed fame. He once said to me, "Very few people know why I received the Nobel Prize but they do know I am famous."

His parents came to New York from pre-war Austria when Rabi was a child. The family lived on the lower East Side, where a quick tongue and intellectual acrobatics were appreciated. I have always felt that Rabi developed his style of arguing early on as a way to protect himself from the local bullies by proving, through clever riposte, that he was much smarter than they and that they would gain more by listening to him than opposing him. His methods in an argument were usually defensive and executed in such a way as to disarm his opponent in the short if not in the long run.

I never met an individual who disliked Rabi personally, although there always were those who disagreed with some of his conclusions. Everyone would admit, however, that the great majority of his works were intrinsically good.

He had cause to hate the Germans and was explicit on this. If he ever visited Germany after World War II, I do not know of it. When the NATO science committee met in Munich and West Berlin in the mid-1960s he asked me to go in his place.

We attended many meetings together in many parts of the globe, often accompanied by Betty and Rabi's wife Helen, and we worked together on many committees. It was always a treat to be in his presence, no matter how much I might disagree with him on one point or another. At some time early in our association he had cast me into a specific mental mold of his own choosing, but very late in his life, perhaps in the 1970s or when we were together at a meeting in Martinique in the 1980s, he made the comment, "You are really a much more complicated person than I ever realized."

Although he was undoubtedly critical of some of Oppenheimer's actions when the latter was in a position of very great influence, he was always loyal to him, particularly during Oppenheimer's most difficult days. He was openly hostile to those who had criticized Oppenheimer, and had a particular, almost pathological, antagonism for Edward Teller.

There is an essentially unlimited number of stories about Rabi, that together would form the basis for a fine anthology. One that I always enjoyed relates to a meeting Rabi had with a Russian scientist. He had become very interested in the Russian's work and had arranged to meet him at a conference. It turned out, however, that Rabi's colleague spoke no English, while Rabi spoke no Russian. The problem was solved when they discovered that they both spoke Yiddish. Their meeting went off exceedingly well.

Soon after World War II Rabi and I were invited to a small conference sponsored by the Office of Naval Research to discuss the desirability of supporting cosmic-ray research. Important discoveries in the field of high energy physics were being made through cosmic-ray studies but the leaders of the ONR were concerned that financial support might be looked at askance by the powers higher up in the Navy. At that time, Rabi was strongly endorsing the development of high energy accelerators and was concerned that money diverted for cosmic-ray research might impede what he regarded as the most promising lead to the future of high energy physics. When one of the cosmic-ray research enthusiasts tried to emphasize that it was relatively inexpensive because much of it could be carried out with simple equipment at the top of a mountain, Rabi responded by saying: "Your argument is unfair. You have left out the cost of the mountain!" The Navy ultimately did support cosmic-ray research once space science became popular a few years later.

Rabi had been an advisor to the Department of State for nearly thirty five years when, late in the 1970s, the Department decided to rotate those posts. He and a group of his admiring friends were invited to attend a special ceremony in one of the meeting rooms at the Department where the Deputy Secretary, Warren Christopher, presented Rabi with an inscribed scroll honoring his distinguished service. Normally, the individual so honored responds with a few words expressing gratitude for the privilege of serving. Rabi, perhaps with tongue in cheek, took the occasion to deliver a lengthy speech recounting his many experiences, the importance of science as a touchstone for international cooperation, and a number of other issues which he deemed important, holding forth for at least a half-hour. Meanwhile the Deputy who had remained standing and had expected to be finished with this duty in a few minutes, shuffled impatiently from one foot to another, powerless to intervene.

Once, when Rabi and I were visiting the Hans Memling Museum in the ancient town of Bruges in Flanders to admire the exquisitely detailed paintings of the Flemish artist, Rabi commented, "When an artist is good, he can be very great." In this way he put into simple words the miraculous quality of great genius: Once it breaks out of the bonds of the commonplace it can reach new, previously unattainable, heights with revolutionary consequences.

Rabi displayed great perception on matters small as well as great. For example, he took considerable exception to the use of three-tined rather than four-tined forks in a table setting. He felt that the former were completely impractical as a working tool and should be eliminated from tableware. One need only test the two types in practice to confirm his observational wisdom.

Rabi had a great instinct for knowing just which elements made for a fine research environment. He had of course been eminently successful in managing his own research laboratory at Columbia University, which was devoted to investigations with the use of molecular beams. But, along with scientific and technical expertise, Rabi stressed the vital importance of guidance and support that must be provided at the administrative level of the institutions where research is conducted. Generally speaking, this support requires the special dedication of one or more individuals who not only understand the research process, but are willing to sacrifice, if that is the proper word to use, some of their own time and ego for the sake of the institutional good.

A major figure at Columbia University whom Rabi much admired, was Dean George B. Pegram (Figure 15.14) — locally known as "The Dean." Pegram was, incidentally, a graduate student when the American Physical Society was created in 1899 and remained close to it throughout his professional life. He had great capabilities himself as a research investigator, but was willing to devote most of his substantial energies to encouraging the work of others at the University. It was he, for example, who persuaded Enrico Fermi to join the Department of Physics at Columbia in 1939. His was also the principal guiding hand for the advancement of scholarly work during Eisenhower's period as President at Columbia University.

Dean Pegram was indefatigable as well as dedicated. Whether on or off

Figure 15.14. Dr. George B. Pegram, center, with Niels Bohr and Harrison M. Randall. Randall was Head of the Physics Department of the University of Michigan and did much to bring that institution into a leading position in the field. (AIP Emilio Segrè Visual Archives.)

campus his thoughts were always on research and the well-being of Columbia University. One day a group of concerned colleagues came to him and suggested that he take a vacation. He thought this advice over carefully. On the following Saturday he got into his automobile and drove from New York City to Syracuse, a distance of perhaps three hundred miles. He spent the evening with a sister there, and then returned to his office on Sunday.

Pegram visited Oak Ridge during our year there. Betty and I had him to dinner at a time when her Aunt Roberta was a houseguest, visiting from her home in Virginia. Pegram had grown up in the same general area in the South and, as the evening progressed, the two discovered that they had many old friends in common. Enlivened by the occasion, Aunt Bertie became once again the southern belle and he the dashing blade.

Rabi was always interested in observing the human scene and making cogent comments based on his observations. For example, he pointed out a unique degree of parallelism between the American immigration pattern of the Puritans in the seventeenth century and that of the Jewish population in the period just before and after 1900. In both cases movements were caused by persecution. Both groups arrived with a strong internal structure already in place, and both came seeking freedom. Most other immigrants to America, at least up to World War I, were motivated as individuals by economic pressure and were primarily seeking better living conditions.

During the winter of 1987–88, I received an invitation to give a series of lectures on topics of my choosing in physics departments at several South African universities. The invitation came from Professor J. S. Vermaak, the Chairman of the Department of Physics at the University of Port Elizabeth in the city of that name on the southeastern coast, east of Cape Town. We had met Professor Vermaak in 1979 when we visited the University as a guest of its Chancellor, the enlightened and scholarly industrialist, Dr. Anton E. Rupert, who served on the Advisory Council of Rockefeller University when I was president. It is traditional at South African universities that the Chancellor, who is essentially the Chairman of the Board of Trustees, be a prominent citizen. Dr. Rupert is one of the industrial leaders who had taken a strong public stand against apartheid. In 1979, through his courtesy and assistance, I had met leading colored and black academic leaders and had also enjoyed visits to the Kimberly Diamond Center, Kruger Park, and a number of universities, including those at Cape Town and Witwatersrand in Johannesburg.

I first went to South Africa in 1965 as part of a trip sponsored by Dr. J. George Harrar, President of the Rockefeller Foundation, who urged me as a member of his board to visit a number of countries in Africa such as Nigeria, Ethiopia, and Kenya, where the Foundation was supporting research projects. To that itinerary, I added Tanzania and South Africa.

This first visit to South Africa in 1965 was motivated in part out of general curiosity (which inspires a great number of foreign trips), but I was also pursuing three scientific interests. The first was to visit the landmark neu-trino experiment which was being conducted twelve thousand feet down one of the Rand mines in Johannesburg. That experiment was a joint effort of scientists in the United States and South Africa, with Professor Fridel Sellschop of the University of the Witwatersrand serving as head of the South African group.

My second purpose was to visit Willard Bascom's team, who were sta-tioned in Cape Town and were searching for diamonds in the offshore gravel off the West Coast that originates in the Orange River. Bascom had played a key role in the early development of the ill-fated Mohole program in the United States. While the Bascom team found a considerable number of dia-monds, very few were of gem quality.

Finally, I had for years had an interest in the development of our knowl-edge of ancestral hominids, and was anxious to learn as much as possible firsthand concerning the works of Raymond A. Dart, R. Broom, and P. V. Tobias in South Africa. This interest had made it possible for me, during that same trip, to visit L. S. B. Leakey in Olduvai Gorge, Tanzania (see Chap-ter 13). Friends in Johannesburg, led by Professor Sellschop, not only pro-vided me with special access to the rooms where the artifacts discovered by Dart, Broom, and Tobias were stored, but also took me to see the exact spot where the skull of the Taung Child had been found and identified by Dart in 1924. Since Dart had become a distinguished Professor at Witwatersrand, it was natural that the artifacts would be kept on that campus. Partly be-cause of the uncertain implications they suggest concerning the origins of man, and partly because of political pressure, the hominid artifacts were placed off-limits as far as the general public was concerned. I was, however, permitted to view them as a special guest.

Attitudes toward ancestral material had changed by 1984; the South Af-rican ban on display was lifted and the major South African pieces were lent to the once-in-a-lifetime exhibit "Ancestors" at the American Museum of Natural History in New York City. This exhibit, incidentally, involved the display of skulls and other bones, some of them as casts, of a wide vari-ety of hominids of various ages from all over the globe. The lack of enlight-enment in the United States regarding our antecedents is, however, indi-cated by the intensive picketing of the New York City museum because of the presence of the prehistoric South African bones. Both the Museum and the City of New York endured much trouble and expense as a result.

During the 1965 visit to South Africa, I also had an opportunity to visit Dr. S. Meiring Naudé at the major national laboratory, the CSIR, and was given a special tour. He had spent a significant period in his early career at the University of Chicago working with the distinguished physicist, Robert S. Mulliken.

When Professor Vermaak's invitation came in 1987, it was agreed that Betty and I would come to South Africa for four weeks (mid-June to mid-July 1989); we would visit some ten of the eighteen universities in the country and two major government laboratories. It was also agreed that I would give any one of four lectures of my choosing at the universities, and participate in the five-day annual conference of the South African Institute of Physics, which would take place in Pretoria during the week of 11 July. I would be prepared to offer comments and suggestions concerning the state of research in solid state physics at the universities.

This program obviously would involve extensive travel by air as well as by automobile. To facilitate matters, Professor Vermaak and his staff arranged with colleagues in the academic network to provide accommodations along the way. The system worked flawlessly. The hospitality offered us everywhere was warm and comfortable beyond description; we even had three-and-a-half days in Kruger Park on our own with a rented car.

The universities visited were, in chronological order: The University of the Witwatersrand and the Rand Afrikaans University in Johannesburg, the University of Port Elizabeth, Rhodes University in Grahamstown, the University of Cape Town and the nearby University of Western Cape, the University of Stellenbosch, the University of Bloemfontein, the University of Natal, in the Northeast, which has campuses at both Durban and Pietermaritzburg, and the University of Pretoria. The first scientifically confirmed specimen of a coelacanth, a living fossil fish, is on exhibit in a preserved state at Rhodes University. The visit to Pretoria also gave me an opportunity to make an informal visit to that most remarkable educational institution, the University of South Africa, which caters specifically to part-time students through class work and correspondence, and had an enrollment at that time of some one hundred thousand.

Incidentally, some reports from the U.S. media in 1989 implied that only white citizens were permitted on the public beaches in South Africa. We stayed at a beach resort hotel convenient to the University in Port Elizabeth. The weekend residents of the hotel included both ethnic blacks and Asian families, all of whom had completely free access to the fine public beach there.

South African universities vary in age and tradition as well as in their academic entrance requirements. By and large the older ones, such as Witwatersrand, Cape Town, and Stellenbosch, require entering students to have a fairly high performance record at the secondary school level. Some of the newer ones, such as that of the Western Cape and the Rand Afrikaans University, are more lenient with respect to admission standards but they require a significant level of performance during the first year if the student is to continue beyond that year. This is not unlike the policy once used by some state universities in the United States. For example, until the large post-World War II generation arrived at their doors, most state universities were relatively easy to enter, but performance in the first year was used as a critical test for continued matriculation.

It will be appreciated that the backgrounds of potential students in South Africa are, on the whole, probably much more varied than those of students in the United States. There are four or five main ethnic groups: white, Cape Colored or mixed (including an admixture of Chinese), Asian Indian, Malaysian and black, and there are important further sub-divisions within these groups. For example, whites may come from an English-, or Afrikaans-, or even German-, speaking background, and the blacks, even those living in urban areas, may have strong links to tribal origins. The languages spoken among the blacks, which exhibit much tribal and ethnic diversity, can be divided into four main groups, twenty three sub-groups, and numerous dialects. People in different sub-groups often cannot converse with others in their main group. A new student at a university founded on modern European standards may face culture shock, as well as intellectual challenges. To compensate for this, it is not uncommon for universities to offer students who have difficulty in their first year an opportunity to repeat that year.

It will come as no surprise that the university faculties joined with many other groups in opposing the principle of apartheid. All of the many university faculty and administrators I spoke with voiced a strong sense of responsibility toward blacks, as well as other non-white ethnic groups, and provided me with literature describing their goals and policies. It was recognized universally that the educational system has a duty to serve the needs of some twenty five million people, from all ethnic groups. When I was there, on average, about twenty to twenty five percent of university students were black and this fraction was growing rapidly.

The two greatest obstacles to expanding educational opportunities for blacks are a great dearth of good teachers at the primary and secondary school levels, and a shortage of public funds for all forms of public services. In 1989, this shortage was undoubtedly exacerbated as a result of the sanctions imposed by other countries on trade with South Africa and the withdrawal of some foreign companies from the region. It is very important, if the educational goals of the country are to be met, that an appropriate number of blacks and other ethnic groups enter into the various fields of education, both elementary and advanced. The problems and needs of South Africa are much too vast for the white population to resolve without the constructive cooperation of other ethnic groups,.particularly at the elementary and secondary levels.

During our visit, the government was taking drastic steps to raise funds for public services. One measure was a significant increase in currency production, which of course brought with it a high rate of inflation—probably fifteen percent per year—reminiscent of that in the United States in the latter part of the 1970s. Also, in addition to a heavily graded income tax, the government levied a thirteen percent sales tax on all purchased items.

Universities which were new in 1979, and which were built primarily for so-called colored students, are now open to all. It is expected that eventually they will become predominantly black.

The University of Natal in Durban was constructing a fine modern medical school and clinic to serve the educational and medical needs of the mainly Zulu black population there. It was a matter of debate led by the black population whether white students would be admitted to the school.

The two national laboratories we visited were the greatly enlarged CSIR Center near Pretoria, then headed by Dr. C. F. Garbers, which is now a general-purpose applied laboratory serving many programs, and the National Accelerator Laboratory at Faure mid-way between Cape Town and Stellenbosch, a notably fine, multi-use facility which serves several purposes.

The National Accelerator Laboratory, which has been under the direction of Dr. D. Reitmann, is a superbly designed and engineered separated-sector cyclotron. It can operate at proton energies up to two hundred MeV and currents of one hundred milliamperes. It designed to serve a variety of uses in medicine and experimental physics as well as in isotope production. A fully equipped hospital for both in- and outpatients is attached to the facility.

Whereas at the time of my visits in 1965 and 1979, universities and laboratories had great freedom to pursue research determined by the personal interests of the investigator, the government was now providing special incentives to conduct research in specific areas considered important to industrial growth. Part of this incentive consists of special funding through the Foundation for Research and Development. University groups that wish to continue pursuing long-range basic research now have more modest budgets and hence a strong incentive to use ingenuity in acquiring apparatus. The University of the Witwatersrand, for example, has raised funds by becoming one of the nation's main suppliers of liquid helium. The University also works in close association with a laboratory devoted to basic research on diamonds.

The South African Institute of Physics, whose activities combine, more or less, those of the American Physical Society and the American Institute of Physics, holds an annual conference at one of the major universities during the winter inter-semester break. As mentioned above, my visit to South Africa was designed to coincide with the thirty fourth annual meeting of the Institute, which was held at the University of Pretoria under the sponsorship of both that university and the University of South Africa. About two hundred physicists from all parts of South Africa, in addition to a half-dozen foreign guests of whom I was one, participated in the five-day meeting. My fellow foreign guests represented Canada, England, Germany, and Israel. Our first day was devoted to reviews of a number of topics in theoretical physics of both specialized and general interest, including condensed matter physics, cryogenics, nuclear and high energy particle physics. The second day focussed on the so-called summer school for advanced students and others who wished to obtain information in specialized areas. That year the lectures dealt mainly with aspects of theoretical nuclear physics.

The last three days of the meeting were devoted to a combination of fifteen-minute papers by investigators, mainly from the universities, and one-hour plenary lectures given by invited speakers. The schedule was arranged so that all participants could attend the plenary lectures. The general atmosphere during these three days reminded me strongly of a typical meeting of the American Physical Society prior to World War II; the attendees formed a fairly close-knit group, most of whom were acquainted with one another, and everyone spent a great deal of time discussing research in the corridors. Most of the lectures were given in English, although Afrikaans has equal standing. Perhaps fifteen percent of the attendees were black, colored, or Asian. Interspersed at various times throughout the meeting were a set of ten half-hour, non-specialist lectures that provided an overview of frontier areas of basic and applied physics.

There are three active scientific academies in South Africa at present. The first is the Royal Society of South Africa, which operates under a charter granted to it by the Royal Society of London. Its president in recent years has been a valued friend, Frank Nabarro. The second is the Afrikaans Academy of Sciences and Art which conducts its business mainly in the Afrikaans language. The third is intended to encourage participation in science and engineering among the black community; this is the Science and Engineering Academy of South Africa. I had the privilege of meeting with its President, Dr. G. S. Sibiya. We agreed that the field of engineering offered a wide range of opportunities for those blacks who are prepared to seek them. This academy has a council, chaired by Professor Sellschop, which is composed of a mixture of white and black specialists in various fields.

Dr. Sibiya visited me in New York City in the summer of 1992 to discuss the further development of his plans. In the meantime, the laws regarding apartheid have been rescinded, but the ultimate decisions on how the country will be governed have not been completed. One hopes that the struggle for political power among the various groups will not lead to a chaotic collapse of the organization of the society.

The ethnic mixture in South Africa is diverse and complex; exceptional caution must therefore be exercised by all who serve as leaders of the various factions in the times ahead, now that apartheid is at an end. The region could easily become the scene of a costly, destructive, internecine struggle from which no one would benefit. Fortunately, it is probably far less likely now than it was a decade ago that foreign countries will attempt to exploit the current situation, fishing in troubled waters for their own benefit. Perhaps over-zealous individuals in the United States, who do not appreciate the complexity of the situation, represent the greatest danger to a peaceful transition.

Chapter Sixteen
Reflections[1]

PATTERNS OF CULTURE

In 1934, a semi-popular book, *Patterns of Culture*, generated much comment. Its author, Professor Ruth Benedict,[2] was a distinguished anthropologist who had been a disciple of Franz Boas. The book focusses on three groups of peoples living in relative isolation who had developed special cultures, remarkably different both from one another and from the cultures that developed in the more advanced interconnected societies of the Mediterranean, Northern Europe, the Near East, and Asia. Two of the groups were American Indians, namely the Kwakiutls of Vancouver Island in the Northwest and the former cliff-dwelling Zuni or Pueblos; the third was a Melanesian group on Dobu, an isolated Pacific island. Other isolated cultures are mentioned in passing.

Benedict's studies of these vastly differing groups led her to one inescapable finding: an isolated, initially primitive people can, if undisturbed, develop any one of a great variety of cultures. There is substantial arbitrariness in the formation of cultures, although any given form will be conditioned to a significant degree by the group's natural environment. In some cases, but by no means all, population pressure can have a profound effect.[3]

---

[1] Additional reflections will be found in a recent book by the author: *The Science Matrix* (Springer-Verlag, New York, 1992).

[2] Ruth Benedict, born June 5, 1887, died September 17, 1948. In this book, she conceived of cultures as "total constructs of intellectual, religious, and aesthetic elements. The term "culture" is used in this special sense here.

[3] Any isolated, normally healthy human community can be expected to have a natural birthrate that permits a population growth of about three percent per year, corresponding to a doubling time of about twenty five years. If such growth is not constrained, the community will have grown by a factor of about one thousand or so in three centuries, putting great pressure on its resources. It must seek new options such as limiting the birthrate, introducing new technology to increase resources, seek new unoccupied lands, or become a militant expanding society displacing other groups from their lands and resources. The lifestyle of those in the Garden of Eden would have changed eventually even without expulsion. The collapse of the Mayan society and the dispersal of the Polynesian peoples to many distant islands are probably related in significant part to such changes.

Among many other observations, Benedict emphasized that what appears to be the desirable norm of behavior in one culture may well be looked upon as undesirable or even pathological in another.

Recognizing the almost infinite variety of possible human cultural patterns, one is tempted to conclude that there are no absolutes. Any one pattern should be regarded as being as acceptable as any other. "Lifestyles" are arbitrary and therefore all have equal merit. Many Americans who were coming of age during the 1960s and 1970s did, in fact, adopt the view that there are no absolutes with respect to human behavior. One mode is in fact as good as another, and each individual is privileged to select his or her own. They generated many slogans deprecating previously accepted standards of civility and other forms of behavior.

On deeper reflection, however, one comes to realize that there may be significant penalties associated with such a viewpoint. If an isolated culture of the type described by Benedict happens to encounter a technically advanced culture, it can expect to survive, if at all, only on the sufferance of the latter. The American Indian cultures she studied survive today only on reservations. Even there, they have had to make extensive adjustments in order to exist within the society outside their reservations.

Similarly, all the technically less-advanced countries of the world must, in one way or another, come to terms with the influence of the highly industrialized societies with which they come into contact. Whether we like it or not, there does appear to be such a thing as cultural evolution on our planet, even if some wish to deny that it represents "progress." The survival of any society or nation depends on its recognizing this evolution. Adjustments must be made to take advantage of the benefits associated with such growth, and to mitigate what may be regarded as its negative features, granting that the good and the bad may be closely intertwined and may often even be inseparable. Coming to terms with the dominant pattern, while admittedly entailing certain disadvantages, usually offers the redeeming benefit of permitting a wide range of lifestyles to coexist—far more than were possible in any of the isolated cultures described by Benedict.

One thing is certain: any group, national or otherwise, which attempts to stay completely removed from the mainstream of the developing world civilization will find it increasingly difficult to maintain its integrity. The rapidly growing world technological culture is far too penetrating to leave untouched for long any significant areas of isolation.

POOR COLUMBUS!

At the time of the quincentennial celebration of Columbus's discovery of America, it was the practice among some groups in the United States to denigrate that great adventure because of the drastic changes it brought to what they regarded as an unspoiled people. Be that as it may, there was a high degree of inevitability both in Columbus's voyage, and in what followed it. Eurasia and the Americas were bound to meet sooner or later as science-based technology advanced in Europe and made global voyages possible. The real wonder is that an Asian civilization did not make this kind of contact earlier.

The Chinese were operating large, technically advanced fleets of ocean-going ships a full century before the time of Columbus's journey and could, in principle, have ventured to the west coast of North America. Their main interest, however, was in India and the Moslem lands in the Near East and Africa, where commercial activity was high. Some of the most challenging voyages out of China were made under the leadership of Admiral Cheng Ho (1371–1433), a captive Mongol general, who sailed fleets into the Indian Ocean, including the Persian Gulf, and down the east coast of Africa, causing excitement that endured for a considerable time in those areas. Internal problems, combined with a change in attitude toward such journeys on the part of the Chinese emperor, brought the voyages to an end.

The initial rewards of ventures eastward into the Pacific Ocean would have been relatively modest, although the long-range consequences for human history could have been momentous, particularly if the Chinese had begun colonizing the west coast of North America.

The Vikings actually did cross the Atlantic five hundred years before Columbus. However, their scope was limited for a variety of reasons, even though they established colonies on relatively uninhabited Iceland and Greenland. Their ships, although technically remarkable, were less suitable than Columbus's for long, commercial voyages. Moreover, their weapons may have been barely a match for those of the war-like North American Indians whom they probably encountered from time to time during their expeditions.

The great tragedy of the peoples of the Americas was their susceptibility to common old-world diseases to which the populations of Europe and Asia had long since adjusted, epidemics having eliminated those with the least tolerance. The great plagues that swept over the Americas in the wake of the European explorers seriously compromised the ability of the native people to deal with the aggressive foreigners. This does not excuse the greed, religious fanaticism, and inhuman exploitation pursued by those who followed Columbus. It is, however, only fair to recognize that Columbus's voyage represented just one more step in the expansion of a form of civilization which continues to the present time. It was another stage in the sequence that presumably began with the trading, sometimes over long distances, of flints, obsidian, amber, and other materials in the Stone Age. Doubtless the early traders exchanged bacteria and viruses as well as goods, as we still do now. The vacationing tourist of today may unwittingly contribute to the spread of infectious diseases, much as Columbus and his colleagues did. The recent opening up of Eastern Europe has exposed Western populations to strains of bacteria which do not respond to most standard antibiotics because of misuse of such agents in the eastern countries.

Although the Chinese exhibited exploratory tendencies from time to time, as is indicated both by their extensive sea voyages in the Indian Ocean, and by the creation earlier of the Silk Roads to foreign areas such as India and the west during the Han Dynasty, most of their contacts with other cultural groups came about either through invasions by barbarians bent on conquest, such as that of Ghengis Khan and his Mongol hordes in the thirteenth century, or through the establishment of business colonies in China by other peoples. There was, for example, a substantial settlement of Jewish merchants in the northern Sung city of Kaifung starting in the tenth century. All such groups, however, were ultimately overpowered by the highly advanced Chinese culture; those who were militant and not defeated in battle were eventually assimilated into the Chinese ways.

One wonders if some invading groups might not have been overwhelmed by the indigenous diseases to which the Chinese had become adapted. Apparently Attila the Hun decided against sacking Rome when, as he traveled down the peninsula, he saw the depredations wrought on his nearly invincible army by local diseases. A large factor in the defeat of the Turkish army in its first attempt to take Vienna in 1529, under the leadership of Suleiman the Magnificent, was the outbreak of epidemics that decimated Turkish forces.

ENVIRONMENTAL MOVEMENTS

Had the most influential Europeans been more sensitive at the time of Columbus, they might have recognized that their culture's expansion and evolution could damage both the people and the environment, in spite of the immediate benefits to Europe in the shape of precious metals and new and useful plants. Unfortunately, Europe's social conscience was hardly developed enough at that time for such ideas to have taken hold. The reasons for this are complex. There was widespread poverty in Europe, and the lure of new riches drove out all other considerations. Moreover, the greatest damage was taking place far away and was affecting a completely alien people.

Both the Spanish conquistadors and the members of the Spanish church who accompanied them to America had recently seen the ultimate struggle with the "infidel" Moslems, and had finally driven them out of Spain. This experience, combined with the forced expulsion of unconverted Jews, who had played an important role in Arabian culture, gave the Spaniards a highly distorted view of what a "civilizing" mission should be, particularly when treasures were there for the taking. Ethnic and religious differences justified any action. A few conscientious priests who had settled in the New World were frankly appalled, and appealed to Rome to exercise more control, but their voices were almost unheard. The rise of Protestantism in Europe claimed almost all of the Papacy's attention and made it more, rather than less, militant.

Fortunately, changing times have brought greater wisdom, at least in the most developed countries. It is difficult to say just when this process began. It was evident to a degree in nineteenth century Europe, when the preservation of forests by governments was already considered essential and the purity of water in communal wells was beginning to be regarded as a major issue in many communities.

By the second half of the twentieth century, conservation and preservation, as well as the safeguarding of the public health, had become matters of major concern for large portions of the population in developed countries.

We now know that the scientific knowledge that has given mankind the benefits of improved health, higher living standards, and easier travel and communications, must also be used both to control the ill effects that may accompany such advances and to rectify some of the damage incurred in the past. Technology is continuously devising procedures to protect our health and safety and the natural beauty and resources of our world. Some observers will argue that the measures we have taken are too little, too late — that we are on the wrong path entirely. Some even claim that the most advanced countries are on the whole less convenient, pleasant, and healthy to live in now than they were at the beginning of the twentieth century or even earlier.

The consequences of science became a matter of great controversy on two other occasions: at the time of Galileo and then in the nineteenth century when Darwin and Wallace advanced the theory of evolution. In both cases, the main instigators of controversy were outside the scientific community. The tensions that surround science today are generated fully as much within as without the community and rest on socio-political attitudes rather than religious orthodoxy. Moreover, a special feature of the current situation is the extent to which segments of the editorial staffs of some professionally popular scientific journals feel the need to display bias by means of editorials or the way in which they deal with material offered for publication. The concept of "political correctness" is not confined to college campuses.

When I speak to those who would roll back the years, I point out to them that their health and physical well-being, and that of their families and friends, are substantially the result of the science-based medical advances of this century. Most such individuals promptly respond that they welcome biomedical research, but feel that physical research and its applications have led us astray. There is a clear separation, as they see it, between "good" science and "evil" science. Unfortunately, no such separation is possible. Most biological advances have been significantly dependent upon those in the physical sciences, including physical chemistry. The days of string and sealing wax research in medicine are long past. Science is, to a large extent, a seamless whole in spite of professional specialization, including the specialized languages and attitudes in different fields. One must accept the whole of the scientific endeavor if one wishes to maximize its benefits.

Efforts to better understand and control the consequences of our actions will never end. However, none of the problems associated with human endeavor seem insoluble within the basic framework of our effort.

Natural forces, both terrestrial and extra-terrestrial, that are beyond our control may, of course, cause serious problems if we are really unlucky. We live on an inherently unstable planet amid many closely neighboring planetary bodies with complex orbits, and in a universe that is violently active.[4]

THE EXTREMISTS

We are currently seeing in the United States a highly desirable interest on the part of the public in both the environment and human health as influenced by the expansion of technology. Some of the most radically extreme attitudes expressed, however, appear to me to be part of a movement whose aim is to overturn much of the gains humanity has made through scientific advances and their applications. These individuals behave as prophets, but more of doom than of salvation. Many elements of their gospel seem aimed at denigrating and misinforming rather than enlightening. Their tactics encourage universal fear of countless hazards that lie on the fringes of the truly demonstrable. These "dangers" include weak electromagnetic fields, the merest traces of ionizing radiation, even at levels well below those arising from natural radiation, and barely measurable quantities of chemical compounds that may cause illness in rats only when consumed in very large amounts. Many of the extremists' claims are couched in terms far removed from the traditional, objective language of good science. And yet their warnings seem solidly authentic to the uninitiated. Such "experts" are quick to demean more temperate voices, and with the help of the media, they frequently create the impression that many activities of the scientific community, both basic and applied, are at best highly irresponsible, and at worst downright immoral.

It seems reasonable to assume that if extremists gain the upper hand, the result, whether intended or not, will be to send national economies into tailspins, and to force great social and political changes as well.

---

[4] It now seems reasonably clear that the large dinosaurs were hastened on their way to extinction as a consequence of events caused by a meteorite crashing to earth, perhaps in the vicinity of the northern Yucatan Peninsula, sixty five million years ago, as was first proposed by Luis Alvarez and his colleagues. The dinosaurs had probably become highly specialized and therefore were particularly vulnerable to any disruption of their environment. More flexible forms of animal and plant life survived, including species of birds which are believed to have emerged from the dinosaur line early in its history. Our detailed knowledge of the motion of potentially hazardous meteorites and comets should be sufficient during the coming century that we will at least have ample warning of such an event recurring. We may even be able to alter trajectories, if space technology is encouraged to advance sufficiently.

Such groups are richly funded in the United States, not infrequently, although by no means exclusively, through the support of sympathetic officials at private foundations and government agencies. Fortunately, the number of such scientists and officers is not yet large. One of them told me that "perceptions are at least as important as facts." Another made the following remarkable admission:

> On the one hand, as scientists, we are ethically bound to the scientific method, in effect promising to tell the truth, the whole truth, and nothing but.... On the other hand, we are not just scientists but human beings as well. ...we need...to capture the public's imagination. That, of course, entails getting loads of media coverage. So we have to offer up scary scenarios, make simplified, dramatic statements, and make little mention of any doubts we might have. ...Each of us has to decide what the right balance is between being effective and being honest.

This attitude is not uncommon at present: the public is to be manipulated rather than objectively informed — and not necessarily for its own good, but in support of the extremists' prejudices.

One might well ask when it became popular for scientists in the United States to distort scientific facts to a degree in order to serve their personal political or social goals. Certainly many scientists hold strong religious or political beliefs; this has been true since the rise of western science in the thirteenth and fourteenth centuries. After all, most European scientists were religious clerics until the time of Galileo in the sixteenth century. (See, for example, the book listed in footnote 1 to this chapter.) Yet they and those who followed, regardless of their political leanings or religious beliefs, held firmly to the pursuit of what they believed to be a scientific truth — excepting, of course, the occasional charlatans, who had little effect on the course of good science.

I believe that, in the United States, the shift in outlook began relatively innocently in the post-World War II period when scientists, somewhat overwhelmed by the not undeserved adulation they had received as a result of their work in World War II, began forming political groups publicly favoring this or that presidential candidate, or social or political cause. This phenomenon became increasingly common, particularly among younger scientists who began their careers in the post-war period. Inevitably, political or social interests began to be intertwined with scientific pursuits.[5] It is too

---

[5] See, for example, "The Crisis in Authority," by A. Hunter Dupree, Brown Alumni Monthly, October 1969 (Volume 70, No. 1) describing the build-up of activist tensions at the University of California at Berkeley several years before the dramatic events of 1964 when the campus became the scene of riots. Joseph A. Schumpeter (1883–1950), the Austrian economist and social commentator who spent his final years (1932–50) at Harvard University, actually suggested that successful free-enterprise societies would be destroyed by their own success. They would engender large numbers of intellectuals who would prefer a socialist system in which they had a strong influence. He also proposed that the ideals which led to the creation of the democratic system may wane and that the system could become primarily a means of selecting leadership. See, for example, his book, *Capitalism, Socialism and Democracy* (Harper and Row, New York, 1950, original edition 1942, courtesy of Paul W. McCracken).

much to hope that the situation will ever return to the relative separation of science and politics found prior to World War II, since scientists now expect to be called upon routinely to serve the public interest. I sincerely hope, however, that the current trend will taper off, and that at least something of traditional objectivity will return to science. One hopes that future generations of scientists will avoid the attitude expressed by Party Member O'Brien in George Orwell's *1984*:[6]

> There is nothing we could not do. Invisibility, levitation—anything. I could float off this floor like a soap bubble if I wished. I do not wish to, because the Party does not wish it. You must get rid of those nineteenth century ideas about the laws of nature. We make the laws of nature.

Be that as it may, the continuing growth of the world's population, as well as the plight of large numbers of that population in both the Third World and developed countries, demand the continued advance of science along traditional, well-proven pathways. Only in this manner will we be able to deal with the pressing problems we shall face in the future. Current attempts to draw firm but distorted conclusions in the absence of an understanding of basic facts will, if successful, seriously damage the attempts of conscientious scientists to play out their essential roles.

Should the integrity of scientific traditions not be maintained by researchers and practitioners, the well-being of humanity is almost certain to regress. Our need to achieve and maintain, for example, good medical services, adequate nutrition, and enhanced communications means we must sustain the highest possible scientific standards. It is clear that the application of good science that resulted in improved agriculture has been the salvation of much of the growing population of the world. Whether our population should continue to grow at its present rate is an issue which must be faced on its own terms.

SPECIAL INTERESTS

In criticizing the extremist aspects of the environmental movement, I am not unaware that governments everywhere have been and always will be subject to pressures exerted by special interest groups of one kind or another. Such activity is an integral part of the life of any society. Moreover, the process tends to have a compounding effect since the activity of one group may stir up controversy and lead to the creation of one or more supporting or opposing groups.

It is no less evident that the fate of a society may depend upon its responses to such pressures. At present, most democratic nations contain many such groups; in the United States, their numbers have increased unceas-

---

[6] George Orwell, 1984 (Harcourt Brace Jovanovich, New York, 1983), p. 218. Courtesy of Michael Fumento. The book was originally published in 1949.

ingly since 1960. They take up positions pro and con on matters such as: the status and rights of school teachers and their students; the appropriate levels of armament; human rights, both national and international; handgun control; the future of space research; the use of animals in biomedical research; and so on. Increasingly, the U.S. political scene resembles that of democratic countries where a large number of splinter parties must find some form of compromise in order to maintain anything resembling an effective government.[7]

I myself have joined such groups on occasion; for example, Scientists and Engineers for Secure Energy, created with Miro Todorovich, which supports the continued development of nuclear power plants, in keeping with our early post-war successes and those of nations such as France and Canada. I have also joined organizations such as the George C. Marshall Institute, created with Robert Jastrow, which encourages research on defensive antiballistic missiles, on space science, and critical studies of factors that could have a major effect on the environment.

It is not at all clear how wide a variety of controversial opinions or "deviationism" a society can sustain and still retain the coherence necessary to play a significant role among its fellow nations. The questions raised regarding the future of California (see Chapter 4) apply equally to the United States as a nation. The country has clearly lost, at least for the present, much of the sense of national unity which it possessed prior to the 1960s. The coming century promises to be a critical one.[8]

There is little need for those engaged in the basic sciences to form a special interest group in support of *general* scientific purposes since the government has many ways of obtaining reasonably sound scientific advice when

---

[7] At the time this is written, the level of controversy in the United States is matched by troubles of at least comparable magnitude in other countries. Even Japan faces major social and economic problems as a result of global changes and the inevitable domestic perturbations that occur as a new generation takes over the stewardship of a country. Should the European countries achieve the level of unity many strive for, and should Eastern Asia achieve its full potential as a commercial dynamo, the United States may be compelled to re-examine its internal differences and, seek its own higher degree of unity.

[8] The population of the United States is currently increasing at the rate of about ten percent per decade, in substantial part as a result of immigration. This corresponds to a doubling time of about 75 years—a rate which presumably could be sustained well into the next century. While the "melting pot" concept which was prevalent in my childhood now tends to be discounted, particularly by those who have a preference for the preservation of ethnic diversity, it is worth observing that the population remains highly mobile, guaranteeing much geographical mixing. Should both population growth and geographical mixing eventually become negligible, it is possible that the various parts of the country will exhibit strong regional differences, leading to pressures for a re-definition of our form of federalism. Such pressures could become particularly acute should languages other than English become official in some regions. The English language increasingly binds the United States together, much as Chinese character writing does the diversified Chinese people.

it so wishes, always allowing for human error. Potential sources include the National Academy of Sciences, which is quasi-independent, the Office of Technology Assessment[9] of the Congress and the many scientific experts within the existing executive offices and agencies, including the Science Board of the National Science Foundation, as well as the Office of Science and Technology and the Office of Management and Budget in the White House.

Generally speaking, whenever the name of an independent special interest group has the word "science" or "scientists" in its title, it is more than likely that its aims are political, economic, and sociological, and have only indirectly to do with the re-affirmation of the well-established traditions of scientific research.

RELIGIOUS MOVEMENTS

The evolution of science-based technology has made life more comfortable for all of us; this ease has inevitably led us to focus on the material benefits of life. Nevertheless, mankind does not live by bread alone. The changes that are occurring will inevitably be accompanied by religious movements both old and new. It is to be hoped that these movements will serve the constructive purpose of furthering harmony and spiritual unification among the peoples of the planet rather than increasing tension — as has too often been the case with many (but not all) such movements.

To me the ideal framework for religious activity is found in most of China. There religious beliefs, many conforming to today's major religions, are viewed as personal matters. Religion and politics are not closely linked. This is probably a form of wisdom learned through long experience. I hope that it will become more nearly universal, particularly to the degree that it encourages the separation of religion and politics.

SCIENCE IN THE UNITED STATES

When I entered the world in 1911, scientific research was far advanced in Europe but had barely begun in the United States, although many people were interested in science on an individual basis. Hundreds, perhaps thousands, of such earnest American scholars had gone to Europe to absorb the

---

[9] When the Office of Technology Assessment was first established in Congress in the mid-1960s, substantially through the efforts of Representative Emilio Q. Daddario who was much interested in the progress of science and technology, its underlying concept was balanced in the sense that fully as much emphasis was to be placed upon the benefits of science and technology as upon negative side effects. In keeping with the changing spirit of the times, however, the office soon turned into a rallying hall for the nay-sayers who were available in abundance. (See Chapter 13.) Many such opponents of this or that form of advancing technology have since learned to earn a good living through such activity.

scientific traditions and knowledge developed in the various fields. In some cases they became involved there in frontier research programs. On returning home many struggled to establish comparable standards and institutions here, as circumstances permitted. This pattern of transfer had actually started in colonial times but had moved ahead slowly because the environment at home was not very receptive.

It is true that the Smithsonian Institution was established in the 1840s, but that was the result of an unsolicited gift to our government by an Englishman—an individual who felt inspired by our form of government and believed that science was destined to become an important force on the western side of the Atlantic. The miracle is that the Congress used the money remarkably well, with few traditional guidelines. It is also true that the National Academy of Sciences was created in 1863; however, its immediate goals were very practical and were specifically related to the Civil War. The Academy probably would have died following the war had not Joseph Henry, the inspired head of the Smithsonian and a high-minded scientist, supported it and given it renewed life, focussing its activities on basic science.

The last quarter of the nineteenth century had seen the beginning of a more formal recognition of institutions devoted to science. For example, the Geological Survey was established in 1879, the Weather Bureau in 1891, and the Bureau of Standards just at the turn of the century. The Johns Hopkins University (1876), the University of Chicago (1891), Stanford University (1890), and The Rockefeller Institute for Medical Research (1901) all emphasized scientific research and education and helped set the stage for a new era.

By 1911, a number of American scientists, such as Joseph Henry, Albert A. Michelson, Henry Rowland, Willard Gibbs, Robert A. Millikan, Ira Remsen, and Thomas H. Morgan, had either carried out or were in the process of initiating world-class research. Our country was no longer a scientific desert.

Twenty years later in 1931, the process of preparation was essentially complete and the United States was ready to assume a leadership role in science—a position which was enhanced during the following decade by the arrival of refugees from the persecutions of Hitler, Stalin, and Mussolini and the stimulation of research during World War II. By the start of the 1930s it was no longer *necessary* for the embryonic scientist to spend a period studying abroad. Many members of my own generation did not visit Europe before the end of World War II.

It should be added that an enormous stimulus to the rapid advance of science in the United States was provided in the 1920s by the National Research Council Fellowships funded by the Rockefeller Foundation. They made it possible for many promising young American scientists to go abroad for two years and work with leading scientists in Europe. They almost invariably returned with a deeper and more comprehensive understanding of frontier issues and problems.

PUBLIC VERSUS PRIVATE SUPPORT

Until World War II, when the federal government enlisted a large percentage of the scientific community in war-related research, much of the support for science came from private sources—individuals, foundations, and, in some cases, enlightened industries. Federal and state support was not entirely negligible; federal laboratories and state universities received modest research funds through official channels in amounts that depended upon the mood of Congress or the Legislators when annual or bi-annual budgets were passed. Private support, however, was not only significant in magnitude, but at that time usually came with fewer stipulations attached, once its general purpose had been determined. Most of the great philanthropists who established foundations prior to World War II, such as James Lick, Andrew Carnegie, John D. Rockefeller, Frederick G. Cottrell, and Alfred P. Sloan, appreciated both the short- and long-term benefits of scientific research. They expected that a significant fraction of the funds dispersed through their philanthropies would support science as part of the life-blood of an advancing society.

Nevertheless the scientific community, not least the component in the physical sciences, rejoiced when, in the wake of World War II, several agencies of the federal government, such as the Office of Naval Research, the Atomic Energy Commission, the National Institutes of Health, and, finally, the National Science Foundation were created or expanded to provide support for science in a systematic way. The application process for government funds now appeared to be more direct and less subject to the delays and uncertainties that were sometimes part of dealing with private philanthropy. Moreover, both the scientific community itself and the complexity of its research programs were increasing rapidly, and the large grants that could be provided through the bounty of government agencies were a great boon to a scientific community eager to play an important role in society.

THE NATIONAL SCIENCE FOUNDATION

Immediately after the war, scientists in all basic fields were hoping for the rapid creation of a National Science Foundation that could support broad-based basic research and that would replace the Office of Scientific Research and Development, created within the White House during World War II by Vannevar Bush, James Conant, and their colleagues to support military research. Unfortunately, neither the scientific community nor the Congress could concur on the details of the form such a foundation should take. Everyone agreed that it should have a Director and an advisory structure, but differences of opinion soon developed over the degree of independence granted to key positions in the organization. Many scientists felt that the advisory board should appoint the director without having to seek the approval of the President. President Truman made it clear, however, that he would veto any bill which denied the President the right to select the Foundation's director.

There were also concerns about the control of any patents that might result from Foundation-supported research. Indeed, many industrial leaders greatly feared that the government would assume control of most patents for basic research and thereby draw the country into a quasi-socialist form of society. As a result, no bill was forthcoming at that time.

In fact, it took five long years and the indirect pressure of the Korean War to bring the creation of a National Science Foundation to the top of the Congressional agenda once again. While many previously disputed issues were no longer of serious concern, the scientific community became alarmed when the possibility arose that those receiving funds might be required to have security clearance, or at least take a formal oath of loyalty. All such questions were finally resolved and President Truman signed the bill establishing the foundation in 1951.

Fortunately, the delay was not as harmful as it might have been. As mentioned earlier, other federal agencies, such as the Department of Defense and the Atomic Energy Commission, had been awarding contracts and grants to the scientific community since immediately after the war, permitting a variety of research programs to proceed without much hindrance, and within a reasonable policy framework. Moreover the National Institutes of Health began funding biomedical research on a significant scale early in the 1950s and in response to a rapidly growing demand. Actually, it was not until nearly a decade after its creation that the National Science Foundation became a truly significant source of major funds.

PROBLEMS WITH FEDERAL FUNDING

In spite of the initial promise of large-scale federal funding, a number of far-seeing individuals expressed a certain concern over the government's playing such an important role in sustaining basic research. Even today, very few members of the Congress have had sufficient experience with research to appreciate the subtle ways in which basic science can lead to revolutionary innovations in technology, or to understand that the best way to support research is to provide funds to the most gifted scientists and then allow them the freedom to determine their own research agendas. Even though a research program may seem to provide purely intellectual gains far removed from any material benefits, the ultimate return on our investment may be far-reaching and eminently practical applications. What may appear on the surface to be mere tinkering with the workings of nature could provide the key to the development of sophisticated technology one or more decades down the road. Interfering with this well-established method will almost certainly benefit neither science nor technology in the long run.

With respect to basic scientific research, Vannevar Bush (Figure 16.1) made a number of clear and forceful statements in his report, "Science the Endless Frontier," written at the end of World War II:

> "Support of basic research in the public and private colleges, universities, and research institutes must leave the internal control of policy, personnel, and the method and scope of the research to the institutions themselves. This is of the utmost importance."

He added:

> "The distinction between applied and pure research is not a hard and fast one, and industrial scientists may tackle specific problems from broad fundamental viewpoints. But it is important to emphasize that there is a perverse law governing research: under the pressure for immediate results, and unless deliberate policies are set up guarding against this, *applied research invariably drives out pure.*

> This moral is clear. It is pure research which deserves and requires special protection and especially assured support."

It takes a combination of genius and artistry to bring about a technological revolution such as we have seen born of the creation and evolution of the transistor and the microchip—the integrated circuit. However, the individuals directly responsible for such great technical advances are usually quite different from those who laid their foundations through prior (unhin-

Figure 16.1. Vannevar Bush, war-time leader of the scientific research community. His wise counsel concerning the support of basic science needs to be reiterated in each generation. (Courtesy of the National Academy of Sciences.)

dered) basic research. There are those who are outstanding at both basic and applied research, but such prodigies are rare indeed. (One, William D. Coolidge, appears in Chapter 7.) Those who influence advances in technology are more often than not to be found in organizations—laboratories, if you will—where invention, development, and production are the principal goals; that is, in industrial institutions.

During the debates in 1993 which led to the cancellation of the superconducting supercollider—a magnificently conceived machine that was designed to probe more deeply into the structure of matter—some members of the House of Representatives displayed quite clearly their lack of fundamental understanding of the scientific base upon which our current civilization rests. Experience over a long period of time has shown that the ultimate fallout of such frontier endeavors usually repays the cost of investment by generating new technology of one form or another. In some cases the returns have been enormous. If the behavior of congress in this case signals a general trend in decline of the level of appreciation of the scientific and technological underpinnings of our present-day world, the United States is in for deep trouble.

PRESSURE FOR APPLIED RESEARCH

Since the second half of the 1960s, when the total budget of the Foundation began to appear significant, and continuing through to the present time, there has been a growing pressure upon the National Science Foundation to devote an ever-increasing effort to applied research (Figure 16.2). I was privileged in the 1960s to witness first-hand the lobbying efforts in the Congress to induce the National Science Foundation to undertake the direct support of applied research, on what initially appeared to be a relatively small scale. The goal seemed innocent at the time. I realize now, a quarter of a century later, that the plan contained the seeds of disaster for both basic and applied science. If Congress wishes to support engineering research, it should either set up a separate agency with guidelines to assure a direct link between the research supported and well-designated applications, or provide additional funds to established, mission-oriented agencies.

If the leaders of the National Science Foundation yield to pressure, perhaps under the threat of a budget freeze or cut, the result will almost certainly be the production of much mediocre science without any significant improvement in the output of new technology. In fact, technological advance will be hindered in the long run. It would be far more useful to provide financial incentives to industry that would inspire more in-house research and development efforts, with the goal of bringing innovative products to market. Scientists and engineers in academic life who have the aptitude and inclination to assist in such developments should be encouraged to serve as consultants to industry.

In fact, the major responsibility for advancing technological development and production in the United States must rest with industry, and the princi-

*"Oh, I give him full credit for inventing fire,
but what's he done since?"*

Figure 16.2. *One is reminded of current Congressional and Parliamentary debates concerning the merits of scientists and scientific research. (Drawing by Kraus, © 1960, 1988 The New Yorker Magazine, Inc., All Rights Reserved, and The Rockefeller University Faculty Club.) The caption reads: "Oh, I give him full credit for inventing fire, but what's he done since?"*

---

pal role of government should be to encourage that development. There are several ways of doing this. One is to provide tax incentives for investment in research and development programs selected by the leadership within a given industry. It is worth remembering that innovative development in the defense industry has been helped substantially, at least in the past, by what are termed Independent Research and Development Grants, based percentage-wise upon the magnitude of the procurement contracts held by a company. The use of this research and development money was determined by mutual agreement between the producing industry and the sponsoring agent, which together looked for ways to promote innovation.

For many decades important exploratory research, akin to basic research, was carried on in laboratories of major consumer-oriented corporations such as Bell Labs, the General Electric Laboratory, the DuPont Experimental Station, and various laboratories at IBM. Changing times and circumstances may require such organizations to place more emphasis on applied research immediately related to practical developments. It is not clear that the smaller "start-up" companies will be able to support significant amounts of basic research and thereby fill in the gap. The ultimate significance of the changes taking place in the corporate world is far from clear but they do bear close watching.

It should be added that, if there is a substantial cutback in the support of basic science, the most capable individuals in the scientific community will do their best to compensate with ingenuity and patience. However, many areas of science such as astronomy and high energy physics, as well as some areas of biology and medicine, will require special funding for more sophisticated equipment if they are to advance. On the whole, there is little doubt that the progress of science will be curtailed to a significant degree.

THE NEED FOR EXTENDED PRIVATE SUPPORT OF SCIENCE

A serious problem remains even if wisdom prevails at each critical juncture and the federal agencies, particularly the National Science Foundation, ultimately do not interfere with basic research at academic institutions. One cannot expect the funds available for basic research from government sources to grow at a substantially greater rate than the gross domestic product. It follows that, percentage-wise, we are probably somewhere near the peak of what we can expect — barring a catastrophic event that requires an extraordinary expansion of one or more fields of research. In fact, substantially less than half of all qualified requests for basic research are funded at present, and even they are not supported in anything like a luxurious manner.

Those most likely to suffer in such a situation are young scientists who have just completed their formal education and are about to take the next important step in the development of their careers. Some will find shelter with a senior investigator under an umbrella of research grants; some will accept junior faculty positions and struggle with heavy teaching loads while carrying out research with limited facilities. A few will receive relatively unrestricted fellowships. And while it is true that the genuinely committed can be expected to face such challenges with fortitude, something very precious may be lost if the path to a stable career is made too steep. Some very capable pre-doctoral students will be turned away from science in their formative years. Others will be forced into channels outside basic research before they have had a reasonable opportunity to assess their own aptitudes or make a fair contribution.

ROLE OF THE PRIVATE FOUNDATIONS

One might hope that somehow the nation's private foundations would be prepared to recognize this problem as a major issue and adjust their programs accordingly. For example, they might either individually or collectively agree to devote a fraction of their annual budgets to support a modern equivalent of the two-year National Research Council Fellowships of the 1920s, focussing on the basic sciences and employing competitive screening. At the present time, an appropriate annual grant would be between fifty and seventy five thousand dollars, tax free. Five to ten thousand such fellowships per annum distributed over the basic sciences could alleviate substantially the problems facing young scientists. Such a program would encourage young science students, and could provide the best and the brightest of the postdoctoral scientists with a much longer period in which to decide what form of career best matches their talents.

If we examine the present status of many private foundations; there appear to be two major obstacles to this plan:

Once the federal government began to fund basic science on a substantial scale in the period between 1945 and 1960, many foundations that had previously regarded the support of science to be among their most important obligations abandoned that policy, except possibly for investigations regarded as "socially relevant" according to some specially selected peer group—often not including scientists. Along with this change came inevitable changes in the personnel of the boards of trustees and the staffs of the foundations. Unfortunately few foundation leaders had the foresight to realize that federal funds devoted to science would eventually peak, and that some crucial problems might remain which could be far better handled by private rather than public organizations.[10]

In addition to this change, which is perhaps understandable in view of the normal limitations of human vision, many foundations fell under the spell of a widely held view which emerged out of the intellectual ferment of the 1960s and 1970s, that our mode of life and our social structure are deeply flawed and require extensive, even radical, changes. These changes should be supported in a spirit of penitence, and from a sense of common guilt. For example, much money is spent on attempts to alleviate the plight of those living in what are termed "the inner cities," even in the absence of firm plans or traditions supported by the leadership, if any, within the inner cities. It would, incidentally, be difficult to recall an instance where a group or society has benefitted from outside help in the absence of strong internal leadership. The resurgence of Western Europe and Japan after World War II, and the "miracles" observed in South Korea and Taiwan depended on such strong internal leadership in addition to some outside support.

Until the right kind of leadership arises within the inner cities and provides directions for development that are reasonably consonant with "outside" culture, the future of the populations trapped in such enclaves seems almost hopeless.

Money is also being used to "save" the environment of our planet, even though many of the most shrilly voiced "problems" are based on emotion rather than solid scientific evidence. Estimates indicate that from one half to one billion dollars is spent annually by private philanthropic sources, much of it from private foundations, on projects that supposedly affect the

[10] Fortunately there still are philanthropists who are inspired to give substantial support to fields such as astronomy and biomedical research through their foundations. Nevertheless, support by governments has inevitably gained the dominant role.

environment. One might think that comparable amounts of money spent to help advance the careers of budding scientists would prove at least as productive.

This is not to say that the environmental issues can be ignored. Many of the worst and most obvious, however, can be resolved without dramatic changes in national productivity by the application of good science, engineering, and common sense. The more profound ones, such as the possibility of global warming or changes in the ozone layer, are not well-understood at present and require continuing observation and analysis along traditional lines of scientific research. Some of the more extreme remedies recommended, such as immediate, radical cut-backs in the use of fossil fuels and halogen-containing compounds with uniquely useful properties, and the discouraging of nuclear power plant development, could well prove not only unnecessary, but could undermine the economy and weaken our ability to take appropriate actions on truly deserving problems.

TIME FOR COURAGE!

It would be easy to be pessimistic when viewing the obstacles that appear to be standing in the way of the continued advance of science in the United States at the present time. One must, however, look at the big picture and take courage. Science has come of age in the United States during the past century; there is widespread appreciation of the benefits it confers on us, it is both a source of enlightenment and a handmaiden to technology. We continue to be enriched on both scores.

Then too, attitudes do change. If I look back on the years since I first gained awareness, it is clear that what might be called the mood of the country has undergone a radical shift every five years or so. The course of a nation somewhat resembles the journey of a vehicle on a curved road, being steered first one way and then another as changing circumstances require. Most people realize that we will not continue to enjoy the improvements in our quality of life that we have come to expect in the twentieth century by abandoning the pursuit of science, even though scientific advances bring problems along with benefits.

Name Index*

A

Ablon, R.E., 315, 345-346
Ablon, R.R., 345
Abrahamson, J.A., 365- 366
Acton, J.E.E.O., 159
Adams, Albert M., xi
Adams, E.P., 48
Adams, R., 204-205, 222, 347
Adrian, E.D., 117
Aigrain, Francine, x, 178, 232, 265,
Aigrain, P.R., 178, 232, 255
Alexopoulos, K., 215, 263
Allen, E.R., 132
Allison, S.K., 154
Almy, G.M., 203-204, 215, 241
Alpert, D., 97, 107, 247
Alvarez, L.W., 18, 381
Amouzegar, J., 335-336
Angello, S.J., 135
Apker, L.W., 87
Armand, L., 260
Armstrong, Elizabeth, 73
Arnold, R.T., 222
Aron, R., x, 264-265
Arwade, Florence, ix
Astbury, W.T., 81
Astor, Mrs. Vincent B., 315
Attila, 379
Avery, O.T., 322

B

Bacher, R.F., 69-70, 82, 246
Bacon, Helen G., 77, 79
Bacon, L.L., ix, 77-79, 269, 292
Bailey, Jean A., 70
Baker, T.S., 127
Baker, W.O., 293, 306, 315, 326, 337
Baltimore, D., 315
Bardeen, J., 63 et seq., 144, 148, 151-152,
218-221
Bardeen, Jane M., 63, 65, 83, 219
Barger, A.C., 354
Barnes, Margaret, 77
Barnes, S.W., 77-78 et seq., 103
Barrett, C.S., 123

Barton, H.A., 223, 227
Bartholdi, F.A., 251
Bascom, W.N., 371
Bassani, F., 209
Baust, F., 4, 18
Bearn, A.G., x, 358
Beberman, M., 211
Becker, J.A., 98
Becker, R., 170
Beethoven, L. Van, 18, 154
Beltran Family, 333-334
Benedict, Ruth F., 376-377
Benitez, J., 306
Bennett, B., 261
Bennett, I.L., 351
Bergstrom, F.W., 33
Beria, L.P., 168
Berkner, L.V., 21, 279-281, 302
Bernardini, G., 203
Bethe, H.A., 58, 69, 82, 87, 98, 137, 167
Bhabbha, H.J., 308-309
Bing, R., 207
Birge, R.T., 40
Bitter, F., 84, 363
Bjerknes, J., 259
Blackett, P.M.S., 287-288, 309
Blackwelder, E., 34
Bleakney, W., 49-50, 94-95, 138-139
Blewett, J.P., 52-53, 71
Blewitt, T.H., 177
Blichfeldt, H.F., 39
Bloch, F., 58, 66, 131, 193
Blodgett, Katherine, 102
Blokhintsev, D.I., 246
Blumberg, S.A., 110
Boas, F., 376
Boeke, Clara, 29
Bogoliubov, N.N., 342-343
Bohr, H., 40
Bohr, N., x, 40, 61, 66, 80, 86, 110, 170, 193,
221, 369
Bollinger, B., x
Booge, J.E., 133, 143
Booker, E.R., 28
Booth, J.W., 141
Borden, B.F., 347
Borelius, G., 215
Born, M., 356

Subject Index

405